Cluster Analysis and Applications

Rudolf Scitovski • Kristian Sabo
Francisco Martínez-Álvarez • Šime Ungar

Cluster Analysis
and Applications

Rudolf Scitovski
Department of Mathematics
University of Osijek
Osijek, Croatia

Kristian Sabo
Department of Mathematics
University of Osijek
Osijek, Croatia

Francisco Martínez-Álvarez
Department of Computer Science
Pablo de Olavide University
Sevilla, Spain

Šime Ungar
Department of Mathematics
University of Osijek
Osijek, Croatia

ISBN 978-3-030-74554-7   ISBN 978-3-030-74552-3   (eBook)
https://doi.org/10.1007/978-3-030-74552-3

This Springer imprint is published by the registered company Springer Nature Switzerland AG
The registered company address is: Gewerbestrasse 11, 6330 Cham, Switzerland

Preface

For several years, parts of the content of this textbook have been used in undergraduate courses in the Department of Mathematics and in the Faculty of Economics at the University of Osijek. Also, some parts have been used at the graduate level in the Faculty of Electrical Engineering and in the Faculty of Mechanical Engineering at the University of Osijek.

The textbook is intended for graduate students and experts using methods of cluster analysis, particularly in mathematics, computer science, electrical engineering, medicine, biology, chemistry, physics, agriculture, civil engineering, economy, and social sciences.

We assume acquaintance with the basics of mathematical analysis, linear algebra, numerical mathematics, and computer programming. Recently, with development of *Big Data platforms* for managing massive amount of data and wide availability of tools for processing these data, the biggest limitation, as is often pointed out in the literature, is the lack of trained experts who are qualified to process and interpret the results. Precisely this is what this textbook is aiming at—to contribute in training of such specialists who will be qualified to work as data analysts or data scientists. According to many web portals, these are the professions which are going to be most sought after in the future.

Numerous illustrative examples are provided to help comprehend the presented material. In addition, every chapter contains exercises of various difficulties—from very simple ones to such that could be presented at seminars and similar projects. It is possible to work out many of these exercises using *Mathematica* or Matlab computation systems.

The content of this textbook, as well as the accompanying *Mathematica*-modules can also usefully serve in practical research.

At the end of the book, we provide a comprehensive bibliography, referencing numerous books and related articles published in scientific journals, giving a necessary overview of the most important and accurate insights into the field, which may serve as a basis for further independent research in this scientific area.

We would like to thank our colleagues who read parts of the manuscript and offered valuable comments, significantly improving this text: M. Briš–Alić,

D. Grahovac, S. Majstorović, R. Manger, T. Marošević, A. Morales-Esteban, I. Slapničar, I. Soldo, and M. Zekić–Sušac. We are also thankful to our colleagues P. Nikić, P. Taler, and J. Valenta for their technical support.

Osijek, Croatia and Sevilla, Spain
February, 2021

Rudolf Scitovski
Kristian Sabo
Francisco Martínez-Álvarez
Šime Ungar

Contents

Chapter 1
Introduction

Grouping data into meaningful clusters is a popular problem omnipresent in the recent scientific literature having practical importance in various applications, such as investigating earthquakes, climate changes (temperature fluctuations, water-levels, ...), in medicine (recognizing shapes, positions, and dimensions of human organs, grouping medical data, ...), in biology and bioinformatics, agriculture, civil engineering, robotics, computer simulations, managing energy resources, quality control, text analysis and classification, in social sciences, etc. Grouping and ranking objects has attracted the interest of not only researchers of different profiles but also decision-makers, such as government and public administration.

We will consider the problem of grouping a data set $\mathcal{A} = \{a_i \in \mathbb{R}^n : i = 1, \ldots, m\}$ complying with certain requirements, while clearly and precisely defining the basic concepts and notions. This is in the literature known as *cluster analysis*. Special attention will be paid to the analysis and implementation of the most important methods for searching for optimal partitions.

A considerable part of the book stems from the results obtained by the authors which were published during the past ten years in scientific journals, or were presented at various conferences.

Let us give a brief overview of the individual chapters.

In Chap. 2, *Representatives*, we introduce the notion of representatives of data sets with one or with several features. In particular, we consider centroids of data sets by applying the least squares distance-like function, and medians of data sets by applying the ℓ_1-metric. In addition, periodic data are analyzed using a particular distance-like function on the unit circle.

The problem of grouping weighted data with one or more features is considered in Chap. 3, *Data clustering*. We introduce the notion of optimal k-partition solving the global optimization problem (GOP)

© The Author(s), under exclusive license to Springer Nature Switzerland AG 2021
R. Scitovski et al., *Cluster Analysis and Applications*,
https://doi.org/10.1007/978-3-030-74552-3_1

$$\underset{\Pi \in \mathcal{P}(\mathcal{A};k)}{\arg\min} \ \mathcal{F}(\Pi), \qquad \mathcal{F}(\Pi) = \sum_{j=1}^{k} \sum_{a \in \pi_j} d(c_j, a),$$

where $\mathcal{P}(\mathcal{A}; k)$ denotes the set of all k-partitions of the set $\mathcal{A}$, with clusters $\pi_1, \ldots, \pi_k$, and their centers $c_1, \ldots, c_k$, and d is some distance-like function. Also, we introduce the minimal distance principle, Voronoi diagrams, and a simple form of the well known k-means algorithm. While discussing application of the least squares distance-like function, we analyze the corresponding dual problem, and while discussing the application of the ℓ_1-metric, we give a detailed analysis of the median of data without weights and the median of data with weights. At the end of this chapter, we consider the objective function

$$F(c_1, \ldots, c_k) = \sum_{i=1}^{m} \min_{1 \le j \le k} d(c_j, a_i).$$

Introducing an appropriate smooth approximation of this function, we prove that it is Lipschitz continuous. In addition, we also show the relationship between the objective functions $\mathcal{F}$ and F.

In Chap. 4, *Searching for an optimal partition*, we list some methods devised to search for optimal partitions. We introduce the objective function $\mathcal{F}$ using the membership matrix, construct the standard k-means algorithm, and prove its basic properties. Also, we progressively introduce the incremental algorithm by using the globally optimizing DIRECT algorithm. We also analyze the agglomerative hierarchical algorithm and the DBSCAN method.

In Chap. 5, *Indexes*, we analyze the choice of a partition with the most appropriate number of clusters using the Calinski–Harabasz, Davies–Bouldin and Dunn indexes, and the Silhouette Width Criterion. Furthermore, we consider the problem of comparing two different partitions using the Rand and Jaccard indexes and also using the Hausdorff distance. This issue comes up while testing various clustering methods on artificial data sets.

Chapter 6, *Mahalanobis data clustering*, deals with Mahalanobis data grouping. First, we introduce the notion of the total least squares line and the covariance matrix. Using covariance matrix of a data set, we define the Mahalanobis distance-like function in the plane and, generally, in $\mathbb{R}^n$. Next, we construct the Mahalanobis k-means algorithm and the Mahalanobis incremental algorithm for searching for an optimal partition with ellipsoidal clusters. In addition, all previously defined indexes are adapted to this case.

In Chap. 7, *Fuzzy clustering problem*, we consider fuzzy clustering. After defining the fuzzy membership matrix, we introduce the Gustafson–Kessel fuzzy c-means algorithm and the appropriate fuzzy incremental algorithm. Again, we adapt to this situation the previously introduced indexes for deciding on partitions with the most appropriate number of ellipsoidal clusters, and we also analyze some other fuzzy indexes as well as the fuzzy variant of the Rand index.

In Chap. 8, *Applications*, we analyze in more detail some applications which were previously investigated in our published papers and which originated from different real-world problems. First, we examine the multiple geometric objects detection problem applied to lines, circles, ellipses, and generalized circles. In particular, we consider the cases when the number of geometric objects is known in advance and when this number is not known in advance. We introduce a new efficient approach to find a partition with the most appropriate number of clusters, using parameters MinPts and ϵ from the DBSCAN algorithm. Furthermore, we examine the possibility to solve this problem using the RANSAC method. We discuss also some other applications, such as determining the seismic zones in an area, the temperature seasons according to temperature fluctuations, and determining the optimal constituencies.

In Chap. 9, *Modules*, we describe functions, algorithms, and data sets used in this textbook, coded as *Mathematica*-modules, and provide links to all these freely available *Mathematica*-modules.

Chapter 2
Representatives

In applied research, it is often necessary to represent a given set of data by a single datum which, in some sense, encompasses most of the features (properties) of the given set. The quantity commonly used is the well known arithmetic mean of the data. For example, a student's grade point average can be expressed by arithmetic mean, but it would not be appropriate to represent the average rate of economic growth during several years in such a way (see [167]).

In order to determine the best representative of a given set, first one has to decide how to measure the distance between points of the set. Of course, one could use some standard metric function, but various applications show (see e.g. [21, 96, 184]) that to measure the distance it is more useful to take a function which does not necessarily satisfy all the properties of a metric function (see [96, 182]).

Definition 2.1 A function $d\colon \mathbb{R}^n \times \mathbb{R}^n \to \mathbb{R}_+$, which satisfies[1]

(i) $d(x, y) = 0 \Leftrightarrow x = y$,
(ii) $x \mapsto d(x, y)$ is continuous on $\mathbb{R}^n$ for every fixed $y \in \mathbb{R}^n$,
(iii) $\lim\limits_{\|x\|\to\infty} d(x, y) = +\infty$ for every fixed $y \in \mathbb{R}^n$,

will be called a ***distance-like function***.

It is readily seen that every ℓ_p metric, $p \geq 1$, is a distance-like function, but an important example is the well known *least squares* (LS) *distance-like function* $d_{LS}(x, y) = \|x - y\|^2$, where $\| \ \|$ is the usual 2-norm.[2]

In general, a distance-like function is neither symmetric nor does it satisfy the triangle inequality. But, as shown by the following lemma, given a finite set of data

---

[1] We use the following notation: $\mathbb{R}_+ = \{x \in \mathbb{R} : x \geq 0\}$ and $\mathbb{R}_{++} = \{x \in \mathbb{R} : x > 0\}$.
[2] If there is no risk of misapprehension, throughout the text we are going to use $\| \ \|$ to denote the Euclidean, i. e. the 2-norm $\| \ \|_2$.

© The Author(s), under exclusive license to Springer Nature Switzerland AG 2021
R. Scitovski et al., *Cluster Analysis and Applications*,
https://doi.org/10.1007/978-3-030-74552-3_2

points[3] $\mathcal{A} = \{a_i = (a_i^1, \ldots, a_i^n) : i = 1, \ldots, m\} \subset \mathbb{R}^n$ with weights $w_1, \ldots, w_m > 0$, there exists a point $c^* \in \mathbb{R}^n$ such that the sum of its weighted d-distances to the points of $\mathcal{A}$ is minimal.

Lemma 2.2 *Let* $\mathcal{A} = \{a_i : i = 1, \ldots, m\} \subset \mathbb{R}^n$ *be a set of data points with weights* $w_1, \ldots, w_m > 0$, *let* $d \colon \mathbb{R}^n \times \mathbb{R}^n \to \mathbb{R}_+$ *be a distance-like function, and let* $F \colon \mathbb{R}^n \to \mathbb{R}_+$ *be the function given by*

$$F(x) = \sum_{i=1}^{m} w_i \, d(x, a_i) \,. \tag{2.1}$$

Then there exists a point $c^* \in \mathbb{R}^n$ *such that*

$$F(c^*) = \min_{x \in \mathbb{R}^n} F(x) \,. \tag{2.2}$$

Proof Since $F(x) \geq 0$, $x \in \mathbb{R}^n$, there exists $F^* := \inf_{x \in \mathbb{R}^n} F(x)$. Let (c_k) be some sequence in $\mathbb{R}^n$ such that $\lim_{k \to \infty} F(c_k) = F^*$. Let us show that the sequence (c_k) is bounded. In order to do this, assume the contrary, i.e. that there exists a subsequence (c_{k_ℓ}) such that $\|c_{k_\ell}\| \to +\infty$. Then, according to the properties (ii) and (iii) from Definition 2.1, it follows that $\lim_{\|c_{k_\ell}\| \to \infty} F(c_{k_\ell}) = +\infty$, contrary to $\lim_{\ell \to \infty} F(c_{k_\ell}) = F^*$. Finally, the sequence (c_k), being bounded, has a convergent subsequence (c_{k_j}), and let c^* be its limit point. Then $F(c^*) = F(\lim_{j \to \infty} c_{k_j}) = \lim_{j \to \infty} F(c_{k_j}) = \lim_{k \to \infty} F(c_k) = F^*$, showing that (2.2) holds true. $\square$

Remark 2.3 Note that for a global minimum point $c^* \in \mathbb{R}^n$ and all $x \in \mathbb{R}^n$,

$$F(x) = \sum_{i=1}^{m} w_i \, d(x, a_i) \geq \sum_{i=1}^{m} w_i \, d(c^*, a_i) = F(c^*) \,, \tag{2.3}$$

and the equality holds true if and only if $x = c^*$, or some other point satisfying (2.4).

Lemma 2.2 enables the following definition:

Definition 2.4 Let $d \colon \mathbb{R}^n \times \mathbb{R}^n \to \mathbb{R}_+$ be a distance-like function. A best representative of the set $\mathcal{A}$ with weights $w_1, \ldots, w_m > 0$, with respect to the distance-like function d, is any point

---

[3] We are going to use lower indexes for elements $a_i \in \mathbb{R}^n$, and the upper indexes for the coordinates of elements in $\mathbb{R}^n$.

$$c^\star \in \underset{x \in \mathbb{R}^n}{\arg\min} \sum_{i=1}^{m} w_i \, d(x, a_i) \, . \qquad (2.4)$$

The notation (2.4) suggests that the best representative might not be unique, i.e. there may exist more best representatives of the set $\mathcal{A}$.

In this chapter we shall consider the two most commonly used representatives of a data set—the *arithmetic mean* and the *median*.

2.1 Representative of Data Sets with One Feature

A set of data without weights and with a single feature is usually interpreted as a finite subset $\mathcal{A} = \{a_1, \ldots, a_m\}$ of real numbers $a_i \in \mathbb{R}, i = 1, \ldots, m$.

The two most frequently used distance-like functions on $\mathbb{R}$ are the *LS distance-like function* and ℓ_1 metric function, also known as the *Manhattan* or *taxicab metric function* (see e.g. [46, 52, 94, 149])

$$d_{LS}(x, y) = (x - y)^2, \qquad \text{[LS distance-like function]}$$

$$d_1(x, y) = |x - y| \, . \qquad \text{[ℓ_1 metric function]}$$

Exercise 2.5 Check whether

$$d_1(x, y) = d_2(x, y) = d_\infty(x, y) = d_p(x, y), \quad p \geq 1, \quad x, y \in \mathbb{R},$$

holds, where d_p is the p-metric on $\mathbb{R}$ (see e.g. [184]).

Exercise 2.6 Show that the function d_{LS} is not a metric function on $\mathbb{R}$, but the function d_1 is a metric on $\mathbb{R}$.

2.1.1 Best LS-Representative

For data without weights and the LS distance-like function, the function (2.1) becomes

$$F_{LS}(x) := \sum_{i=1}^{m} (x - a_i)^2 \, , \qquad (2.5)$$

and because it is a convex function and $F_{LS}'(c_{LS}^\star) = 0$ and $F_{LS}''(x) = 2m > 0$ for all $x \in \mathbb{R}$, it attains its global minimum at the unique point

$$c_{LS}^{\star} = \arg\min_{x \in \mathbb{R}} \sum_{i=1}^{m} d_{LS}(x, a_i) = \frac{1}{m} \sum_{i=1}^{m} a_i . \qquad (2.6)$$

Hence, the best LS-representative of the set $\mathcal{A} \subset \mathbb{R}$ is the ordinary *arithmetic mean*[4], and it has the property (cf. Remark 2.3) that the sum of squared deviations to the given data is minimal:

$$\sum_{i=1}^{m} (x - a_i)^2 \geq \sum_{i=1}^{m} (c_{LS}^{\star} - a_i)^2, \qquad (2.7)$$

where the equality holds true for $x = c_{LS}^{\star}$.

As a measure of dispersion of the data set $\mathcal{A}$ around the arithmetic mean $c_{LS}^{\star}$, in statistics literature [24, 141] one uses the *variance of data*

$$s^2 = \frac{1}{m-1} \sum_{i=1}^{m} (c_{LS}^{\star} - a_i)^2, \quad m > 1. \qquad (2.8)$$

The number s is called the *standard deviation*.

Example 2.7 Given the set $\mathcal{A} = \{2, 1.5, 2, 2.5, 5\}$, its arithmetic mean is $c_{LS}^{\star} = 2.6$.

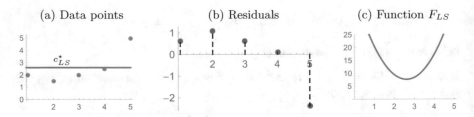

Fig. 2.1 Arithmetic mean of the set $\mathcal{A} = \{2, 1.5, 2, 2.5, 5\}$[5]

Figure 2.1a shows the data and the arithmetic mean $c_{LS}^{\star}$, Fig. 2.1b shows the so-called *residuals* (the numbers $c_{LS}^{\star} - a_i$), and Fig. 2.1c depicts the graph of the

---

[4]The problem of finding the best LS-representative of a data set occurs in the literature as the *least squares principle*, which was proposed in 1795 by German mathematician Carl Friedrich Gauss (1777–1855) while investigating the movements of celestial bodies, published in *Teoria Motus Corporum Coelestium in Sectionibus Conicis Solem Ambientium*, Perthes and Besser, Hamburg, 1809. One should also mention that in 1805 French mathematician Adrien-Marie Legendre (1752–1833) was the first one to publish an algebraic procedure for the least squares method.

[5]The set $\mathcal{A} = \{2, 1.5, 2, 2.5, 5\}$ is in fact a multiset, see Remark 3.6, and Fig. 2.1 actually depicts the set $\widetilde{\mathcal{A}} = \{(1, 2), (2, 1.5), (3, 2), (4, 2.5), (5, 5)\}$.

function F_{LS}. Note that the graph is a parabola and $F_{LS}(c^\star_{LS}) = 7.7$. What is the variance and what is the standard deviation of this set?

What would happen if there were an *outlier* (strongly jutting datum) among the data? How would it effect the best LS-representative (arithmetic mean) of the set $\mathcal{A}$? What would be the result if $a_5 = 5$ were changed to $a_5 = 10$?

Exercise 2.8 Let $c^\star_{LS}$ be the arithmetic mean of the set $\mathcal{A} = \{a_1, \ldots, a_m\} \subset \mathbb{R}$ without weights. Show that

$$\sum_{i=1}^{m} (c^\star_{LS} - a_i) = 0.$$

Check this property for data in Example 2.7.

Exercise 2.9 Let $\mathcal{A} = \{a_1, \ldots, a_p\}$ and $\mathcal{B} = \{b_1, \ldots, b_q\}$ be two disjoint sets in $\mathbb{R}$, and let $a^\star_{LS}$ and $b^\star_{LS}$ be their arithmetic means. Show that the arithmetic mean of the union $\mathcal{C} = \mathcal{A} \cup \mathcal{B}$ equals

$$c^\star_{LS} = \frac{p}{p+q} a^\star_{LS} + \frac{q}{p+q} b^\star_{LS}.$$

Check the formula in several examples. What would the generalization of this formula for r mutually disjoint sets $\mathcal{A}_1, \ldots, \mathcal{A}_r$ containing $p_1, \ldots, p_r$ elements respectively look like?

2.1.2 Best ℓ_1-Representative

For data without weights and the ℓ_1 metric, the function (2.1) becomes

$$F_1(x) := \sum_{i=1}^{m} |x - a_i|. \tag{2.9}$$

The next lemma shows that if $\mathcal{A}$ is a set of distinct real numbers, the function F_1 attains its global minimum at the median of $\mathcal{A}$ (see e.g. [144, 184]). The case when some data may be equal will be considered in Sect. 2.1.3.

Lemma 2.10 *Let $\mathcal{A} = \{a_i \in \mathbb{R} : i = 1, \ldots, m\}$ be a set of distinct data points. The function F_1 given by (2.9) attains its global minimum at the median of the set $\mathcal{A}$.*

Proof Without loss of generality, we may assume that $a_1 < a_2 < \cdots < a_m$. Note that F_1 is a convex piecewise linear function (see Fig. 2.2c) and therefore it can attain its global minimum at a single point in $\mathcal{A}$ or at all points between two points in $\mathcal{A}$.

For $x \in (a_k, a_{k+1})$ we have

$$F_1(x) = \sum_{i=1}^{k}(x - a_i) - \sum_{i=k+1}^{m}(x - a_i) = (2k - m)x - \sum_{i=1}^{k} a_i + \sum_{i=k+1}^{m} a_i,$$

$$F_1'(x) = 2k - m.$$

Thus, the function F_1 decreases on intervals (a_k, a_{k+1}) for $k < \frac{m}{2}$ and increases for $k > \frac{m}{2}$.

Therefore, we have to consider two cases:

- If m is odd, i.e. $m = 2p + 1$, the function F_1 attains its global minimum at the middle datum a_p;
- If m is even, i.e. $m = 2p$, the function F_1 attains its global minimum at every point of the interval $[a_p, a_{p+1}]$.

Hence, the best ℓ_1-representative of the set $\mathcal{A} \subset \mathbb{R}$ is the *median* of $\mathcal{A}$.   □

Note that the median of a set $\mathcal{A}$ may be either a set (a segment of real numbers) or a single real number. If the median of $\mathcal{A}$ is a set it will be denoted by Med $\mathcal{A}$ and its elements by med $\mathcal{A}$. The number med $\mathcal{A}$ has the property (cf. Remark 2.3) that the sum of its absolute deviations to all data is minimal, i.e.

$$\sum_{i=1}^{m} |x - a_i| \geq \sum_{i=1}^{m} |\text{med}\,\mathcal{A} - a_i|,$$

and the equality holds true if and only if $x = \text{med}\,\mathcal{A}$.[6]

Example 2.11 Given the data set $\mathcal{A} = \{2, 1.5, 2, 2.5, 5\}$, its median is med $\mathcal{A} = 2$. What is the sum of absolute deviations?

Figure 2.2a shows the data and the median $c_1^\star$, Fig. 2.2b shows the residuals (numbers $c_1^\star - a_i$), and Fig. 2.2c depicts the graph of the function F_1. Note that F_1 is a convex piecewise linear function and that $F_1(c_1^\star) = 4$.

How would the median of this set change if data contained an outlier? What would be the median if the datum $a_5 = 5$ were replaced by $a_5 = 10$, and what if it were replaced by $a_5 = 100$?

---

[6]The problem of finding the best ℓ_1-representative of a data set appears in the literature as the *least absolute deviations principle*, LAD, ascribed to Croatian scholar Josip Ruđer Bošković (1711–1787), who posed it in 1757 in his article [25]. Due to complicated calculations, for a long time this principle was neglected in comparison to the Gauss least squares principle. Not until modern computers came about did this take an important place in scientific research, in particular because of its robustness: In contrast to the Gauss least squares principle, this principle ignores the outliers (strongly jutting data) in data sets. Scientific conferences devoted to ℓ_1 methods and applications are regularly held in Swiss city Neuchâtel, and the front page of the conference proceedings shows the Croatian banknote depicting the portrait of Josip Ruđer Bošković [43].

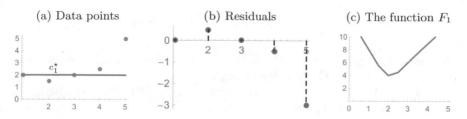

Fig. 2.2 Median of the set $\mathcal{A} = \{2, 1.5, 2, 2.5, 5\}$[7]

To find the median of a set $\mathcal{A}$, first one has to sort the elements. Then, if the number of elements is odd, median is the middle element, and if the number of elements is even, median is any number between the two middle elements. For example,[8]

$$\mathrm{Med}\{3,\ 1,\ 4,\ 5,\ 9\} = \{4\},$$

$$\mathrm{Med}\{-1,\ 1,\ -2,\ 2,\ -5,\ 5,\ -9,\ 9\} = [-1, 1],$$

but med$\{3,\ 1,\ 4,\ 5,\ 9\} = 4$ and med$\{-1,\ 1,\ -2,\ 2,\ -5,\ 5,\ -9,\ 9\} \in [-1, 1]$.

Remark 2.12 Note that the median of a data set $\mathcal{A}$ can always be chosen among the elements of $\mathcal{A}$ itself. This means that median, as the best ℓ_1 representative of a set, can always be an element of that set, contrary to the case of the arithmetic mean as the best LS-representative. In some applications this fact might be useful.

Note also that a half of elements of the set $\mathcal{A}$ are placed to the left, and the other half to the right of the median of $\mathcal{A}$ (see e.g. [153, 187]).

As a measure of dispersion of a data set $\mathcal{A}$ around the median, in statistics literature [140, 141] one uses the *Median of Absolute Deviations from Median* (MAD):

$$\mathrm{MAD}\,\mathcal{A} = 1.483 \operatorname*{med}_{i=1,\ldots,m} |a_i - \operatorname*{med}_{j=1,\ldots,m} a_j|. \tag{2.10}$$

Example 2.13 The relative magnitudes of elements of the set

$$\mathcal{A} = \{9.05,\ 2.83,\ 3.00,\ 3.16,\ 4.12,\ 3.00,\ 3.50\}$$

can be better compared after mapping the set $\mathcal{A}$ to the unit interval $[0, 1]$ using the linear map

---

[7]See the footnote on page 8 and Remark 3.6.

[8]The median of a set can be obtained using *Mathematica* instruction `Median[]`. If the median of the given set happens to be an interval, the instruction `Median[]` will give the midpoint of that interval.

$$\varphi(x) = \frac{x - a}{b - a}, \qquad \text{where } a = \min \mathcal{A}, \ b = \max \mathcal{A}. \tag{2.11}$$

We get $\varphi(\mathcal{A}) = \{1., 0., 0.027, 0.053, 0.207, 0.027, 0.108\}$, and it is readily seen that $a_1 \in \mathcal{A}$ is by far the largest element in $\mathcal{A}$.

Following [140], this can be ascertained more exactly by first using (2.10) to find MAD $= 0.489$ and define the new set

$$\widetilde{\mathcal{A}} = \{\tilde{a}_i = |a_i - \underset{j=1,\ldots,m}{\mathrm{med}} \ a_j| / \mathrm{MAD} : a_i \in \mathcal{A}\}$$

$$= \{12.04, \ 0.67, \ 0.33, \ 0, \ 1.96, \ 0.33, \ 0.69\}.$$

The element $a_i \in \mathcal{A}$ for which $\tilde{a}_i > 2.5$ is considered, according to [140], to be an outlier. So, in our example, only the element $a_1 = 9.05$ is an outlier in $\mathcal{A}$.

In statistics literature [24, 141], median of a set $\mathcal{A}$ is related to the *first quartile* (the element of $\mathcal{A}$ placed at ¼ of the sorted data) and the *third quartile* (the element of $\mathcal{A}$ placed at ¾ of the sorted data). What are the first and third quartiles of the data set in previous example?

2.1.3 Best Representative of Weighted Data

In practical applications, it is sometimes necessary to equip the data with some weights. In this way, we associate with each datum its impact or the frequency of occurrence. For example, to find the student's average grade in the exams he passed, the data set is $\{2, 3, 4, 5\}$ and weights are the frequencies of occurrence of each grade.[9]

As with data without weights, one can prove that the function

$$F_{LS}(x) = \sum_{i=1}^{m} w_i (x - a_i)^2$$

attains its global minimum at the unique point

$$c_{LS}^{\star} = \underset{x \in \mathbb{R}}{\arg\min} \sum_{i=1}^{m} w_i \, d_{LS}(x, a_i) = \frac{1}{W} \sum_{i=1}^{m} w_i \, a_i, \quad W = \sum_{i=1}^{m} w_i \,,$$

which we call the *weighted arithmetic mean* [96, 144].

In case of ℓ_1 metric function, the function (2.1) looks like

---

[9]In Croatia at all levels of education the grading scale runs from 1 (insufficient) to 5 (excellent).

$$F_1(x) = \sum_{i=1}^{m} w_i \, |x - a_i|, \tag{2.12}$$

and it attains its global minimum at the *weighted median* $\text{Med}_i(w_i, a_i)$ of the set $\mathcal{A}$, as shown by the following lemma.

Lemma 2.14 ([144]) *Let $a_1 < \cdots < a_m$ be a set of data points with weights $w_1, \ldots, w_m > 0$, and let $I = \{1, \ldots, m\}$. Denote*

$$J := \{v \in I : \sum_{i=1}^{v} w_i \leq \sum_{i=v+1}^{m} w_i\},$$

and for $J \neq \emptyset$, denote $v_0 = \max J$. Then:

(i) *If $J = \emptyset$, (i. e. $w_1 > \sum\limits_{i=2}^{m} w_i$), then the minimum of F_1 is attained at the point $\alpha^{\star} = a_1$.*

(ii) *If $J \neq \emptyset$ and $\sum\limits_{i=1}^{v_0} w_i < \sum\limits_{i=v_0+1}^{m} w_i$, then the minimum of F_1 is attained at the point $\alpha^{\star} = a_{v_0+1}$.*

(iii) *If $J \neq \emptyset$ and $\sum\limits_{i=1}^{v_0} w_i = \sum\limits_{i=v_0+1}^{m} w_i$, then the minimum of F_1 is attained at every point $\alpha^{\star}$ in the segment $[a_{v_0}, a_{v_0+1}]$.*

Proof Notice that on each interval

$$(-\infty, a_1), \ [a_1, a_2), \ldots, [a_{m-1}, a_m), \ [a_m, \infty)$$

F is a linear function with slopes of these linear functions being consecutively d_v, $v = 0, \ldots, m$, where

$$d_0 = -\sum_{i=1}^{m} w_i,$$

$$d_v = \sum_{i=1}^{v} w_i - \sum_{i=v+1}^{m} w_i = d_{v-1} + w_v + w_{v+1}, \quad v = 1, \ldots, m-1,$$

$$d_m = \sum_{i=1}^{m} w_i.$$

If $J = \emptyset$, then $2\sum_{i=1}^{v} w_i - \sum_{i=1}^{m} w_i > 0$ for every $v = 1, \ldots, m$, and $d_0 < 0 < d_v, v = 1, \ldots, m$. It follows that the function F_1 is strongly decreasing on $(-\infty, a_1)$ and strongly increasing on $(a_1, +\infty)$. Therefore the minimum of F_1 is attained for $\alpha^{\star} = a_1$.

If $J \neq \emptyset$, note that $v_0 = \max\{v \in I : d_v \leq 0\}$. Since $d_{v+1} - d_v = 2w_{v+1} > 0$, $d_0 < 0$, and $d_m > 0$, the sequence (d_v) is increasing and satisfies

$$d_0 < d_1 \ldots < d_{v_0} \leq 0 < d_{v_0+1} < \ldots < d_m. \qquad (2.13)$$

If $d_{v_0} < 0$, i.e. $2\sum_{i=1}^{v_0} \omega_i < \sum_{i=1}^{m} \omega_i$, from (2.13) it follows that F_1 is strongly decreasing on $(-\infty, a_{v_0+1})$ and strongly increasing on $(a_{v_0+1}, +\infty)$. Therefore the minimum of F_1 is attained for $\alpha^* = a_{v_0+1}$.

If $d_{v_0} = 0$, i.e. $2\sum_{i=1}^{v_0} w_i = \sum_{i=1}^{m} w_i$, from (2.13) it follows that F_1 is strongly decreasing on $(-\infty, a_{v_0})$, constant on $[a_{v_0}, a_{v_0+1}]$, and strongly increasing on $(a_{v_0+1}, +\infty)$. Therefore the minimum of F_1 is attained at every point α^* in the segment $[a_{v_0}, a_{v_0+1}]$.   □

Hence, the best ℓ_1-representative of a weighted data set is the *weighted median* $\underset{i}{\mathrm{Med}}(w_i, a_i)$. Note that weighted median can be a set (a segment of real numbers) or a single real number. Weighted median $\underset{i}{\mathrm{med}}(w_i, a_i)$ is any number with the property that the sum of weighted absolute deviations to all data is minimal, i.e.

$$\sum_{i=1}^{m} w_i \, |x - a_i| \geq \sum_{i=1}^{m} w_i \, |\underset{j}{\mathrm{med}}(w_j, a_j) - a_i|,$$

and the equality holds true for $x = \underset{j}{\mathrm{med}}(w_j, a_j)$ (see also Remark 2.3).

The next corollary shows that Lemma 2.10 is just a special case of Lemma 2.14.

Corollary 2.15 *Let $a_1 < a_2 < \cdots < a_m$, $m > 1$, be a set of data points with weights $w_1 = \cdots = w_m = 1$. Then:*

(i) *If m is odd ($m = 2k + 1$), then the minimum of the function F_1 is attained at the point $\alpha^* = a_{k+1}$;*

(ii) *If m is even ($m = 2k$), the minimum of the function F_1 is attained at every point α^* of the segment $[a_k, a_{k+1}]$.*

Proof First, note that in this case the set J from Lemma 2.14 is always nonempty.

Let $m = 2k + 1$. According to Lemma 2.14(*ii*),

$$v_0 = \max\{v \in I : 2v - m \leq 0\} = \max\{v \in I : v \leq k + \tfrac{1}{2}\} = k,$$
$$d_{v_0} = d_k = 2k - m = 2k - 2k - 1 < 0,$$

and therefore $\alpha^* = a_{k+1}$.

Let $m = 2k$. According to Lemma 2.14(*iii*),

$$v_0 = \max\{v \in I : 2v - m \leq 0\} = \max\{v \in I : v = k\} = k,$$
$$d_{v_0} = d_k = 2k - m = 2k - 2k = 0.$$

It follows that the minimum of the function F_1 is attained at every point α^* of the segment $[a_k, a_{k+1}]$.   □

In general, determining the weighted median is a very complicated numerical procedure. For this purpose, there are numerous algorithms in the literature [72].

Example 2.16 Weighted median of a set $\mathcal{A} \subset \mathbb{R}$ with weights being positive integers can be determined similarly as for median of a set without weights. First we sort the elements of the set $\mathcal{A}$. Next we form the *multiset*[10] where each element of $\mathcal{A}$ is repeated according to its weight, and then we take the middle element of that multiset. A set $\mathcal{A} = \{a_1, \ldots, a_m\}$ with weights $\mathrm{w} := \{w_1, \ldots, w_m\}$ will be called a *weighted set*, and its median will be denoted by $\mathrm{med}_\mathrm{w} \mathcal{A}$. For example, the weighted median of the set $\mathcal{A} = \{3, 1, 4, 5, 9\}$ with weights $\mathrm{w} = \{3, 1, 3, 2, 2\}$ is the middle element of the multiset $\{\{1, 3, 3, 3, 4, 4, 4, 5, 5, 9, 9\}\}$.[10]

In our example, the weighted median of the weighted set $\mathcal{A}$ with weights w is $\mathrm{med}_\mathrm{w} \mathcal{A} = 4$. What is the first and third quartile of the weighted set $\mathcal{A}$?

2.1.4   Bregman Divergences

Let us consider yet another class of distance-like functions which is important in applications. Following [96, 182] we introduce the following definition.

Definition 2.17 Let $D \subseteq \mathbb{R}$ be a convex set (i. e. an interval) and let $\phi \colon D \to \mathbb{R}_+$ be a strictly convex continuously differentiable function on $\mathrm{int}\, D \neq \emptyset$. The function $d_\phi \colon D \times \mathrm{int}\, D \to \mathbb{R}_+$ defined by

$$d_\phi(x, y) := \phi(x) - \phi(y) - \phi'(y)(x - y) \qquad (2.14)$$

is called the **Bregman divergence**.

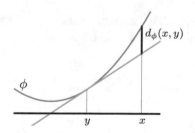

Fig. 2.3 Geometric meaning of Bregman divergence

It is not difficult to see that such a function is indeed a distance-like function. Geometrically, for a given $x \in D$, $d_\phi(x, y)$ represents the difference between the

---

[10]See Remark 3.6.

value $\phi(x)$ and the value at x of the linear function whose graph is the tangent line at the point $(y, \phi(y))$ (see Fig. 2.3). In the past 10 years, distance-like functions of this kind have been intensely investigated and applied in operations research, information theory, nonlinear analysis, machine learning, wireless sensor network, etc. (see e.g. [101, 136]).

Exercise 2.18 Show that for $\phi(x) := x^2$, $x \in \mathbb{R}$, the Bregman divergence becomes the LS distance-like function.

Example 2.19 Bregman divergence for $\phi\colon \mathbb{R}_{++} \to \mathbb{R}_+$, $\phi(x) := -\ln x$, is known as the *Itakura–Saito divergence* given by

$$d_{IS}(x, y) = \tfrac{x}{y} - \ln \tfrac{x}{y} - 1. \tag{2.15}$$

Let us find the corresponding best representative of the set $\mathcal{A} = \{a_i \in \mathbb{R}_{++} : i = 1, \ldots, m\}$. In this case, the function (2.1) becomes

$$F(x) = \sum_{i=1}^{m} d_{IS}(x, a_i) = \sum_{i=1}^{m} \left(\tfrac{x}{a_i} - \ln \tfrac{x}{a_i} - 1 \right).$$

Since $F'(x) = \sum_{i=1}^{m} \tfrac{1}{a_i} - \tfrac{m}{x}$, the point $c_{IS}^\star = m \left(\sum_{i=1}^{m} \tfrac{1}{a_i} \right)^{-1}$ is the unique stationary point, and since $F''(x) = \tfrac{m}{x^2} > 0$, F is a convex function and $c_{IS}^\star$ is its only point of global minimum. Note that $c_{IS}^\star$ is the harmonic mean of the set $\mathcal{A}$.

Example 2.20 With the convention $0 \cdot \ln 0 = 0$, Bregman divergence for $\phi\colon \mathbb{R}_+ \to \mathbb{R}_+$, $\phi(x) := x \ln x$, is known as the *Kullback–Leibler divergence* given by

$$d_\phi(x, y) = x \ln \tfrac{x}{y} - x + y. \tag{2.16}$$

Let us find the corresponding best representative of the set $\mathcal{A} = \{a_i \in \mathbb{R}_+ : i = 1, \ldots, m\}$. In this case, the function (2.1) becomes

$$F(x) = \sum_{i=1}^{m} d_{KL}(x, a_i) = \sum_{i=1}^{m} \left(x \ln \tfrac{x}{a_i} - x + a_i \right).$$

Since $F'(x) = \sum_{i=1}^{m} \ln \tfrac{x}{a_i}$, the point $c_{KL}^\star = \sqrt[m]{\prod_{i=1}^{m} a_i}$ is the unique stationary point, and since $F''(x) = \tfrac{m}{x} > 0$ for all $x \in \mathbb{R}_{++}$, F is a convex function, and the point $c_{KL}^\star$ is its only point of global minimum. Note that $c_{KL}^\star$ is the geometric mean of the set $\mathcal{A}$.

Note that, in general, the Bregman divergence $d_\phi\colon D \times \text{int}\, D \to \mathbb{R}_+$ is not a symmetric function. It is not difficult to see that the function $D_\phi\colon \text{int}\, D \times D \to \mathbb{R}_+$ defined by $D_\phi(x, y) := d_\phi(y, x)$ is also a distance-like function.

Assume that the function ϕ is twice continuously differentiable and find the representative of the data set $\mathcal{A} = \{a_i \in D : i = 1, \ldots, m\}$ using the distance-like function D_ϕ. This boils down to finding the global maximum of the function

$$F(x) = \sum_{i=1}^{m} D_\phi(x, a_i) = \sum_{i=1}^{m} d_\phi(a_i, x) = \sum_{i=1}^{m} \big(\phi(a_i) - \phi(x) - \phi'(x)(a_i - x)\big).$$

Since the function ϕ is strictly convex, and $F'(x) = \phi''(x) \sum_{i=1}^{m} (x - a_i) = 0$, the corresponding representative $c^\star$ is the mean$(\mathcal{A}) = \frac{1}{m} \sum_{i=1}^{m} a_i$, and it is independent of the choice of the function ϕ.

The above distance-like functions can be generalized to data sets with corresponding weights $w_i > 0$ (see e.g. [96, 182]).

Exercise 2.21 Let $\mathcal{A} = \{a_i \in \mathbb{R}_{++} : i = 1, \ldots, m\}$ be a data set with corresponding weights $w_i > 0$. Show that using the Itakura–Saito divergence, the weighted center of the set $\mathcal{A}$ is

$$c_{IS} = W \left(\sum_{i=1}^{m} w_i \frac{1}{a_i} \right)^{-1}, \quad W = \sum_{i=1}^{m} w_i.$$

Exercise 2.22 Let $\mathcal{A} = \{a_i \in \mathbb{R}_{++} : i = 1, \ldots, m\}$ be a data set with corresponding weights $w_i > 0$. Show that using the Kullback–Leibler divergence, the weighted center of the set $\mathcal{A}$ is

$$c_{KL} = \left(\prod_{i=1}^{m} a_i^{w_i} \right)^{\frac{1}{W}}, \quad W = \sum_{i=1}^{m} w_i.$$

Remark 2.23 The above distance-like functions can be generalized to data sets with $n \geq 1$ features (see e.g. [96, 182]). Let $D \subseteq \mathbb{R}^n$ be a convex set and let $\phi : D \to \mathbb{R}_+$ be a strictly convex continuously differentiable function on int $D \neq \emptyset$. The function $d_\phi : D \times \text{int } D \to \mathbb{R}_+$ defined by

$$d_\phi(x, y) = \phi(x) - \phi(y) - \nabla\phi(y)(x - y)^T$$

is called the *Bregman divergence*.

2.2   Representative of Data Sets with Two Features

A set with two features without weights is usually interpreted as a finite set $\mathcal{A} = \{a_i = (x_i, y_i) : i = 1, \ldots, m\} \subset \mathbb{R}^2$, and geometrically it can be visualized as a finite set of points in the plane. In the next section, we give a short historical overview of looking for a best representative of a data set with two features and possible applications.

2.2.1 Fermat–Torricelli–Weber Problem

Let $A, B, C \in \mathbb{R}^2$ be three non-collinear points in the plane (see Fig. 2.4).

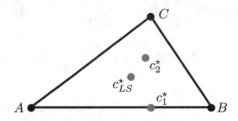

Fig. 2.4 Fermat's problem

The *Fermat's problem* consists in finding the point $c_2^\star \in \mathbb{R}^2$ with the property that the sum of its Euclidean, i. e. ℓ_2-distances to the vertices of the triangle $\triangle ABC$ is minimal. The point $c_2^\star$ is called the *geometric median* of the points A, B, C, and it can be obtained (see e.g. [113]) as the intersection of the so-called Simpson's lines (see Fig. 2.5a) or as the intersection of the so-called Torricelli's circles (see Fig. 2.5b). The same problem can also be treated for a different distance-like functions: in the sense of physics—the *Torricelli's problem*, and in the sense of econometrics—the *Weber's problem* (see e.g. [46]).

The point $c_{LS}^\star \in \mathbb{R}^2$ (see Fig. 2.4), with the property that the sum of its LS-distances (i.e. the sum of squared Euclidean distances) to vertices of the triangle $\triangle ABC$ is minimal is called the *centroid* or the *Steiner point* (this is related to the center of mass in physics), and it is obtained as the intersection of medians of the triangle, i. e. line segments joining vertices to the midpoints of opposite sides.

The point $c_1^\star \in \mathbb{R}^2$ (see Fig. 2.4), with the property that the sum of its ℓ_1-distances to the vertices of the triangle $\triangle ABC$ is minimal is called the *median* of the set $\{A, B, C\}$.

In general, one can consider a finite set of points in $\mathbb{R}^n$ and an arbitrary distance-like function d (see e.g. [153]). The problem of finding best d-representative has many applications in various fields: telecommunication (optimal antenna coverage problem, discrete network location), public sector (optimal covering problem), economy (optimal location of consumer centers), hub location problems, robotics, optimal assignation problems, hourly forecast of natural gas consumption problem, etc. [46, 115, 150].

Exercise 2.24 Given the triangle $\triangle ABC$ with vertices $A = (0, 0)$, $B = (6, 0)$, and $C = (4, 3)$, find the vertices A_1, B_1, C_1 of the equilateral triangles constructed on the sides of triangle $\triangle ABC$, and find the intersection of line segments joining the points A–A_1, B–B_1, and C–C_1 as in Fig. 2.5a.

Solution: $A_1 = (7.598, 3.232)$, $B_1 = (-0.598, 4.964)$, $C_1 = (3., -5.196)$; Geometric median: $c_2^\star = (3.833, 1.630)$.

Exercise 2.25 Given the triangle $\triangle ABC$ with vertices $A = (0, 0)$, $B = (6, 0)$, and $C = (4, 3)$, find the vertices A_1, B_1, C_1 of the equilateral triangles constructed on the sides of triangle $\triangle ABC$, construct the circumcircles of these triangles, and find the intersection of these circles as in Fig. 2.5b.

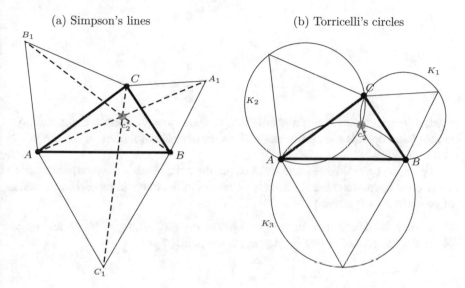

(a) Simpson's lines   (b) Torricelli's circles

Fig. 2.5 Fermat's problem

Solution: $A_1 = (7.598, 3.232)$, $B_1 = (-0.598, 4.964)$, $C_1 = (3., -5.196)$; $K_1 = ((5.866, 2.077), 2.082)$; $K_2 = ((1.134, 2.655), 2.887)$; $K_3 = ((3., -1.732), 3.464)$; Geometric median: $c_2^\star = (3.833, 1.630)$.

2.2.2 Centroid of a Set in the Plane

Let $\mathcal{A} = \{a_i = (x_i, y_i) : i = 1, \ldots, m\} \subset \mathbb{R}^2$ be a set without weights in the plane. The *centroid* $c_{LS}^\star$ of the set $\mathcal{A}$ is the solution to the optimization problem

$$\underset{c \in \mathbb{R}^2}{\arg \min} \sum_{i=1}^{m} d_{LS}(c, a_i), \qquad (2.17)$$

where $d_{LS}(a, b) = d_2^2(a, b) = \|a - b\|^2$. The point $c_{LS}^\star$ is the point at which the function

$$F_{LS}(x, y) = \sum_{i=1}^{m} \|c - a_i\|^2 = \sum_{i=1}^{m} \left((x - x_i)^2 + (y - y_i)^2 \right), \quad c = (x, y)$$

attains its global minimum. $F_{LS}(x, y)$ is the sum of squared Euclidian, i.e. ℓ_2-distances from the point $c = (x, y)$ to the points $a_i \in \mathcal{A}$. From (2.7) it follows that

$$F_{LS}(x, y) = \sum_{i=1}^{m} \left((x - x_i)^2 + (y - y_i)^2 \right) \geq \sum_{i=1}^{m} (\bar{x} - x_i)^2 + \sum_{i=1}^{m} (\bar{y} - y_i)^2, \quad (2.18)$$

where

$$\bar{x} = \frac{1}{m} \sum_{i=1}^{m} x_i , \qquad \bar{y} = \frac{1}{m} \sum_{i=1}^{m} y_i ,$$

and the equality holds true if and only if $x = \bar{x}$ and $y = \bar{y}$. Therefore, the solution to the global optimization problem (2.17) is the centroid of the set $\mathcal{A}$, i.e. the point $c_{LS}^\star = (\bar{x}, \bar{y})$.

Hence, the centroid of a finite set $\mathcal{A}$ of points in the plane is the point whose first and second coordinates are the arithmetic means of the first and second coordinates of points in $\mathcal{A}$, respectively.

Example 2.26 Check that for the given points $a_1 = (0, 0)$, $a_2 = (6, 0)$, and $a_3 = (4, 3)$, the centroid of the set $\{a_1, a_2, a_3\}$ is the point $c_{LS}^\star = (\frac{10}{3}, 1)$.

2.2.3 Median of a Set in the Plane

Median of a set of points $\mathcal{A} = \{a_i = (x_i, y_i) : i = 1, \ldots, m\} \subset \mathbb{R}^2$ without weights is a solution to the optimization problem

$$\arg\min_{c \in \mathbb{R}^2} \sum_{i=1}^{m} d_1(c, a_i) . \quad (2.19)$$

This is every point at which the function

$$F_1(x, y) = \sum_{i=1}^{m} \|c - a_i\|_1 = \sum_{i=1}^{m} (|x - x_i| + |y - y_i|), \quad c = (x, y)$$

attains the global minimum. $F_1(x, y)$ is the sum of ℓ_1-distances from $c = (x, y)$ to the points $a_i \in \mathcal{A}$. From (2.9) it follows that

$$F_1(x, y) = \sum_{i=1}^{m} (|x - x_i| + |y - y_i|) \geq \sum_{i=1}^{m} |\operatorname*{med}_k x_k - x_i| + \sum_{i=1}^{m} |\operatorname*{med}_k y_k - y_i| ,$$

$$(2.20)$$

and the equality holds true if and only if $x = \operatorname*{med}_{k} x_k$ i $y = \operatorname*{med}_{k} y_k$. Therefore, the solution to the global optimization problem (2.19) is a median of the set $\mathcal{A}$, which is a point

$$(\operatorname*{med}_{k} x_k, \operatorname*{med}_{k} y_k). \tag{2.21}$$

Hence, the median of a finite set $\mathcal{A}$ of points in the plane is any point whose first and second coordinates are medians of the first and second coordinates of points in $\mathcal{A}$, respectively.

Example 2.27 Check that for the three points $A_1 = (0, 0)$, $A_2 = (6, 0)$, and $A_3 = (4, 3)$, median of the set $\{A_1, A_2, A_3\}$ is the unique point $c_1^\star = (4, 0)$.

Example 2.28 Median of the set $\mathcal{A} = \{(1, 1), (1, 3), (2, 2), (3, 1), (3, 4), (4, 3)\} \subset \mathbb{R}^2$ is any point in the square $[2, 3] \times [2, 3]$ (see Fig. 2.6), since median of the first coordinates of the data is $\operatorname{med}\{1, 1, 2, 3, 3, 4\} \in [2, 3]$, and median of the second coordinates is $\operatorname{med}\{1, 3, 2, 1, 4, 3\} \in [2, 3]$.

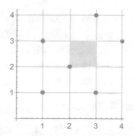

Fig. 2.6 Median of the set $\mathcal{A} = \{(1, 1), (1, 3), (2, 2), (3, 1), (3, 4), (4, 3)\}$

Exercise 2.29 Change the position of just one point of the set $\mathcal{A}$ in previous example in such a way that median becomes a single point, a segment, or a rectangle.

2.2.4   Geometric Median of a Set in the Plane

Geometric median $c^\star$ of the set $\mathcal{A} = \{a_i = (x_i, y_i) : i = 1, \ldots, m\} \subset \mathbb{R}^2$ without weights is the solution to the global optimization problem

$$c^\star = \arg\min_{c \in \mathbb{R}^2} \sum_{i=1}^{m} \|c - a_i\|. \tag{2.22}$$

The point $c^\star$ is the point at which the function

$$F_2(x, y) = \sum_{i=1}^{m} \|c - a_i\| = \sum_{i=1}^{m} \sqrt{(x - x_i)^2 + (y - y_i)^2}, \quad c = (x, y) \tag{2.23}$$

attains the global minimum. $F_2(x, y)$ is the sum of ℓ_2-distances between the point $c = (x, y) \in \mathbb{R}^2$ and points $a_i \in \mathcal{A}$, and in this case the variables x and y cannot be separated. Therefore, the solution to the global optimization problem (2.22) cannot be written down explicitly.

Example 2.30 In order to find the geometric median of the set of three points $a_1 = (0, 0)$, $a_2 = (6, 0)$, and $a_3 = (4, 3)$, one has to solve the following optimization problem:

$$\underset{(x,y)\in\mathbb{R}^2}{\arg\min}\ F_2(x, y),$$

$$F_2(x, y) = \sqrt{x^2 + y^2} + \sqrt{(x - 6)^2 + y^2} + \sqrt{(x - 4)^2 + (y - 3)^2}.$$

Using *Mathematica* computation system, we can solve this optimization problem like this: first define the function

```
In[1]:= F2[x_, y_]:= Sqrt[x^2 + y^2] + Sqrt[(x-6)^2 + y^2]
                   + Sqrt[(x-4)^2 + (y-3)^2]
```

We can try to solve our problem as a global optimization problem using the *Mathematica*-module `NMinimize[]` [195], but because this module sometimes finds only a local minimum, in such cases we can try to solve the problem as a local optimization problem using the *Mathematica*-module `FindMinimum[]` with some good initial approximation (x_0, y_0) close to the solution:

```
In[2]:= FindMinimum[F2[x, y], {x, 1}, {y, 2}]
```

With the initial approximation $(x_0, y_0) = (1, 2)$ we obtain $c_2^* = (3.833, \ 1.630)$.

Remark 2.31 The best known algorithm for searching for the geometric median by solving the optimization problem (2.22) is the *Weiszfeld's algorithm* (see [84, 165, 194]). This is an iterative procedure which arose as a special case of simple-iteration method for solving systems of nonlinear equations (see e.g. [1, 128, 179]).

First we find partial derivatives of the function (2.23) and make them equal to zero:

$$\frac{\partial F_2}{\partial x} = \sum_{i=1}^{m} \frac{x - x_i}{\|c - a_i\|} = x \sum_{i=1}^{m} \frac{1}{\|c - a_i\|} - \sum_{i=1}^{m} \frac{x_i}{\|c - a_i\|} = 0,$$

$$\frac{\partial F_2}{\partial y} = \sum_{i=1}^{m} \frac{y - y_i}{\|c - a_i\|} = y \sum_{i=1}^{m} \frac{1}{\|c - a_i\|} - \sum_{i=1}^{m} \frac{y_i}{\|c - a_i\|} = 0,$$

which can be written as

$$x = \Phi(x, y), \quad y = \Psi(x, y), \tag{2.24}$$

where

$$\Phi(x, y) = \frac{\sum_{i=1}^{m} \frac{x_i}{\|c-a_i\|}}{\sum_{i=1}^{m} \frac{1}{\|c-a_i\|}}, \quad \Psi(x, y) = \frac{\sum_{i=1}^{m} \frac{y_i}{\|c-a_i\|}}{\sum_{i=1}^{m} \frac{1}{\|c-a_i\|}}. \tag{2.25}$$

After choosing an initial approximation (x_0, y_0) from the convex hull of the set $\mathcal{A}$, $(x_0, y_0) \in \text{conv}(\mathcal{A})$, the system (2.24) can be solved by successive iteration method

$$x_{k+1} = \Phi(x_k, y_k), \quad y_{k+1} = \Psi(x_k, y_k), \quad k = 0, 1, \ldots \tag{2.26}$$

Example 2.32 Let the data set $\mathcal{A} \subset \mathbb{R}^2$ be defined like this:

```
In[1]:= SeedRandom[13]
  sig = 1.5; m = 50; cen = {4,5};
  podT = Table[cen + RandomReal[NormalDistribution[0, sig], {2}],
          {i, m}];
  podW = RandomReal[{0,1}, m];
  Show[Table[ListPlot[{podT[[i]]},
      PlotStyle -> {PointSize[podW[[i]]]/20], Gray}], {i,m}],
      PlotRange -> {{0,8},{0,8}}, AspectRatio > Automatic]
```

Each datum in Fig. 2.7 is equipped with weight according to the point-size (small disc) representing the datum. Check that the centroid is the point $c_{LS}^\star = (4.151, 4.676)$, median $c_1^\star = (4.350, 4.750)$, and the geometric median $c_2^\star = (4.251, 4.656)$.

Exercise 2.33 Let the set $\mathcal{A} = \{(x_i, y_i) \in \mathbb{R}^2 : i = 1, \ldots, 10\}$ be given by the table

i	1	2	3	4	5	6	7	8	9	10
x_i	9	6	8	1	1	4	4	3	9	10
y_i	5	5	5	2	5	8	1	8	8	4

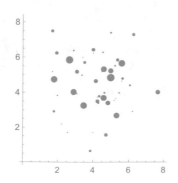

Fig. 2.7 Set $\mathcal{A}$ of weighted data

Depict the set $\mathcal{A}$ in the coordinate plane, and find its centroid, median, and geometric median.

Hint: Use the following *Mathematica*-program:

```
In[1]:= SeedRandom[2]
        A = RandomInteger[{1, 10}, {10, 2}]
            ListPlot[A, ImageSize -> Small]
                Print["Centroid = ", Mean[A]]
                Print["Median = ", Median[A]]
    Psi[x_, y_] := Sum[Norm[{x, y} - A[[i]]], {i, Length[A]}]
                Print["Geometric median:"]
                    NMinimize[Psi[x, y], {x, y}]
```

Solution: $c_{LS}^{\star} = (5.5,\ 5.1)$, $c_1^{\star} = (5,5)$, $c_2^{\star} = (6,5)$.

2.3   Representative of Data Sets with Several Features

In practical applications, data can have more than one or two features as was mentioned at the beginning of the Introduction, where we listed several such examples. Since the number of features represents the dimension of data, it will be necessary to find representatives also for data of arbitrary dimension.

We want to find a point in $\mathbb{R}^n$ which represents, as good as possible, a given set of points $\mathcal{A} = \{a_i = (a_i^1, \ldots, a_i^n) \in \mathbb{R}^n : i = 1, \ldots, m\}$ without weights.

In case of LS distance-like function, the best representative of the set $\mathcal{A}$ is its *centroid (barycenter)*[11]

$$c_{LS}^{\star} = \arg\min_{c \in \mathbb{R}^n} \sum_{i=1}^{m} d_{LS}(c, a_i) = \arg\min_{c \in \mathbb{R}^n} \sum_{i=1}^{m} \|c - a_i\|^2 = \tfrac{1}{m} \sum_{i=1}^{m} a_i ,$$

and the corresponding minimizing function is

$$F_{LS}(c) = \sum_{i=1}^{m} \|c - a_i\|^2 .$$

In case of ℓ_1 metric function, a best representative of the set $\mathcal{A}$ is its *median*

$$c_1 = \operatorname*{med}_i a_i = \left(\operatorname*{med}_i a_i^1, \ldots, \operatorname*{med}_i a_i^n \right) \in \operatorname{Med}\mathcal{A} = \arg\min_{c \in \mathbb{R}^n} \sum_{i=1}^{m} \|c - a_i\|_1 ,$$

---

[11] Recall that $\|\ \|$ denotes the Euclidean, i. e. ℓ_2-norm.

and the corresponding minimizing function is

$$F_1(c) = \sum_{i=1}^{m} \|c - a_i\|_1.$$

2.3.1 Representative of Weighted Data

Let $\mathcal{A}$ be the set of data points with weights $w_1, \ldots, w_m > 0$. If d is the LS distance-like function, the best representative of the set $\mathcal{A}$ with weights $w_1, \ldots, w_m > 0$ is its *weighted centroid (barycenter)*

$$c_{LS}^{\star} = \arg\min_{c \in \mathbb{R}^n} \sum_{i=1}^{m} w_i \, d_{LS}(c, a_i) = \arg\min_{c \in \mathbb{R}^n} \sum_{i=1}^{m} w_i \, \|c - a_i\|^2 = \tfrac{1}{W} \sum_{i=1}^{m} w_i \, a_i,$$

i. e.

$$c_{LS}^{\star} = \left(\tfrac{1}{W} \sum_{i=1}^{m} w_i \, a_i^1, \ldots, \tfrac{1}{W} \sum_{i=1}^{m} w_i \, a_i^n \right) \qquad \text{[coordinate-wise]}, \qquad (2.27)$$

where $W = \sum_{i=1}^{m} w_i$, and the corresponding minimizing function is

$$F_{LS}(c) = \sum_{i=1}^{m} w_i \, \|c - a_i\|^2. \qquad (2.28)$$

If d is the ℓ_1 metric function, a best representative of the set $\mathcal{A}$ with weights $w_1, \ldots, w_m > 0$ is its *weighted median*

$$c_1^{\star} = \operatorname{med}_i(w_i, a_i) = \left(\operatorname{med}_i(w_i, a_i^1), \ldots, \operatorname{med}_i(w_i, a_i^n) \right) \in \operatorname{Med} \mathcal{A}$$

$$= \arg\min_{c \in \mathbb{R}^n} \sum_{i=1}^{m} w_i \, \|c - a_i\|_1, \qquad (2.29)$$

and the corresponding minimizing function is

$$F_1(c) = \sum_{i=1}^{m} w_i \, \|c - a_i\|_1. \qquad (2.30)$$

Namely,

$$
F_1(c) = \sum_{i=1}^{m} w_i \, \|c - a_i\|_1 = \sum_{i=1}^{m} w_i \left(\sum_{k=1}^{n} |c^k - a_i^k| \right)
$$

$$
= \sum_{k=1}^{n} \left(\sum_{i=1}^{m} w_i \, |c^k - a_i^k| \right) = \sum_{k=1}^{n} \sum_{i=1}^{m} w_i \, |c^k - a_i^k|
$$

$$
\geq \sum_{k=1}^{n} \sum_{i=1}^{m} w_i \, |\underset{j}{\mathrm{med}}(w_j, a_j^k) - a_i^j| = \sum_{i=1}^{m} \sum_{k=1}^{n} w_i \, |\underset{j}{\mathrm{med}}(w_j, a_j^k) - a_i^k|
$$

$$
= \sum_{i=1}^{m} w_i \, \|c_1^\star - a_i\|_1 = F(c_1^\star),
$$

where $\underset{j}{\mathrm{med}}(w_j, a_j^k)$ is the weighted median of data $\{a_1^k, \ldots, a_m^k\}$ with weights $w_1, \ldots, w_m > 0$.

Exercise 2.34 Show, similarly as was shown for the weighted median, that the function F_{LS} given by (2.28) attains its global minimum at the weighted centroid $c_{LS}^\star$ given by (2.27).

2.4   Representative of Periodic Data

Often it is the case that one has to find the best representative of a data set describing events which are periodic in nature, and this is indeed frequently discussed in the literature. For instance, air temperatures at certain measuring point during a year, water levels of a river at certain measuring place, seismic activities in specific area over some extended period of time, the illuminance i.e. the measure of the amount of light during a day, etc., are examples of such phenomena. Mathematically speaking, one has to deal with data sets on a circle. Namely, if we represent such a data set on the real line, as we did before, then the data corresponding to the beginning and the end of the same year, for example, would appear far apart, although they belong to the same year period. Therefore, one has to define a distance-like function also for such data sets and find the center of such data.

Example 2.35 Let $t_i \in \mathcal{A}$ represent the position of the small clock hand on a clock with 12 marks (see Fig. 2.8a). Distances in $\mathcal{A}$ will be measured as the *elapsed time* from the moment t_1 to t_2:

$$
d(t_1, t_2) = \begin{cases} t_2 - t_1, & \text{if } t_1 \leq t_2 \\ 12 + (t_2 - t_1), & \text{if } t_1 > t_2 \end{cases}.
$$

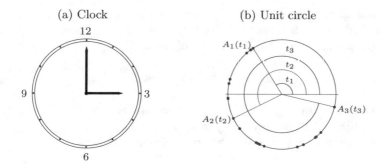

Fig. 2.8 Data set on a circle

For example, $d(2, 7) = 5$, but $d(7, 2) = 12 + (-5) = 7$. Note that this function is not symmetric.

Example 2.36 Let $t_i \in \mathcal{A}$ represent the position of the small clock hand on a clock with 12 marks (see Fig. 2.8a). Define the function measuring the distances on $\mathcal{A}$ as the *length of the time interval* from moment t_1 to t_2:

$$
d(t_1, t_2) = \begin{cases} |t_2 - t_1|, & \text{if } |t_2 - t_1| \le 6 \\ 12 - |t_2 - t_1|, & \text{if } |t_2 - t_1| > 6 \end{cases}.
$$

For example, $d(2, 9) = 12 - 7 = 5$ and $d(2, 7) = 7 - 2 = 5$. Check whether this defines a metric function on the set $\mathcal{A}$.

2.4.1   Representative of Data on the Unit Circle

In general, let (T_i, w_i), $i = 1, \ldots, m$, be a data set where T_i denotes the moment over $M \ge 1$ successive years during which the event we investigate occurred, and let $w_i > 0$ denote the intensity of the event at the moment T_i. The time moments T_i can denote days (for example, for water levels of a river at some point), hours (air temperatures at some place), or seconds (earthquake moments). We want to identify the moment at which this event is most notable. See [45, 106] for various aspects and applications of such data.

If the moments $T_1, \ldots, T_m$ were considered as simple time series, then the data from, say, the beginning of a year to the end of the same year would be far apart, although they belong to the same season, i. e. time of the year. Therefore, to each year we allot an interval of length 2π, and to a sequence of M successive years the interval $[0, 2\pi M]$. In this way the sequence $T_1, \ldots, T_m$ is transformed into the sequence $T'_1, \ldots, T'_m \in [0, 2\pi M]$.

In our discussion, important is only the moment of the year, and not the particular year in which the event occurred. Therefore, instead of the sequence (T_i') we define the new sequence $t_i \in [0, 2\pi]$, $i = 1, \ldots, m$, where

$$t_i = 2\pi\, T_i' \pmod{2\pi}, \quad i = 1, \ldots, m, \tag{2.31}$$

(the remainder of dividing $2\pi\, T_i'$ by 2π). The number $t_i \in [0, 2\pi]$ represents the moment which is $t_i/2\pi$-th part of a year apart from January 1.

Using the sequence (2.31) we define the following data set:

$$\mathcal{A} = \{a(t_i) = (\cos t_i,\ \sin t_i) \in \mathbb{R}^2 : i = 1, \ldots, m\} \subset K, \tag{2.32}$$

where $K = \{(x, y) \in \mathbb{R}^2 : x^2 + y^2 = 1\}$ is the unit circle.

In the following lemma we define a metric on the unit circle and prove its basic properties (see also [87, 106])

Lemma 2.37 *Let* $K = \{a(t) = (\cos t, \sin t) \in \mathbb{R}^2 : t \in [0, 2\pi]\}$ *be the unit circle in the plane. The function* $d_K : K \times K \to \mathbb{R}_+$ *defined by*

$$d_K(a(t_1), a(t_2)) = \begin{cases} |t_1 - t_2|, & \text{if } |t_1 - t_2| \le \pi, \\ 2\pi - |t_1 - t_2|, & \text{if } |t_1 - t_2| > \pi, \end{cases} \tag{2.33}$$

is a metric on K *and can equivalently be defined as*

$$d_K(a(t_1), a(t_2)) = \pi - \big|\, |t_1 - t_2| - \pi \,\big|, \quad t_1, t_2 \in [0, 2\pi]. \tag{2.34}$$

Proof It is straightforward to see that (2.33) and (2.34) are equivalent definitions, and that the function d_K is symmetric. Let us show that d_K is a metric on K.

First we show that $d_K(a(t_1), a(t_2)) \ge 0$ for all $t_1, t_2 \in [0, 2\pi]$. Let $t_1, t_2 \in [0, 2\pi]$. Then

$$0 \le |t_1 - t_2| \le 2\pi \ \Rightarrow\ -\pi \le |t_1 - t_2| - \pi \le \pi \ \Rightarrow\ \big|\, |t_1 - t_2| - \pi \,\big| \le \pi$$

$$\Rightarrow\ d_K(a(t_1), a(t_2)) = \pi - \big|\, |t_1 - t_2| - \pi \,\big| \ge 0.$$

Next we show that $d_K\big(a(t_1), a(t_2)\big) = 0$ if and only if $a(t_1) = a(t_2)$: If $a(t_1) = a(t_2)$, then either $t_1 = t_2$ or $|t_1 - t_2| = 2\pi$. In both cases $d_K(a(t_1), a(t_2)) = 0$. Conversely, if $d_K(a(t_1), a(t_2)) = 0$, then

$$\pi = \big|\, |t_1 - t_2| - \pi \,\big|. \tag{2.35}$$

If $|t_1 - t_2| \leq \pi$, then from (2.35) it follows that $\pi = \pi - |t_1 - t_2|$, and hence $t_1 = t_2$, thus $a(t_1) = a(t_2)$. If $|t_1 - t_2| \geq \pi$, then from (2.35) it follows that $\pi = |t_1 - t_2| - \pi$, i.e. $|t_1 - t_2| = 2\pi$, which is possible if and only if $a(t_1) = a(t_2)$.

Finally, $d_K(a(t_1), a(t_2)) \leq d_K(a(t_1), a(t_3)) + d_K(a(t_3), a(t_2))$ for all $t_1, t_2, t_3 \in K$. The equality holds if $a(t_3)$ lies on the arc between $a(t_1)$ and $a(t_2)$. Otherwise, the strict inequality holds true. □

Using the metrics (2.33), we define the *best representative* of the set $\mathcal{A}$ on the unit circle, as follows:

Definition 2.38 The best representative of the set $\mathcal{A} = \{a(t_i) \in K : i = 1, \ldots, m\}$ and weights $w_1, \ldots, w_m > 0$, with respect to the metric d_K defined by (2.33), is the point $c^\star(t^\star) = (\cos t^\star, \sin t^\star) \in K$, where

$$
t^\star = \underset{\tau \in [0, 2\pi]}{\arg\min} \sum_{i=1}^{m} w_i \, d_K(a(\tau), a(t_i)), \quad a(\tau) = (\cos \tau, \sin \tau) \in K, \tag{2.36}
$$

i.e. $t^\star \in [0, 2\pi]$ is the point at which the function $\Phi \colon [0, 2\pi] \to \mathbb{R}_+$ defined by

$$
\Phi(\tau) = \sum_{i=1}^{m} w_i \, d_K(a(\tau), a(t_i)) \tag{2.37}
$$

attains its global minimum.

Note that the function Φ does not have to be either convex or differentiable, and generally it may have several local minima. Therefore, this becomes a complex global optimization problem. In order to solve (2.36) one can apply the optimization algorithm DIRECT [69, 89, 154].

Example 2.39 Let $t_1, \ldots, t_m$ be a random sample from Gaussian normal distribution $\mathcal{N}(4, 1.2)$, and let $\mathcal{A} = \{a(t_i) = (\cos t_i, \sin t_i) \in \mathbb{R}^2 : i = 1, \ldots, m\}$ with weights $w_i > 0, i = 1, \ldots, m$ be a data set. The set $\mathcal{A}$ is depicted as black points in Fig. 2.9a, and the function $\tau \mapsto d_K(a(\tau), a(t_1))$ and the corresponding function Φ are shown in Fig. 2.9b,c.

2.4.2 Burn Diagram

In order to graphically represent periodic events, it is appropriate to use *Burn diagram* (see e.g. [166]). In the Burn diagram, points are represented as $T = r(\cos t, \sin t)$, where (r, t) are the polar coordinates, i.e. t is the angle (in radians) between the x-axis and the radius vector of T, and r is the distance from T to the origin.

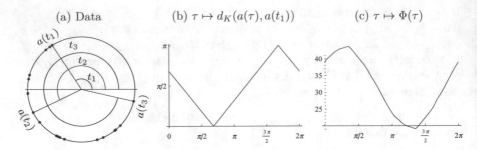

Fig. 2.9 Data and distances on the unit circle

Example 2.40 Figure 2.10a shows earthquake positions in the wider area of Osijek since 1880. Points in the Burn diagram (Fig. 2.10b) identify individual earthquakes, where the distance to the origin represents the year when the earthquake happened, the position on the circle reflects the day of the year, and the size of the point (small disc) corresponds to the magnitude. Figure 2.10 shows that the last stronger earthquake in close vicinity of Osijek happened by the end of winter 1922 (November 24, at 2:15:40), located at geographic position (18.8, 45.7) (close to village Lug, some twenty kilometers to the northeast of Osijek).

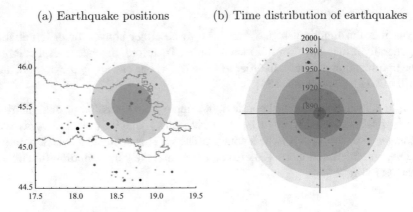

Fig. 2.10 Earthquakes in the wider area of Osijek since 1880

Chapter 3
Data Clustering

Definition 3.1 Let $\mathcal{A}$ be a set with $m \geq 2$ elements. A partition of the set $\mathcal{A}$ into $1 \leq k \leq m$ disjoint nonempty subsets $\pi_1, \ldots, \pi_k$, i.e.

$$\bigcup_{j=1}^{k} \pi_j = \mathcal{A}, \quad \pi_r \cap \pi_s = \emptyset, \ r \neq s, \quad |\pi_j| \geq 1, \quad j = 1, \ldots, k, \tag{3.1}$$

is called a **k-partition** of the set $\mathcal{A}$ and will be denoted by $\Pi = \{\pi_1, \ldots, \pi_k\}$. The elements of a partition are called **clusters**, and the set of all partitions of $\mathcal{A}$ containing k clusters satisfying (3.1) will be denoted by $\mathcal{P}(\mathcal{A}; k)$.

Whenever we are going to talk about a partition of some set $\mathcal{A}$ we will always assume that it consists of subsets as described in Definition 3.1.

Theorem 3.2 *The number of all partitions of the set $\mathcal{A}$ consisting of k clusters is equal to the Stirling number of the second kind*

$$|\mathcal{P}(\mathcal{A}; k)| = \frac{1}{k!} \sum_{j=1}^{k} (-1)^{k-j} \binom{k}{j} j^m. \tag{3.2}$$

In the proof of Theorem 3.2 we will use the well known *inclusion–exclusion principle* (see e.g. [76, p. 156]) written in the following form.

Lemma 3.3 (Inclusion–Exclusion Formula) *Let $X_1, \ldots, X_k$ be subsets of a finite set X. The number of elements of X not belonging to any of the subsets $X_1, \ldots, X_k$ equals*

$$\left| \bigcap_{i=1}^{k} X_i^{\complement} \right| = |X| - \sum_{1 \leq i \leq k} |X_i| + \sum_{1 \leq i < j \leq k} |X_i \cap X_j| - \cdots + (-1)^k |X_1 \cap \cdots \cap X_k|,$$

where $X_i^{\complement}$ denotes the complement $X \setminus X_i$.

© The Author(s), under exclusive license to Springer Nature Switzerland AG 2021
R. Scitovski et al., *Cluster Analysis and Applications*,
https://doi.org/10.1007/978-3-030-74552-3_3

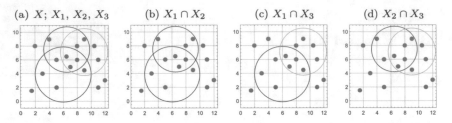

Fig. 3.1 Number of elements of the set X not belonging to any of the subsets X_1, X_2, X_3

Instead of a proof of Lemma 3.3, let us look at an example. Consider the set X containing 16 elements and its three subsets: X_1 (7 elements inside the red circle), X_2 (7 elements inside the blue circle), and X_3 (8 elements inside the green circle), as shown in Fig. 3.1a. The intersections $X_1 \cap X_2$ and $X_1 \cap X_3$ have 4 elements each, and the intersection $X_2 \cap X_3$ has 5 elements (see Fig. 3.1b,c,d). Finally, the intersection $X_1 \cap X_2 \cap X_3$ has 3 elements (see Fig. 3.1a). Therefore,

$$|X_1^C \cap X_2^C \cap X_3^C| = |X| - (|X_1|+|X_2|+|X_3|) + (|X_1 \cap X_2|+|X_1 \cap X_3|+|X_2 \cap X_3|) -$$
$$- |X_1 \cap X_2 \cap X_3|$$
$$= 16 - (7+7+8) + (4+4+5) - 3 = 4.$$

Proof of Theorem 3.2 Without loss of generality, let $\mathcal{A} = \{1, \ldots, m\}$, and let $\Pi^{(k)} = \{\pi_1, \ldots, \pi_k\}$ be its k-partition where $\pi_j \subset \mathcal{A}$ are disjoint nonempty subsets of $\mathcal{A}$. The number of all such partitions is $|\mathcal{P}(\mathcal{A}; k)|$, and define the functions $f: \mathcal{A} \to J, J = \{1, \ldots, k\}$ by

$$f(x) = j, \quad \text{for } x \in \pi_j.$$

The number of these functions equals $|\mathcal{P}(\mathcal{A}; k)|$, and by permuting these k sets we obtain the number of all surjections from $\mathcal{A}$ onto J:

$$k! \, |\mathcal{P}(\mathcal{A}; k)|. \tag{3.3}$$

On the other hand, the number of all surjections from $\mathcal{A}$ onto J equals the number of all functions from $\mathcal{A}$ to J, minus the number of those functions which are not surjective.

Let $X = J^{\mathcal{A}}$ be the set of all functions from $\mathcal{A}$ to J. The number of all such functions is $|X| = k^m$—the number of ways of selecting k, not necessarily distinct items from a collection of m items.

A function from $\mathcal{A}$ to J is not surjective if:

1. Its image does not contain one element of J. The set X_i of all functions whose image does not contain the element $i \in J$ consists of exactly $(k-1)^m$ functions

(the number of ways of selecting $k - 1$, not necessarily distinct items from a collection of m items), and the set $\bigcup_{1 \le i \le k} X_i$ of all functions missing exactly one element of J consists of $\binom{k}{1}(k - 1)^m$ functions;

2. Its image does not contain two distinct elements of J. The set $\bigcup_{1 \le i < j \le k}(X_i \cap X_j)$ of all such functions contains $\binom{k}{2}(k - 2)^m$ elements (the number of ways of selecting $k - 2$, not necessarily distinct items from a collection of m items, for every pair of distinct elements of J); etc.

A function from $\mathcal{A}$ to J is surjective if and only if it does not belong to any of the sets $X_1, \ldots, X_k$, i.e. if and only if it belongs to the set $\bigcap_{i=1}^{k} X_i^{\complement}$. Using Lemma 3.3 we obtain the following number of all surjective functions from $\mathcal{A}$ to J:

$$\left| \bigcap_{i=1}^{k} X_i^{\complement} \right| = |X| - \sum_{1 \le i \le k} |X_i| + \sum_{1 \le i < j \le k} |X_i \cap X_j| - \cdots + (-1)^k |X_1 \cap \cdots \cap X_k|$$

$$= k^m - \binom{k}{1}(k - 1)^m + \binom{k}{2}(k - 2)^m - \cdots + (-1)^k \binom{k}{k}(k - k)^m$$

$$= \sum_{j=0}^{k}(-1)^j \binom{k}{j}(k - j)^m \qquad\qquad [s := k - j]$$

$$= \sum_{s=k}^{0}(-1)^{k-s} \binom{k}{k - s}s^m \qquad\qquad [\text{for } s = 0,\ s^m = 0]$$

$$= \sum_{s=1}^{k}(-1)^{k-s} \binom{k}{k - s}s^m \qquad\qquad \left[\text{since } \binom{n}{r} = \binom{n}{n-r}\right]$$

$$= \sum_{s=1}^{k}(-1)^{k-s} \binom{k}{s}s^m \overset{[j:=s]}{=} \sum_{j=1}^{k}(-1)^{k-j} \binom{k}{j}j^m.$$

Using (3.3) we obtain (3.2), proving the theorem.   □

Example 3.4 Theorem 3.2 gives in particular:

$$\text{for } k = 2: \quad |\mathcal{P}(\mathcal{A}; 2)| = \tfrac{1}{2}(2^m - 2) = 2^{m-1} - 1,$$

$$\text{for } k = 3: \quad |\mathcal{P}(\mathcal{A}; 3)| = \tfrac{1}{2}\left(1 - 2^m + 3^{m-1}\right).$$

The number of all k-partitions of a set $\mathcal{A}$, as described in Definition 3.1, can be rather huge. Table 3.1 shows the approximate number of all k-partitions of the set $\mathcal{A}$ for $m = 5, 10, 50, 1200, 10^6$, and $k = 2, 3, 4, 5, 6, 8, 10$.

Table 3.1 Approximate numbers of all k-partitions for various numbers $m = |\mathcal{A}|$ and numbers $k = 2, 3, 4, 5, 6, 8, 10$ of clusters

m	$k = 2$	$k = 3$	$k = 4$	$k = 5$	$k = 6$	$k = 8$	$k = 10$
5	15	25	10	1	–	–	–
10	511	9330	34105	42525	22827	750	1
50	10^{15}	10^{23}	10^{29}	10^{33}	10^{36}	10^{41}	10^{44}
1200	10^{361}	10^{572}	10^{721}	10^{837}	10^{931}	10^{1079}	10^{1193}
10^6	10^{301030}	10^{477120}	10^{602058}	10^{698968}	10^{778148}	10^{903085}	$10^{1000000}$

Example 3.5 Consider the set $\mathcal{A} \subset \mathbb{R}^2$, shown in Fig. 4.10a, containing $m = 1200$ elements. Table 3.1 shows the approximate number of all its k-partitions with $k = 2, 3, 4, 5, 6, 8$, and 10 clusters.

Provided that one defines the criterion that the *better* partition is the one whose clusters are *more compact* and *better separated*, one could ask the question of defining the *globally optimal (i.e. the best) partition*.

3.1   Optimal k-Partition

Let $\mathcal{A}$ be a set of $m \geq 2$ elements with $n \geq 1$ features. Since each feature is usually expressed by a number, one can, without loss of generality, always regard a set $\mathcal{A}$ with $n \geq 1$ features as a subset of $\mathbb{R}^n$, $\mathcal{A} \subset \mathbb{R}^n$. For example, consider a group of 100 high school students with respect to their gender and height. Assigning the number 0 to males and 1 to females, and expressing heights in centimeters, one can identify this set of students with a subset of $\mathbb{R}^2$.

Remark 3.6 What was said in the previous paragraph is in fact not precisely correct. Consider again the group $\mathcal{A}$ of 100 high school students with two assigned features: gender and height. The gender is denoted by either 0 or 1, and the height is measured in centimeters. It is reasonable to assume that no student is shorter than 160 cm and taller than 200 cm, hence the range of heights is 40 cm. By the Dirichlet's *pigeonhole principle*, some students of the same gender have to be of the same height. Therefore, the set $\mathcal{A}$, regarded as a subset of $\mathbb{R}^2$, consists of at most 80 points, not 120, so some points *have to be counted more than once*, i.e. we are dealing not with an ordinary set but with a **multiset**. There is not yet a standard notation for multisets. One which does occur in papers is $\mathcal{A} = \{\{a_1, \ldots, a_m\}\}$, but we will not bother much with it and, although somehow abusing the notation, just use $\mathcal{A} = \{a_1, \ldots, a_m\}$ and continue to call it a set. But we will keep in mind this fact. For example, the multiset $\mathcal{B} = \{\{1, 3, 1, 3, 3, 2, 3\}\}$ is as a set, just the set $\{1, 2, 3\}$. Multisets can in fact be regarded simply as weighted sets, i.e. as a set $\mathcal{A} = \{a_1, \ldots, a_m\}$ with corresponding weights $w_1, \ldots, w_m$, which are positive integers. In our example, the multiset $\mathcal{B}$ is, as a weighted set, the set $\{1, 2, 3\}$ with weights 2, 1, and 4.

If we defined some distance-like function $d\colon \mathbb{R}^n \times \mathbb{R}^n \to \mathbb{R}_+$, we could have defined a *measure of compactness* and of *good separation* of clusters in a partition $\Pi = \{\pi_1, \ldots, \pi_k\}$ of the set $\mathcal{A} \subset \mathbb{R}^n$, as follows:

1. Find the center $c_j \in \arg\min\limits_{x \in \mathbb{R}^n} \sum\limits_{a_i \in \pi_j} d(x, a_i)$ in every cluster π_j.

2. For every cluster π_j determine its total *dispersion* (the sum of distances from points of π_j to the center c_j) $\mathcal{F}(\pi_j) = \sum\limits_{a_i \in \pi_j} d(c_j, a_i)$.

3. The sum $\sum\limits_{j=1}^{k} \mathcal{F}(\pi_j)$ defines a measure of compactness and of good separation of clusters in the partition Π, and represents an objective function in this optimization problem (see (3.5)).

Figure 4.10 shows some partitions with $k = 2, 3, 4, 5, 6, 7$, and 8 clusters and the corresponding values of the LS objective function. Notice that enlarging the number of clusters decreases the objective function value. For example, Fig. 4.10c shows one of many (see Table 3.1) 3-partitions of the set $\mathcal{A}$. For this partition the objective function $\mathcal{F}_{LS}$ attains the value 16 830. It is plausible to ask whether this is the best 3-partition or could one find a 3-partition with a smaller objective function value? In general, we could pose at least some of the following questions:

1. Are the said objective functions the most appropriate ones for this example?
2. What is the most appropriate number of clusters in a partition?
3. Do the partitions shown in Fig. 4.10 have the smallest objective function values among all possible partitions with those numbers of clusters?

From the previous example, it is evident that the answers to the above questions will not be easy ones. The question of the choice of the objective function, as well as of the appropriate number of clusters in a partition, depends on the previous statistical analysis of the data. For objective functions in this textbook, we are mostly going to use the LS-distance-like function and the ℓ_1 metric function. In Chap. 5 we will deal with the choice of the most appropriate partition with spherical clusters, and in Sect. 6.5 with the choice of the most appropriate fuzzy-partition.

It needs to be said that the problem of finding an optimal partition is an NP-hard problem [184], of a non-convex optimization of, in general, a non-differentiable function of several variables, which, in most cases, does have a substantial number of stationary points. In general, it will not be possible to carry out the search for an optimal partition by searching the whole set $\mathcal{P}(\mathcal{A}; k)$. In the present textbook we are going to deal with finding the optimal partition with spherical clusters in Chap. 4, finding the optimal partition with ellipsoidal clusters in Sect. 6.4, and finding the optimal fuzzy-partition in Chap. 7.

In general, given some distance-like function $d\colon \mathbb{R}^n \times \mathbb{R}^n \to \mathbb{R}_+$, where $\mathbb{R}_+ = [0, +\infty)$ (see Chap. 2), to each cluster $\pi_j \in \Pi$ one can associate its center

$$c_j \in \arg\min_{x \in \mathbb{R}^n} \sum_{a \in \pi_j} d(x, a). \tag{3.4}$$

The quality of the partition determined by the value of the objective function $\mathcal{F}\colon \mathcal{P}(\mathcal{A}; k) \to \mathbb{R}_+$ is usually defined to be the sum over all clusters of the sums of distances from the points of clusters to their centers. The **globally optimal k-partition** (k-GOPart) is then considered to be the solution to the following **global optimization problem** (GOP):

$$\arg \min_{\Pi \in \mathcal{P}(\mathcal{A};k)} \mathcal{F}(\Pi), \qquad \mathcal{F}(\Pi) = \sum_{j=1}^{k} \sum_{a \in \pi_j} d(c_j, a). \tag{3.5}$$

Theorem 3.7 *Increasing the number of clusters in an optimal partition does not increase the value of the objective function $\mathcal{F}$.*[1]

For the proof of this theorem see page 64.

3.1.1 Minimal Distance Principle and Voronoi Diagram

The minimal distance principle (see Algorithm 3.10 or formula (3.39)) is closely related to the so-called *Voronoi diagram* or *Dirichlet tessellation* (see e.g. [6, 127, 195]).

Let d be the usual Euclidean metric in the plane $\mathbb{R}^2$, and let us first consider the case of $k = 2$ clusters in the plane with centers c_1 and c_2. All elements $a \in \mathcal{A} \subset \mathbb{R}^2$ lying on the perpendicular bisector $\sigma(c_1, c_2)$ of the line segment $\overline{c_1 c_2}$ are equally distant from the centers c_1 and c_2. The line bisector $\sigma(c_1, c_2)$ is perpendicular to the segment $\overline{c_1 c_2}$ and divides the plane $\mathbb{R}^2$ into two half-planes—*Voronoi regions*:

$$VR(c_1) = \{x \in \mathbb{R}^2 : d(c_1, x) < d(c_2, x)\},$$

$$VR(c_2) = \{x \in \mathbb{R}^2 : d(c_1, x) > d(c_2, x)\}.$$

The perpendicular bisector $\sigma(c_1, c_2)$ represents the Voronoi diagram of the set of centers $\{c_1, c_2\}$ (see Fig. 3.2a).

In the case of $k = 3$ clusters in the plane with centers c_1, c_2, and c_3, the perpendicular bisector $\sigma(c_1, c_2)$ of the straight line segment $[c_1, c_2]$ defines two half-planes $M(c_1, c_2)$ and $M(c_2, c_1)$, the perpendicular bisector $\sigma(c_1, c_3)$ of the segment $[c_1, c_3]$ defines the half-planes $M(c_1, c_3)$ and $M(c_3, c_1)$, and the perpendicular bisector $\sigma(c_2, c_3)$ of the segment $[c_2, c_3]$ defines the half-planes

---

[1]Notice that $\mathcal{F}(\Pi)$ in (3.5) is in fact defined as $\sum_{\pi \in \Pi} \mathcal{F}(\pi)$ and NOT as $\bigcup_{\pi \in \Pi} \mathcal{F}(\pi)$. Namely, in mathematics in general, if $f\colon X \to Y$ is some function and $A \subseteq X$, then $f(A)$ is, as a rule, defined as $f(A) = \{f(x) : x \in A\} = \bigcup \{f(x)\}$, the latter being usually loosely written as $\bigcup_{x \in A} f(x)$.

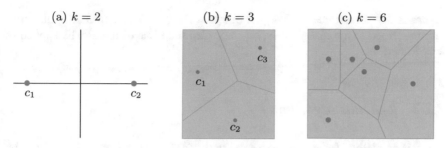

Fig. 3.2 Minimal distance principle and Voronoi diagram

$M(c_2, c_3)$ and $M(c_3, c_2)$. Voronoi regions with centers c_1, c_2, and c_3 are defined as

$$VR(c_1) = M(c_1, c_2) \cap M(c_1, c_3),$$
$$VR(c_2) = M(c_2, c_1) \cap M(c_2, c_3),$$
$$VR(c_3) = M(c_3, c_1) \cap M(c_3, c_2),$$

and the Voronoi diagram of the centers c_1, c_2, c_3 is defined as (see Fig. 3.2b)

$$V(c_1, c_2, c_3) = \left(\overline{VR(c_1)} \cap \overline{VR(c_2)}\right) \bigcup \left(\overline{VR(c_1)} \cap \overline{VR(c_3)}\right) \bigcup \left(\overline{VR(c_2)} \cap \overline{VR(c_3)}\right),$$

where $\overline{VR(c_1)}$ denotes the (topological) closure of $VR(c_1)$, and similarly for other regions.

In general, for k clusters with centers $c_1, \ldots, c_k$, the Voronoi regions are defined as

$$VR(c_j) = \bigcap_{s \neq j} M(c_j, c_s), \quad j = 1, \ldots, k,$$

and the Voronoi diagram is defined as the union of intersections of the closures of Voronoi regions

$$V(c_1, \ldots, c_k) = \bigcup_{s \neq j} \overline{VR(c_j)} \cap \overline{VR(c_s)}.$$

Note that the cluster $\pi(c_j)$, obtained by the minimal distance principle (3.39), lies in the Voronoi region $VR(c_j)$ bounded by the Voronoi diagram.

Figure 3.2c, generated by *Mathematica* computation system, shows the Voronoi diagram for six centers.

Exercise 3.8 Define and draw Voronoi diagrams in case of ℓ_1 and ℓ_∞ metric functions.

Exercise 3.9 Determine the Voronoi diagram for $k = 3$ by considering the circle circumscribed to the triangle $\triangle(c_1, c_2, c_3)$. Can such line of thought be applied for $k > 3$ also?

3.1.2   k-means Algorithm I

There is no method for successfully solving the GOP (3.5). Nevertheless, there exists the well known k-means algorithm giving locally optimal solution which heavily depends on the choice of initial approximation. Choosing an initial partition $\Pi^{(0)}$, the k-means algorithm finds in finitely many steps a locally optimal partition. The algorithm is usually set up in two steps which are iteratively successively repeated until the new partition does not differ from the previous one.

Algorithm 3.10 (*k*-means Algorithm I)

Step A :   (assignment step). Given a finite subset $\mathcal{A} \subset \mathbb{R}^n$ and k distinct points $z_1, \ldots, z_k \in \mathbb{R}^n$, apply the minimal distance principle to determine clusters π_j, $j = 1, \ldots, k$, to get the partition $\Pi = \{\pi_1, \ldots, \pi_k\}$,

$$\pi_j := \pi_j(z_j) = \{a \in \mathcal{A} : d(z_j, a) \le d(z_s, a) \text{ for all } s = 1, \ldots, k\}.$$

Step B :   (update step). For the given partition $\Pi = \{\pi_1, \ldots, \pi_k\}$ of the set $\mathcal{A}$ determine cluster centers $c_j \in \arg\min_{x \in \mathbb{R}^n} \sum_{a \in \pi_j} d(x, a)$, $j = 1, \ldots, k$, and calculate the objective function value $\mathcal{F}(\Pi)$ according to (3.5).
Set $z_j = c_j$, $j = 1, \ldots, k$.

Remark 3.11 It might happen that in Step A some elements $a \in \mathcal{A}$ lie on the border between two or several clusters. The decision as to which cluster should such elements be designated can drastically influence the further course of the iterative process (see [160]). An example of such a situation occurs in the problem of defining optimal electoral districts (see Sect. 8.4). Almost always it becomes necessary to divide the electorate of a city into two or several electoral districts (in Croatia this is the case with the city of Zagreb). We are going to consider this problem later, when discussing fuzzy clustering of data in Chap. 7.

The usual convention for simple clustering of data is to put the datum, which occurs on the border of two or several clusters, into the first cluster in order.

Example 3.12 Let us determine, using the k-means algorithm, the LS-optimal 3-partition of the set $\mathcal{A} = \{0, 2, 4, 8, 9, 10, 12, 16\}$ starting with the initial partition $\Pi^{(0)} = \{\{0, 2, 4\}, \{8, 9\}, \{10, 12, 16\}\}$ (Fig. 3.3).

Table 3.2 Searching for LS-optimal 3-partition of the set $\{0, 2, 4, 8, 9, 10, 12, 16\}$

Iteration	π_1	π_2	π_3	c_1	c_2	c_3	$\mathcal{F}_{LS}(\Pi)$
0	$\{0, 2, 4\}$	$\{8, 9\}$	$\{10, 12, 16\}$	2	8.5	12.67	27.17
1	*$\{0, 2, 4\}$*	*$\{8, 9, 10\}$*	*$\{12, 16\}$*	2	9	14	18
2	$\{0, 2, 4\}$	$\{8, 9, 10\}$	$\{12, 16\}$	2	9	14	18

Results of Step A are typeset in italic and of Step B in slanted sans serif font

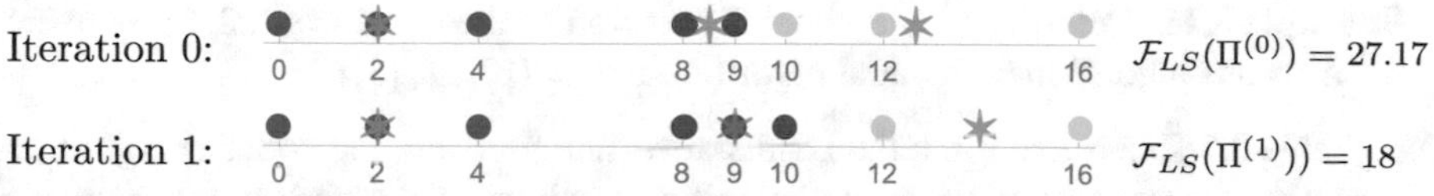

Fig. 3.3 Searching for LS optimal 3-partition of the set $\{0, 2, 4, 8, 9, 10, 12, 16\}$

Exercise 3.13 Using the k-means algorithm find the ℓ_1-optimal 3-partition of the set from Example 3.12 starting with the same initial partition.

The following theorem shows that the sequence of objective function values obtained by the k-means algorithm is monotonically decreasing (see also Theorem 4.8).

Theorem 3.14 *Let $\mathcal{A} \subset \mathbb{R}^n$ be a set, $d: \mathbb{R}^n \times \mathbb{R}^n \to \mathbb{R}_+$ a distance-like function, and $\mathcal{F}$ the objective function given by (3.5). Applying the k-means algorithm, the value of objective function $\mathcal{F}$ is not going to increase in any step.*

Proof Let $\Pi^{(t)} = \{\pi_1^{(t)}, \ldots, \pi_k^{(t)}\}$ be a partition with centers $c^{(t)} = \{c_1^{(t)}, \ldots, c_k^{(t)}\}$ and $\mathcal{F}(\Pi^{(t)})$ be the corresponding objective function value.

Applying Step A (the minimal distance principle) to the set $\mathcal{A}$ with centers $c^{(t)}$, we obtain the new partition $\Pi^{(t+1)} = \{\pi_1^{(t+1)}, \ldots, \pi_k^{(t+1)}\}$ satisfying

$$\mathcal{F}(\Pi^{(t)}) = \sum_{j=1}^k \sum_{a \in \pi_j^{(t)}} d(c_j^{(t)}, a) \overset{\text{(Step A)}}{\geq} \sum_{j=1}^k \sum_{a \in \pi_j^{(t+1)}} d(c_j^{(t)}, a).$$

Next, applying Step B to each cluster $\pi_j^{(t+1)}$ (to determine new centers $c_j^{(t+1)}$), we obtain

$$\sum_{j=1}^k \sum_{a \in \pi_j^{(t+1)}} d(c_j^{(t)}, a) \overset{\text{(Step B)}}{\geq} \sum_{j=1}^k \sum_{a \in \pi_j^{(t+1)}} d(c_j^{(t+1)}, a) =: \mathcal{F}(\Pi^{(t+1)}).$$

Therefore, $\mathcal{F}(\Pi^{(t)}) \geq \mathcal{F}(\Pi^{(t+1)})$. $\qquad\qquad\square$

Table 3.3 Searching for the LS-optimal 2-partition of the set $\mathcal{A} = \{0, 2, 3\}$

Iteration	π_1	π_2	c_1	c_2	$\mathcal{F}_{LS}(\Pi)$
1	{0,2}	{3}	1	3	2
2	{0,2}	{3}	1	3	2

Example 3.15 Determine LS-optimal 2-partition of the set $\mathcal{A} = \{0, 2, 3\}$ by using the k-means algorithm with initial partition $\Pi^{(0)} = \{\{0, 2\}, \{3\}\}$.

Table 3.3 shows that using the LS-distance-like function, the k-means algorithm sometimes cannot improve even the initial partition. However, in the previous example, a better partition is $\Pi^\star = \{\{0\}, \{2, 3\}\}$ because $\mathcal{F}_{LS}(\Pi^\star) = 0.5$. This simple example shows that the k-means algorithm gives a locally optimal partition. Choosing another initial partition we might have obtained the k-GOPart. Give it a try!

In addition to aforementioned shortcomings of the k-means algorithm that it heavily depends on the initial partition and that it produces only some locally optimal partition, as in Example 3.15, one should also point out yet another limitation: during the iterative process it may happen that some clusters become empty sets, i.e. that the number of clusters decreases (see Example 4.14).

3.2   Clustering Data with One Feature

Let $\mathcal{A}$ be a set of $m \geq 2$ elements with one feature. As we remarked on page 34, such a set can be considered as a subset of $\mathbb{R}$, i.e. $\mathcal{A} = \{a_1, \ldots, a_m\} \subset \mathbb{R}$. The set $\mathcal{A}$ should be grouped into $1 \leq k \leq m$ clusters $\pi_1, \ldots, \pi_k$ conforming with Definition 3.1. For example, days of a year can be clustered into three clusters according to mean daily temperatures (mdt) expressed in °C: cluster of cold days (mdt 8°C and below), cluster of mild days (mdt 15 ± 6°C), and cluster of warm days (mdt 22°C and above). According to the named feature, we are going to represent every element $a \in \mathcal{A}$ with a real number which we will also denote by a. Therefore, from now on, we are going to assume that $\mathcal{A} = \{a_1, \ldots, a_m\} \subset \mathbb{R}$ is a multiset[2] of data, i.e. some elements may appear multiple times in $\mathcal{A}$. So, in our example, the multiset $\mathcal{A}$ would have 365 elements, all being, say, 8, 15, or 22, i.e. as a set $\mathcal{A} = \{8, 15, 22\}$.

Given a distance-like function $d \colon \mathbb{R} \times \mathbb{R} \to \mathbb{R}_+$, one can associate with every cluster $\pi_j \in \Pi$ its center c_j as follows:

$$c_j \in \arg\min_{x \in \mathbb{R}} \sum_{a \in \pi_j} d(x, a), \quad j = 1, \ldots, k. \tag{3.6}$$

---

[2]See Remark 3.6.

Furthermore, if we define the objective function $\mathcal{F}: \mathcal{P}(\mathcal{A}; k) \rightarrow \mathbb{R}_+$ on the set $\mathcal{P}(\mathcal{A}; k)$ of all k-partitions of $\mathcal{A}$, by

$$\mathcal{F}(\Pi) = \sum_{j=1}^{k} \sum_{a \in \pi_j} d(c_j, a), \tag{3.7}$$

then the search for an optimal k-GOPart is done by solving the following optimization problem:

$$\underset{\Pi \in \mathcal{P}(\mathcal{A};k)}{\arg \min} \ \mathcal{F}(\Pi). \tag{3.8}$$

Note that the k-GOPart will have the property that the sum of *dispersions* (the sum of deviations) of cluster elements to its center is minimal. In this way, we attempt to obtain as good the inner compactness and separation between clusters as possible.

Remark 3.16 The number of all k-partitions of a set $\mathcal{A}$ with m elements can be rather huge (see Table 3.1). But in the case of data with one feature, $\mathcal{A} \subset \mathbb{R}$, it is obvious that the optimal partition can be expected among partitions where the clusters follow one another. This means that all elements of cluster π_2 are on the right hand side of cluster π_1, all elements of cluster π_3 are on the right hand side of cluster π_2, etc. (see [148, p. 161]). The number of such partitions is considerably smaller as shown by the following proposition (see also Table 3.4).

Table 3.4   The number of k-partitions of a set $\mathcal{A} \subset \mathbb{R}$ whose clusters follow one another

$\binom{m-1}{k-1}$	$k = 2$	$k = 3$	$k = 4$	$k = 5$	$k = 6$	$k = 8$	$k = 10$
$m = 10$	9	36	84	126	126	36	1
$m = 30$	29	406	3 654	23 751	118 755	1 560 780	10 015 005
$m = 50$	49	1 176	18 424	211 876	1 906 884	85 900 584	2 054 455 634

Proposition 3.17 *Let $\mathcal{A} = \{a_i \in \mathbb{R} : i = 1, \ldots, m\}$. The number of all k-partitions of the set $\mathcal{A}$ whose clusters $\pi_1, \ldots, \pi_k$ follow one another equals*

$$\binom{m-1}{k-1}. \tag{3.9}$$

Proof Without loss of generality, assume $\mathcal{A} = \{1, \ldots, m\}$.

Obviously the smallest element of cluster π_1 has to be $1 \in A$, and the largest element of cluster π_k has to be $m \in A$. Denote the largest elements of clusters $\pi_1, \ldots, \pi_{k-1}$ by $r_1, \ldots, r_{k-1}$. These numbers satisfy $1 \leq r_1 < r_2 < \cdots < r_{k-1} < m$. Therefore, the question about the number of all k-partitions of the set A whose clusters follow one another boils down to the question of number of elements of the set

$$S = \{(r_1, \ldots, r_{k-1}) \in A^{k-1} : 1 \leq r_1 < r_2 < \cdots < r_{k-1} < m\},$$

i.e. the number of all subsets of the set $\{1, \ldots, m-1\}$ with $k-1$ elements. And this is the number of $(k-1)$-combinations of a set with $m-1$ elements.   □

3.2.1   Application of the LS-Distance-like Function

Let $\Pi = \{\pi_1, \ldots, \pi_k\}$ be a k-partition of the set $A \subset \mathbb{R}$, and $d_{LS} \colon \mathbb{R} \times \mathbb{R} \to \mathbb{R}_+$, defined by $d_{LS}(x, y) = (x - y)^2$, be the LS-distance-like function. The centers of clusters $\pi_1, \ldots, \pi_k$ are called *centroids* and are determined as follows:

$$c_j = \arg\min_{x \in \mathbb{R}} \sum_{a \in \pi_j} (x - a)^2 = \frac{1}{|\pi_j|} \sum_{a \in \pi_j} a, \quad j = 1, \ldots, k, \tag{3.10}$$

and the objective function (3.7) is defined by

$$\mathcal{F}_{LS}(\Pi) = \sum_{j=1}^{k} \sum_{a \in \pi_j} (c_j - a)^2. \tag{3.11}$$

Example 3.18 Given the set $A = \{2, 4, 8, 10, 16\}$, find all 3-partitions of A satisfying Definition 3.1 and whose clusters follow one another. Determine also the corresponding centroids and values of the objective function $\mathcal{F}_{LS}$.

Table 3.5 All 3-partitions of $A = \{2, 4, 8, 10, 16\}$ whose clusters follow one another

π_1	π_2	π_3	c_1	c_2	c_3	$\mathcal{F}_{LS}(\Pi)$	$\mathcal{G}(\Pi)$
{2}	{4}	{8,10,16}	2	4	11.33	$0 + 0 + 34.67 = 34.67$	$36 + 16 + 33.33 = 85.33$
{2}	{4,8}	{10,16}	2	6	13	$0 + 8 + 18 = 26$	$36 + 8 + 50 = 94$
{2}	{4,8,10}	{16}	2	7.33	16	$0 + 18.67 + 0 = 18.67$	$36 + 1.33 + 64 = 101.33$
{2,4}	{8}	{10,16}	3	8	13	$2 + 0 + 18 = 20$	$50 + 0 + 50 = 100$
{2,4}	{8,10}	{16}	3	9	16	$2 + 2 + 0 = 4$	$50 + 2 + 64 = 116$
{2,4,8}	{10}	{16}	4.67	10	16	$18.67 + 0 + 0 = 18.67$	$33.33 + 4 + 64 = 101.33$

According to the Stirling formula (3.2), the number of all 3-partitions of the set $\mathcal{A}$ is 25. But the number of 3-partitions of the same set with clusters following one another is only $\binom{5-1}{3-1} = \frac{4!}{2!\cdot2!} = 6$, see Table 3.5. From this table we see that the LS-optimal 3-partition is $\Pi^\star = \{\{2, 4\}, \{8, 10\}, \{16\}\}$, the one in the fifth row, where the objective function $\mathcal{F}_{LS}$ attains its (global) minimum $\mathcal{F}_{LS}(\Pi^\star) = 4$, and $\Pi^\star$ is therefore the LS-3-GOPart.

Exercise 3.19 What is the number of all 3-partitions, and the number of 3-partitions with clusters following one another, of the set $\mathcal{A} = \{1, 4, 5, 8, 10, 12, 15\}$? Write down all 3-partitions with clusters following one another and find the LS-optimal one.

Solution: The number of all partitions is 301, and the number of partitions with clusters following one another is 15. The LS-optimal 3-partition is $\Pi^\star = \{\{1, 4, 5\}, \{8, 10\}, \{12, 15\}\}$, and $\mathcal{F}(\Pi^\star) = \frac{91}{6} \approx 15.1667$.

3.2.2 The Dual Problem

The following lemma shows that applying the LS-distance-like function, the dispersion of the set $\mathcal{A}$ about its center c equals the sum of dispersions of clusters π_j, $j = 1, \ldots, k$, about their centers c_j, $j = 1, \ldots, k$, and the weighted sum of squared distances between c and c_j, where the weights are determined by the size of sets π_j.

Lemma 3.20 *Let* $\mathcal{A} = \{a_1, \ldots, a_m\}$ *be a data set, let* $\Pi = \{\pi_1, \ldots, \pi_k\}$ *be its k-partition with clusters* $\pi_1, \ldots, \pi_k$, *and let*

$$c = \frac{1}{m} \sum_{i=1}^{m} a_i, \qquad c_j = \frac{1}{|\pi_j|} \sum_{a \in \pi_j} a, \quad j = 1, \ldots, k. \qquad (3.12)$$

Then

$$\sum_{i=1}^{m} (c - a_i)^2 = \mathcal{F}_{LS}(\Pi) + \mathcal{G}(\Pi), \qquad (3.13)$$

where

$$\mathcal{F}_{LS}(\Pi) = \sum_{j=1}^{k} \sum_{a \in \pi_j} (c_j - a)^2, \qquad (3.14)$$

$$\mathcal{G}(\Pi) = \sum_{j=1}^{k} |\pi_j| (c_j - c)^2. \qquad (3.15)$$

Proof Notice that for c_j we have $\sum_{a_i \in \pi_j} (c_j - a_i) = 0$. Using this, for every $x \in \mathbb{R}$ we have

$$\sum_{a_i \in \pi_j} (x - a_i)^2 = \sum_{a_i \in \pi_j} \left((x - c_j) + (c_j - a_i) \right)^2$$

$$= \sum_{a_i \in \pi_j} (x - c_j)^2 + 2 \sum_{a_i \in \pi_j} (x - c_j)(c_j - a_i) + \sum_{a_i \in \pi_j} (c_j - a_i)^2$$

$$= |\pi_j| (x - c_j)^2 + \sum_{a_i \in \pi_j} (c_j - a_i)^2,$$

i.e.

$$\sum_{a_i \in \pi_j} (x - a_i)^2 = \sum_{a_i \in \pi_j} (c_j - a_i)^2 + |\pi_j| (c_j - x)^2, \quad j = 1, \ldots, k. \tag{3.16}$$

If we put $c = \frac{1}{m} \sum_{i=1}^{m} a_i$ in (3.16) instead of x and sum all equations, we obtain (3.13).

$\square$

The objective function $\mathcal{F}_{LS}$ occurred naturally in formula (3.13), and this formula shows that the total dispersion of elements of the set $\mathcal{A}$ about its centroid c, can be described as the sum of two objective functions $\mathcal{F}_{LS}$ and $\mathcal{G}$.

In particular, the LS-optimal 3-partition of the set $\mathcal{A}$ in Example 3.18 is $\Pi^\star = \{\{2, 4\}, \{8, 10\}, \{16\}\}$, for which $\mathcal{F}(\Pi^\star) = 4$ (see Table 3.5). The obvious question is: what is $\mathcal{G}(\Pi^\star)$ in this example?

To answer this question, let us expand Table 3.5 by adding values of the function $\mathcal{G}$ for each partition (the rightmost part of the table). Notice that the sum $\mathcal{F}_{LS}(\Pi) + \mathcal{G}(\Pi)$ is constant and equals $\sum_{i=1}^{m} (c - a_i)^2 = 120$, which is in accordance with (3.13), and the maximal value of the function $\mathcal{G}$ is attained precisely at the LS-optimal 3-partition $\Pi^\star$ for which the objective function $\mathcal{F}_{LS}$ attains its minimal value.

Is this accidental?

To answer this question, let us first try to solve the following example which considers a similar problem. For this we need some foreknowledge from calculus (see e.g. [78, 85]).

Example 3.21 Let $\varphi, \psi \in C^2(\mathbb{R})$ be two functions such that $\varphi(x) + \psi(x) = \kappa$ for some constant $\kappa \in \mathbb{R}$. The function φ attains its local minimum at $x_0 \in \mathbb{R}$ if and only if the function ψ attains at x_0 its local maximum, and $\varphi(x_0) = \kappa - \psi(x_0)$.

If $\varphi'(x_0) = 0$, then $\psi'(x_0) = 0$, and vice versa. Also, if $\varphi''(x_0) > 0$, then $\psi''(x_0) < 0$, and vice versa. Therefore, we have

⋄ $x_0 \in \arg\min\limits_{x\in\mathbb{R}} \varphi(x)$ if and only if $x_0 \in \arg\max\limits_{x\in\mathbb{R}} \psi(x)$;

⋄ $\min\limits_{x\in\mathbb{R}} \varphi(x) = \kappa - \max\limits_{x\in\mathbb{R}} \psi(x)$,   i.e. $\varphi(x_0) = \kappa - \psi(x_0)$.

Check whether the two functions $\varphi(x) = x^2 - 1$ and $\psi(x) = -x^2 + 3$ satisfy these properties. Draw their graphs in the same coordinate system. Try to come up by yourself with another example of a pair of functions φ, ψ satisfying the said properties.

The next theorem follows directly from Lemma 3.20 [176].

Theorem 3.22 *Using the notation from Lemma 3.20, there exists a partition* $\Pi^\star \in \mathcal{P}(\mathcal{A}; k)$ *such that*

(i) $\Pi^\star \in \underset{\Pi\in\mathcal{P}(\mathcal{A};k)}{\arg\min}\ \mathcal{F}_{LS}(\Pi) = \underset{\Pi\in\mathcal{P}(\mathcal{A};k)}{\arg\max}\ \mathcal{G}(\Pi),$

(ii) $\underset{\Pi\in\mathcal{P}(\mathcal{A};k)}{\min}\ \mathcal{F}_{LS}(\Pi) = \mathcal{F}_{LS}(\Pi^\star)$ *and* $\underset{\Pi\in\mathcal{P}(\mathcal{A};k)}{\max}\ \mathcal{G}(\Pi) = \mathcal{G}(\Pi^\star),$

where $\mathcal{G}(\Pi^\star) = \sum\limits_{i=1}^{m}(c - a_i)^2 - \mathcal{F}_{LS}(\Pi^\star).$

This means that in order to find the LS-optimal partition, instead of minimizing the function $\mathcal{F}_{LS}$ given by (3.11), one can maximize the dual function $\mathcal{G}$:

$$\underset{\Pi\in\mathcal{P}(\mathcal{A};k)}{\arg\max}\ \mathcal{G}(\Pi), \qquad \mathcal{G}(\Pi) = \sum_{j=1}^{k}|\pi_j|(c_j - c)^2. \tag{3.17}$$

The optimization problem (3.17) is called the ***dual problem*** with respect to the optimization problem $\arg\min\limits_{\Pi\in\mathcal{P}(\mathcal{A};k)} \mathcal{F}_{LS}(\Pi)$.

One can say that the LS-optimal partition has the property that the sum of dispersions of cluster elements (sum over all clusters of the sums of LS-distances from cluster elements to their centroids) is minimal, and at the same time the centroids of clusters are distant from each another as much as possible. In this way, one achieves the best inner compactness and best separation between clusters.

3.2.3   Least Absolute Deviation Principle

Let $\Pi = \{\pi_1, \ldots, \pi_k\}$ be a k-partition of the set $\mathcal{A} \subset \mathbb{R}$, and $d_1 \colon \mathbb{R} \times \mathbb{R} \to \mathbb{R}_+$, defined by $d_1(x, y) = |x - y|$, be the ℓ_1 metric function. The centers $c_1, \ldots, c_k$ of clusters $\pi_1, \ldots, \pi_k$ are determined by

$$c_j = \text{med}(\pi_j) \in \text{Med}(\pi_j) = \arg\min_{x\in\mathbb{R}} \sum_{a\in\pi_j} |x - a|, \quad j = 1, \ldots, k, \tag{3.18}$$

Table 3.6 Partitions of the set $\mathcal{A}$ with clusters following one another

π_1	π_2	π_3	c_1	c_2	c_3	$\mathcal{F}_1(\Pi)$
$\{2\}$	$\{4\}$	$\{8,10,16\}$	2	4	10	$0+0+8=8$
$\{2\}$	$\{4,8\}$	$\{10,16\}$	2	6	13	$0+4+6=10$
$\{2\}$	$\{4,8,10\}$	$\{16\}$	2	8	16	$0+6+0=6$
$\{2,4\}$	$\{8\}$	$\{10,16\}$	3	8	13	$2+0+6=8$
$\{2,4\}$	$\{8,10\}$	$\{16\}$	3	9	16	$2+2+0=4$
$\{2,4,8\}$	$\{10\}$	$\{16\}$	4	10	16	$6+0+0=6$

and the objective function (3.7) is defined by

$$\mathcal{F}_1(\Pi) = \sum_{j=1}^{k} \sum_{a \in \pi_j} |c_j - a|. \tag{3.19}$$

If one uses (3.20) from Exercise 3.25 then, in order to calculate the objective function (3.19), one does not need to know the centers of clusters (3.18), which speeds up the calculation process.

Example 3.23 Let $\mathcal{A} = \{2, 4, 8, 10, 16\}$ be the set as in Example 3.18. We want to find all of its 3-partitions satisfying Definition 3.1 with clusters following one another.

In addition, we want to determine the corresponding cluster centers and the values of the objective function $\mathcal{F}_1$ using the ℓ_1 metric function and then find the globally ℓ_1-optimal 3-partition.

The number of all 3-partitions with clusters following one another is $\binom{m-1}{k-1} = 6$ and, as shown in Table 3.6, the ℓ_1-optimal 3-partition is $\Pi^\star = \{\{2, 4\}, \{8, 10\}, \{16\}\}$ since the objective function $\mathcal{F}_1$ defined by (3.19) attains at $\Pi^\star$ its lowest value (global minimum). Hence, partition $\Pi^\star$ is the ℓ_1-GOPart.

Exercise 3.24 Among all partitions of the set $\mathcal{A} = \{1, 4, 5, 8, 10, 12, 15\}$ from Exercise 3.19, find the ℓ_1-optimal 3-partition.

Exercise 3.25 Let $\mathcal{A} = \{a_1, \ldots, a_m\}$ be a finite increasing sequence of real numbers. Prove the following:

$$\sum_{i=1}^{m} |a_i - \mathrm{med}(\mathcal{A})| = \sum_{i=1}^{\lceil \frac{m}{2} \rceil} (a_{m-i+1} - a_i), \tag{3.20}$$

where $\lceil x \rceil$ equals x if x in an integer, and $\lceil x \rceil$ is the smallest integer larger than x if x is not an integer.[3] For example, $\lceil 20 \rceil = 20$, whereas $\lceil 20.3 \rceil = 21$.

---

[3]In *Mathematica* computation system, $\lceil x \rceil$ is obtained by `Ceiling[x]`, and the greatest integer smaller or equal to x, $\lfloor x \rfloor$, by `Floor[x]`.

3.2.4 Clustering Weighted Data

Let $\mathcal{A} = \{a_1, \ldots, a_m\} \subset \mathbb{R}$ be a data set of real numbers, and to each datum $a_i \in \mathcal{A}$ a corresponding weight $w_i > 0$ is assigned. For example, in [155, Example 3.8], where the authors analyze the problem of high water levels of the river Drava at Donji Miholjac, data are days of the year and weights are the measured water levels.

In the case of weighted data the objective function (3.7) becomes

$$\mathcal{F}(\Pi) = \sum_{j=1}^{k} \sum_{a_i \in \pi_j} w_i \, d(c_j, a_i), \tag{3.21}$$

where

$$c_j \in \arg\min_{x \in \mathbb{R}} \sum_{a_i \in \pi_j} w_i \, d(x, a_i), \quad j = 1, \ldots, k. \tag{3.22}$$

In particular, when applying the LS-distance-like function, the centers c_j of clusters π_j are weighted arithmetic means of data in π_j

$$c_j = \frac{1}{\kappa^j} \sum_{a_i \in \pi_j} w_i \, a_i, \qquad \kappa^j = \sum_{a_i \in \pi_j} w_i, \tag{3.23}$$

and when applying the ℓ_1 metric function, the centers c_j of clusters π_j are weighted medians of data in π_j [144, 187]

$$c_j = \underset{a_i \in \pi_j}{\text{med}} (w_i, a_i) \in \text{Med}(w, \mathcal{A}). \tag{3.24}$$

Example 3.26 Let us again consider the set $\mathcal{A} = \{1, 4, 5, 8, 10, 12, 15\}$ from Exercise 3.19. Assign to each but the last datum the weight 1 and to the last datum the weight 3. Now the LS-optimal 3-partition becomes $\Pi^\star = \{\{1, 4, 5\}, \{8, 10, 12\}, \{15\}\}$ with centroids $\frac{10}{3}$, 10, and 15, and the objective function value $\mathcal{F}(\Pi^\star) = \frac{50}{3} \approx 16.667$.

In order to determine the centers of clusters when applying the ℓ_1 metric function, one has to know how to calculate the weighted median of data. As mentioned in Sect. 2.1.3, this might turn out not to be a simple task. If the weights are integers, the problem can be reduced to finding the usual median of data (see Example 2.16). If the weights are not integers, then by multiplying with some appropriate number and taking approximations, one can reduce the weights to integers.

Exercise 3.27 Find the ℓ_1-optimal 3-partition of the set $\mathcal{A}$ from the previous example with all weights being equal to 1 and in the case when the weights are assigned as in the previous example.

Exercise 3.28 Write down formulas for the centroid of the set $\mathcal{A}$, and for the objective functions $\mathcal{F}$ and $\mathcal{G}$ for the data set $\mathcal{A}$ with weights $w_1, \ldots, w_m > 0$.

Solution: $\mathcal{G}(\Pi) = \sum_{j=1}^{k} \left(\sum_{\pi_j} w_s \right)(c_j - c)^2$, where $\sum_{\pi_j} w_s$ is the sum of weights of all elements from the cluster π_j.

3.3 Clustering Data with Two or Several Features

Let $\mathcal{A}$ be a set of $m \geq 2$ elements with $n \geq 2$ features. As already said, such a set can be regarded as a subset of $\mathbb{R}^n$, i.e. $\mathcal{A} = \{a_i = (a_i^1, \ldots, a_i^n) \in \mathbb{R}^n : i = 1, \ldots, m\}$. The set $\mathcal{A}$ should be grouped in accordance with Definition 3.1 into $1 \leq k \leq m$ disjoint nonempty clusters. For example, elements of the set $\mathcal{A} \subset \mathbb{R}^2$ from Example 3.5 have two features—the abscissa and the ordinate, and the elements can be grouped into 2, 3, 4, 5, 6, 7, 8, or more clusters (see Fig. 4.10).

Let $\Pi \in \mathcal{P}(\mathcal{A}; k)$ be a partition of the set $\mathcal{A}$. Given a distance-like function $d \colon \mathbb{R}^n \times \mathbb{R}^n \to \mathbb{R}_+$, to each cluster $\pi_j \in \Pi$ one can assign its center c_j in the following way:

$$c_j \in \arg\min_{x \in \mathbb{R}^n} \sum_{a \in \pi_j} d(x, a), \quad j = 1, \ldots, k. \tag{3.25}$$

Analogously to the case of data with one feature, if we define the objective function $\mathcal{F} \colon \mathcal{P}(\mathcal{A}; k) \to \mathbb{R}_+$ on the set $\mathcal{P}(\mathcal{A}; k)$ of all partitions of the set $\mathcal{A}$ consisting of k clusters, by

$$\mathcal{F}(\Pi) = \sum_{j=1}^{k} \sum_{a \in \pi_j} d(c_j, a), \tag{3.26}$$

then we search for the optimal k-partition by solving the following GOP:

$$\arg\min_{\Pi \in \mathcal{P}(\mathcal{A};k)} \mathcal{F}(\Pi). \tag{3.27}$$

Note that the optimal k-partition has the property that the dispersion (the sum of d-distances of the cluster elements to their centers) is minimal. In this way, we attempt to achieve as good the inner compactness of clusters as possible.

3.3.1 Least Squares Principle

Let $\Pi = \{\pi_1, \ldots, \pi_k\}$ be a partition of the set $\mathcal{A} = \{a_i = (a_i^1, \ldots, a_i^n) \in \mathbb{R}^n : i = 1, \ldots, m\}$. The centers $c_1, \ldots, c_k$ of clusters $\pi_1, \ldots, \pi_k$ for the LS-distance-like function $d_{LS} \colon \mathbb{R}^n \times \mathbb{R}^n \to \mathbb{R}_+$, defined as $d_{LS}(a, b) = \|a - b\|^2$, are called *centroids* and are obtained as follows:

$$c_j = \underset{x \in \mathbb{R}^n}{\arg \min} \sum_{a \in \pi_j} \|x - a\|^2 = \frac{1}{|\pi_j|} \sum_{a \in \pi_j} a$$

$$= \left(\frac{1}{|\pi_j|} \sum_{a \in \pi_j} a^1, \dots, \frac{1}{|\pi_j|} \sum_{a \in \pi_j} a^n \right), \qquad j = 1, \dots, k, \tag{3.28}$$

where $\sum_{a \in \pi_j} a^\ell, \ell = 1, \dots, n$, denotes the sum of ℓ-th components of all elements in cluster π_j. In this case the objective function (3.26) is defined by

$$\mathcal{F}_{LS}(\Pi) = \sum_{j=1}^{k} \sum_{a \in \pi_j} \|c_j - a\|^2. \tag{3.29}$$

Example 3.29 Let $\mathcal{A} = \{a_1 = (1, 1), a_2 = (3, 1), a_3 = (3, 2), a_4 = (2, 2)\}$ be a set in the plane. The number of its 2-partitions is $|\mathcal{P}(\mathcal{A}; 2)| = 2^{4-1} - 1 = 7$, and all partitions are listed in Table 3.7. Let us find the optimal 2-partition.

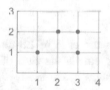

Fig. 3.4 The set $\mathcal{A} \subset \mathbb{R}^2$

The elements of $\mathcal{A}$ have two features—the abscissa and the ordinate, so the set $\mathcal{A}$ can be simply recorded as $\mathcal{A} = \{a_i = (x_i, y_i) \in \mathbb{R}^2 : i = 1, \dots, 4\}$ and depicted in the plane (see Fig. 3.4).

According to (3.2) the set $\mathcal{A}$ has 7 different 2-partitions. Let $\Pi = \{\pi_1, \pi_2\}$ be one of them. Its centroids are defined by

$$c_1 = \frac{1}{|\pi_1|} \sum_{a \in \pi_1} a, \qquad c_2 = \frac{1}{|\pi_2|} \sum_{a \in \pi_2} a,$$

and the corresponding LS-objective function is

$$\mathcal{F}_{LS}(\Pi) = \sum_{a \in \pi_1} \|c_1 - a\|^2 + \sum_{a \in \pi_2} \|c_2 - a\|^2.$$

Table 3.7 Partitions, centers, and values of objective functions $\mathcal{F}_{LS}$ and $\mathcal{G}$ from Example 3.29

π_1	π_2	c_1	c_2	$\mathcal{F}_{LS}(\Pi)$	$\mathcal{G}(\Pi)$
$\{(1,1)\}$	$\{(2,2),(3,1,(3,2)\}$	$(1,1)$	$(2.67,1.67)$	$0+1.33=1.33$	$1.82+0.60=2.42$
$\{(3,1)\}$	$\{(1,1),(2,2),(3,2)\}$	$(3,1)$	$(2.00,1.67)$	$0+2.67=2.67$	$0.81+0.27=1.08$
$\{(3,2)\}$	$\{(1,1),(2,2),(3,1)\}$	$(3,2)$	$(2.0,1.3)$	$0+2.67=2.67$	$0.81+0.27=1.08$
$\{(2,2)\}$	$\{(1,1),(3,1),(3,2)\}$	$(2,2)$	$(2.3,1.3)$	$0+3.33=3.33$	$0.31+0.10=0.42$
$\{(1,1),(3,1)\}$	$\{(2,2),(3,2)\}$	$(2,1)$	$(2.5,2.0)$	$2+0.5=2.5$	$0.625+0.625=1.25$
$\{(1,1),(3,2)\}$	$\{(2,2),(3,1)\}$	$(2,1.5)$	$(2.5,1.5)$	$2.5+1.0=3.5$	$0.125+0.125=0.25$
$\{(1,1),(2,2)\}$	$\{(3,1),(3,2)\}$	$(1.5,1.5)$	$(3.0,1.5)$	$1+0.5=1.5$	$1.125+1.125=2.25$

So the value of the objective function $\mathcal{F}_{LS}$ is obtained by adding the sum of LS-distances of elements of cluster π_1 to its centroid c_1, and the sum of LS-distances of elements of cluster π_2 to its centroid c_2.

Table 3.7 lists all partitions of the set $\mathcal{A}$, the centroids of respective clusters, and values of objective function $\mathcal{F}_{LS}$. As one can see, the LS-optimal partition is $\Pi^\star = \{\{(1,1)\}, \{(2,2),(3,1),(3,2)\}\}$, since $\mathcal{F}_{LS}$ attains the global minimum at it (see also Fig. 3.4).

3.3.2 The Dual Problem

The next lemma shows that in the case of LS-distance-like function, dispersion of the set $\mathcal{A}$ about its center c equals the sum of dispersions of all clusters π_j, $j = 1, \ldots, k$, about their centers c_j, $j = 1, \ldots, k$, and the weighted sum of squared distances between the center c and cluster centers c_j, where the weights are determined by sizes of sets π_j.

Lemma 3.30 *Let $\mathcal{A} = \{a_i \in \mathbb{R}^n : i = 1, \ldots, m\}$ be a data set, $\Pi = \{\pi_1, \ldots, \pi_k\}$ some k-partition with clusters $\pi_1, \ldots, \pi_k$, and let*

$$c = \frac{1}{m} \sum_{i=1}^{m} a_i, \qquad c_j = \frac{1}{|\pi_j|} \sum_{a_i \in \pi_j} a_i, \quad j = 1, \ldots, k \tag{3.30}$$

be the centroid of the set $\mathcal{A}$ and centroids of clusters $\pi_1, \ldots, \pi_k$, respectively. Then

$$\sum_{i=1}^{m} \|c - a_i\|^2 = \mathcal{F}_{LS}(\Pi) + \mathcal{G}(\Pi), \tag{3.31}$$

where

$$\mathcal{F}_{LS}(\Pi) = \sum_{j=1}^{k} \sum_{a_i \in \pi_j} \|c_j - a_i\|^2, \tag{3.32}$$

$$\mathcal{G}(\Pi) = \sum_{j=1}^{k} |\pi_j| \|c_j - c\|^2. \tag{3.33}$$

Proof Note first that c_j satisfies the arithmetic mean property

$$\sum_{a_i \in \pi_j} (c_j - a_i) = 0. \tag{3.34}$$

For an arbitrary $x \in \mathbb{R}^n$ we have

$$\sum_{a_i \in \pi_j} \|x - a_i\|^2 = \sum_{a_i \in \pi_j} \|(x - c_j) + (c_j - a_i)\|^2$$

$$= \sum_{a_i \in \pi_j} \|x - c_j\|^2 + 2 \sum_{a_i \in \pi_j} \langle x - c_j, c_j - a_i \rangle + \sum_{a_i \in \pi_j} \|c_j - a_i\|^2.$$

Since $\sum\limits_{a_i \in \pi_j} \langle x - c_j, c_j - a_i \rangle = \langle x - c_j, \sum\limits_{u_i \in \pi_j} (c_j - a_i) \rangle \overset{(3.34)}{=} 0$, from the previous equality we obtain

$$\sum_{a_i \in \pi_j} \|x - a_i\|^2 = \sum_{a_i \in \pi_j} \|c_j - a_i\|^2 + |\pi_j| \|c_j - x\|^2, \quad j = 1, \ldots, k. \tag{3.35}$$

Substituting $c = \frac{1}{m} \sum\limits_{i=1}^{m} a_i$ for x into (3.35) and adding all equations, we obtain (3.31). $\qquad\square$

The objective function $\mathcal{F}_{LS}$ occurs in (3.31) naturally, and this formula shows that the total dispersion of elements of $\mathcal{A}$ about its centroid c can be expressed as the sum of two objective functions $\mathcal{F}_{LS}$ and $\mathcal{G}$.

As in Sect. 3.2.2, using Lemma 3.30, one can show that the following theorem holds true [42, 176].

Theorem 3.31 *Using the notation as in Lemma 3.30, there exists a partition $\Pi^\star \in \mathcal{P}(\mathcal{A}; k)$ such that*

(i) $\Pi^\star \in \underset{\Pi \in \mathcal{P}(\mathcal{A};k)}{\arg\min} \ \mathcal{F}_{LS}(\Pi) = \underset{\Pi \in \mathcal{P}(\mathcal{A};k)}{\arg\max} \ \mathcal{G}(\Pi),$

(ii) $\underset{\Pi \in \mathcal{P}(\mathcal{A};k)}{\min} \ \mathcal{F}_{LS}(\Pi) = \mathcal{F}_{LS}(\Pi^\star)$ *and* $\underset{\Pi \in \mathcal{P}(\mathcal{A};k)}{\max} \ \mathcal{G}(\Pi) = \mathcal{G}(\Pi^\star),$

where $\mathcal{G}(\Pi^\star) = \sum\limits_{i=1}^{m} \|c - a_i\|^2 - \mathcal{F}_{LS}(\Pi^\star).$

This means that in order to find the LS-optimal partition, instead of minimizing the function $\mathcal{F}_{LS}$ defined by (3.32), one can solve the problem of maximizing the function $\mathcal{G}$

$$\underset{\Pi \in \mathcal{P}(\mathcal{A};k)}{\arg\max} \ \mathcal{G}(\Pi), \qquad \mathcal{G}(\Pi) = \sum_{j=1}^{k} |\pi_j| \, \|c_j - c\|^2. \tag{3.36}$$

The optimization problem (3.36) is called the **dual problem** for the optimization problem $\arg\min_{\Pi \in \mathcal{P}(\mathcal{A};k)} \mathcal{F}_{LS}(\Pi)$.

One can say that the LS-optimal partition has the property that the sum of dispersion of cluster elements (the sum over all clusters of sums of LS-distances between cluster elements and respective centroids) is minimal, and at the same time the clusters are maximally separated. In this way one achieves the best inner compactness and separation between clusters.

Example 3.32 In Example 3.29 one can consider also the corresponding dual problem.

In particular, in this case the formula (3.31) becomes

$$\sum_{i=1}^{m} \|c - a_i\|^2 = \left(\sum_{a \in \pi_1} \|c_1 - a\|^2 + \sum_{a \in \pi_2} \|c_2 - a\|^2 \right) + \left(|\pi_1| \, \|c_1 - c\|^2 + |\pi_2| \, \|c_2 - c\|^2 \right),$$

and the dual optimization problem (3.36) becomes

$$\underset{\Pi \in \mathcal{P}(\mathcal{A};k)}{\arg\max} \ \mathcal{G}(\Pi), \qquad \mathcal{G}(\Pi) = |\pi_1| \, \|c_1 - c\|^2 + |\pi_2| \, \|c_2 - c\|^2.$$

The rightmost part of Table 3.7 shows the values of the dual objective function $\mathcal{G}$ for each 2-partition. As can be seen, $\mathcal{G}$ attains its maximal value at the partition $\Pi^\star = \{\{(1,1)\}, \{(2,2), (3,1), (3,2)\}\}$, the same one at which $\mathcal{F}_{LS}$, given by (3.29), attains the minimal value.

Example 3.33 The set $\mathcal{A} = \{a_i = (x_i, y_i) : i = 1, \dots, 8\} \subset \mathbb{R}^2$ is given by the following table:

i	1	2	3	4	5	6	7	8
x_i	1	4	4	4	7	8	8	10
y_i	3	5	7	9	1	6	10	8

Applying the LS-distance-like function for 2-partitions

$$\Pi_1 = \{\{a_1, a_2, a_5\}, \ \{a_3, a_4, a_6, a_7, a_8\}\},$$

$$\Pi_2 = \{\{a_1, a_2, a_3, a_5\}, \ \{a_4, a_6, a_7, a_8\}\},$$

(a) Partition Π_1 (b) Partition Π_2

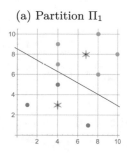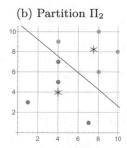

Fig. 3.5 Two partitions of the set $\mathcal{A}$ from Example 3.33

depicted in Fig. 3.5 with clusters colored blue and red respectively, determine the centroids and corresponding values of objective functions $\mathcal{F}_{LS}$ and $\mathcal{G}$, and based on this, identify the partition being closer to the optimal one.

For the partition Π_1 we obtain $c_1 = (4, 3)$, $c_2 = (6.8, 8)$, $\mathcal{F}_{LS}(\Pi_1) = 26 + 38.8 = 64.8$, and $\mathcal{G}(\Pi_1) = 61.575$, and for partition Π_2, $c_1 = (4, 4)$, $c_2 = (7.5, 8.25)$, $\mathcal{F}_{LS}(\Pi_2) = 38 + 27.75 = 65.75$, and $\mathcal{G}(\Pi_2) = 60.625$. Therefore, the 2-partition Π_1 is closer to the LS-optimal partition. Check, applying the *Mathematica*-module WKMeans[], whether this is the globally LS-optimal 2-partition. Note (formula (3.2)) that in this case, in total there are $2^7 - 1 = 127$ different 2 partitions.

Exercise 3.34 Let the set $\mathcal{A} = \{a_i = (x_i, y_i) : i = 1, \ldots, m\}$, depicted in Fig. 3.6, be given by the following table:

i	1	2	3	4	5	6	7	8	9	10	11	12
x_i	1	2	4	4	5	6	7	8	8	8	9	10
y_i	3	1	5	9	7	1	5	2	6	10	4	8

Determine at which of the two 3-partitions shown in Fig. 3.6 does the LS-objective function $\mathcal{F}_{LS}$, given by (3.29), attain the smaller value.

(a) Partition Π_1 (b) Partition Π_2

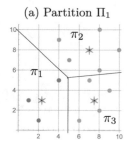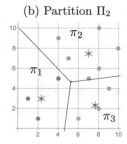

Fig. 3.6 Comparison of the two partitions in Exercise 3.34

Solution:

$$\Pi_1 = \{\{a_1, a_2, a_3\}, \ \{a_4, a_5, a_9, a_{10}, a_{12}\}, \ \{a_6, a_7, a_8, a_{11}\}\} \qquad \text{(Fig. 3.6a)}$$

$$\Pi_2 = \{\{a_1, a_2, a_3\}, \ \{a_4, a_5, a_7, a_9, a_{10}, a_{12}\}, \ \{a_6, a_8, a_{11}\}\} \qquad \text{(Fig. 3.6b)}$$

$$\Pi_1: \quad c_1 = (2.33, 3), \ c_2 = (7, 8), \ c_3 = (7.5, 3);$$

$$\mathcal{F}_{LS}(\Pi_1) = 12.67 + 34 + 15 = 61.67; \quad \mathcal{G}(\Pi_1) = 127.25,$$

$$\Pi_2: \quad c_1 = (2.33, 3), \ c_2 = (7, 7.5), \ c_3 = (7.67, 2.33);$$

$$\mathcal{F}_{LS}(\Pi_2) = 12.67 + 41.5 + 9.33 = 63.5; \quad \mathcal{G}(\Pi_2) = 125.42.$$

Hence, smaller value of LS-objective function $\mathcal{F}_{LS}$ (and larger value of the dual function $\mathcal{G}$) is attained at the 3-partition Π_1, and therefore it is closer to the LS-optimal partition. Try, using the *Mathematica*-module WKMeans[] with various initial partitions, to find a better 3-partition.

3.3.3 Least Absolute Deviation Principle

Let $\mathcal{A} \subset \mathbb{R}^n$ be a set, $\Pi = \{\pi_1, \ldots, \pi_k\}$ some k-partition, and $d_1 : \mathbb{R}^n \times \mathbb{R}^n \to \mathbb{R}_+$, given by $d_1(x, y) = \|x - y\|_1$, the ℓ_1 metric function. The centers $c_1, \ldots, c_k$ of clusters $\pi_1, \ldots, \pi_k$ are determined by

$$c_j = \mathrm{med}(\pi_j) = \left(\underset{a \in \pi_j}{\mathrm{med}}\, a^1, \ldots, \underset{a \in \pi_j}{\mathrm{med}}\, a^n \right) \in \mathrm{Med}(\pi_j)$$

$$= \left(\underset{a \in \pi_j}{\mathrm{Med}}\, a^1, \ldots, \underset{a \in \pi_j}{\mathrm{Med}}\, a^n \right) = \arg\min_{x \in \mathbb{R}^n} \sum_{a \in \pi_j} \|x - a\|_1, \qquad (3.37)$$

where $\underset{a \in \pi_j}{\mathrm{med}}\, a^s$, $s = 1, \ldots, n$, denotes the median of s-th components of elements of π_j. The ℓ_1 metric objective function is, in this case, defined as

$$\mathcal{F}_1(\Pi) = \sum_{j=1}^{k} \sum_{a \in \pi_j} \|c_j - a\|_1. \qquad (3.38)$$

Exercise 3.35 Show that using the ℓ_1 metric function, the globally optimal 2-partition from Example 3.29 is $\{\{(1, 1), (3, 2)\}, \{(2, 2), (3, 1)\}\}$ with cluster centers being $c_1 = (2, 1.5)$ and $c_2 = (2.5, 1.5)$, and the value of objective function $\mathcal{F}_1$ being 5.

Exercise 3.36 Use the least absolute deviation principle to the partitions in Exercise 3.34.

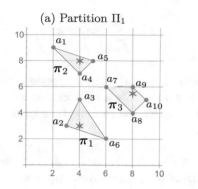
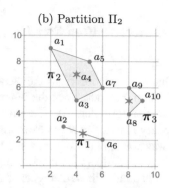

Fig. 3.7 Comparison of the two partitions in Exercise 3.37

Example 3.37 The set $\mathcal{A} = \{a_i = (x_i, y_i) \in \mathbb{R}^2 : i = 1, \ldots, 10\}$ is given by the following table:

i	1	2	3	4	5	6	7	8	9	10
x_i	2	3	4	4	5	6	6	8	8	9
y_i	9	3	5	7	8	2	6	4	6	5

Determine at which of the following two 3-partitions does the ℓ_1-objective function (3.38) attain a smaller value.

$$\Pi_1 = \{\{a_2, a_3, a_6\}, \{a_1, a_4, a_5\}, \{a_7, a_8, a_9, a_{10}\}\} \qquad \text{(Fig. 3.7a)}$$

$$\Pi_2 = \{\{a_2, a_6\}, \{a_1, a_3, a_4, a_5, a_7\}, \{a_8, a_9, a_{10}\}\}. \qquad \text{(Fig. 3.7b)}$$

The following table lists ℓ_1-centers of clusters and values of the objective function for both partitions. One can see that Π_1 is the better partition because the objection function attains a smaller value at Π_1 than at Π_2.

	c_1	c_2	c_3	$\mathcal{F}_1$
Π_1	(4, 3)	(4, 8)	(8, 5.5)	$(1+2+3)+(3+1+1)+(2.5+1.5+0.5+1.5) = 17$
Π_2	(4.5, 2.5)	(4, 7)	(8, 5)	$(2+2)+(4+2+0+2+3)+(1+1+1) = 18$

Remark 3.38 In a similar way as in Sect. 3.2.4, where we have considered the clustering problem for one-dimensional weighted data, we could proceed in the case of two- and more-dimensional data.

3.4 Objective Function $F(c_1, \ldots, c_k) = \sum_{i=1}^{m} \min_{1 \leq j \leq k} d(c_j, a_i)$

The objective function from the GOP (3.5) is not suitable for applying standard optimization methods, since the independent variable is a partition. Therefore, we are going to reformulate GOP (3.5) in such a way that the objective function becomes an ordinary function of several variables.

As was noted, in the k-means algorithm 3.10 we are going, for the given centers $c_1, \ldots, c_k \in \mathbb{R}^n$, to split the set $\mathcal{A}$ into k clusters $\pi(c_1), \ldots, \pi(c_k)$,[4] such that cluster π_j contains those elements of the set $\mathcal{A}$ which are closest to the center c_j, so that for every $a_i \in \mathcal{A}$ we have

$$a_i \in \pi_j(c_j) \Leftrightarrow d(c_j, a_i) \leq d(c_s, a_i) \text{ for all } s = 1, \ldots, k. \tag{3.39}$$

In addition, one has to take care that each element of $\mathcal{A}$ belongs to a single cluster. This principle, which we call *minimal distance principle*, results in a partition $\Pi = \{\pi_1, \ldots, \pi_k\}$ with clusters $\pi_1, \ldots, \pi_k$.

The problem of finding the optimal partition of the set $\mathcal{A}$ can therefore be reduced to the following GOP (see also [176]):

$$\arg\min_{c \in \mathbb{R}^{kn}} F(c), \qquad F(c) = \sum_{i=1}^{m} \min_{1 \leq j \leq k} d(c_j, a_i), \tag{3.40}$$

where $F \colon \mathbb{R}^{kn} \to \mathbb{R}_+$, and $c \in \mathbb{R}^{kn}$ is the concatenation of vectors $c_1, \ldots, c_k$. The function F is non-negative, symmetric, non-differentiable, non-convex, but Lipschitz-continuous. For a survey of most popular methods for finding an optimal partition of the set $\mathcal{A}$, see [184].

The following theorem shows that, when using the LS-distance-like function, the function F in (3.40) is indeed Lipschitz-continuous. Similarly, in [149] it is shown that this function is Lipschitz-continuous also in the case of ℓ_1 metric function. This is an important property of the function F since it opens the possibility to solve GOP (3.40) by using the well known global optimization algorithm DIRECT (see Sect. 4.1).

Theorem 3.39 *Let* $\mathcal{A} = \{a_i \in \mathbb{R}^n : i = 1, \ldots, m\} \subset \Delta$, *where* $\Delta = \{x \in \mathbb{R}^n : \alpha^i \leq x^i \leq \beta^i\}$ *for some* $\alpha = (\alpha^1, \ldots, \alpha^n)$ *and* $\beta = (\beta^1, \ldots, \beta^n) \in \mathbb{R}^n$. *The function* $F \colon \Delta^k \to \mathbb{R}_+$ *defined by*

$$F(c) = \sum_{i=1}^{m} \min_{j=1,\ldots,k} \|c_j - a_i\|^2$$

is Lipschitz-continuous.

---

[4]Notice that the cluster $\pi(c_j)$ depends on neighboring clusters and that the notation $\pi(c_j)$ refers to the fact that the cluster $\pi(c_j)$ is associated with the center c_j.

In order to prove this theorem, we are going to approximate the function F up to an $\epsilon > 0$ by a differentiable (smooth) function F_ϵ. To do this, we will need some theoretical preparations.

Lemma 3.40 *The function $\psi : \mathbb{R}^n \to \mathbb{R}$, defined by $\psi(x) = \ln(e^{x^1} + \cdots + e^{x^n})$, is a convex function.*

Proof One has to show that for all $x, y \in \mathbb{R}^n$ and $\lambda \in [0, 1]$

$$\psi\big(\lambda x + (1 - \lambda)\, y\big) \le \lambda \, \psi(x) + (1 - \lambda)\, \psi(y), \tag{3.41}$$

i.e.

$$\psi(\alpha x + \beta y) \le \alpha \, \psi(x) + \beta \, \psi(y), \tag{3.42}$$

where $\alpha, \beta > 0$ are such that $\alpha + \beta = 1$.

Let the numbers $p := \frac{1}{\alpha}$ and $q := \frac{1}{\beta}$ be such that $\frac{1}{p} + \frac{1}{q} = 1$. Since $\alpha + \beta = 1$, one of the numbers α, β has to be smaller than 1, hence one of the numbers p, q has to be larger than 1. Let $x = (x^1, \ldots, x^n)$ and $y = (y^1, \ldots, y^n) \in \mathbb{R}^n$. Applying the Hölder inequality[5] (see [30, 177]) to vectors

$$a = (e^{\alpha x^1}, \ldots, e^{\alpha x^n}) \text{ and } b = (e^{\beta y^1}, \ldots, e^{\beta y^n}) \in \mathbb{R}^n,$$

we obtain

$$|\langle a, b \rangle| \le \Big(\sum_{i=1}^{n} (e^{\alpha x^i})^p \Big)^{1/p} \Big(\sum_{i=1}^{n} (e^{\beta y^i})^q \Big)^{1/q},$$

i.e.

$$\sum_{i=1}^{n} e^{\alpha x^i + \beta y^i} \le \Big(\sum_{i=1}^{n} e^{x^i} \Big)^{\alpha} \Big(\sum_{i=1}^{n} e^{y^i} \Big)^{\beta}.$$

By taking the logarithm we obtain the required inequality (3.41).   □

---

[5]For two vectors $a, b \in \mathbb{R}^n$, and real numbers p and q such that $\frac{1}{p} + \frac{1}{q} = 1$, $p > 1$, the Hölder inequality states that $\sum\limits_{i=1}^{n} |a^i b^i| \le \|a\|_p \|b\|_q$, i.e.

$$\sum_{i=1}^{n} |a^i b^i| \le \Big(\sum_{i=1}^{n} |a^i|^p \Big)^{1/p} \Big(\sum_{i=1}^{n} |b^i|^q \Big)^{1/q}.$$

In particular, for $p = q = 2$ this becomes the well known Cauchy–Schwarz–Bunyakovsky inequality.

Corollary 3.41 *Let $A \in \mathbb{R}^{n \times n}$ be a square matrix, $b \in \mathbb{R}^n$ a vector, and $\psi : \mathbb{R}^n \to \mathbb{R}$ defined by $\psi(x) = \ln(e^{x^1} + \cdots + e^{x^n})$. Then $\Phi(x) := \psi(Ax + b)$ is a convex function.*

Proof As in the proof of previous lemma, it suffices to show that for arbitrary $x, y \in \mathbb{R}^n$ and $\alpha, \beta > 0$ such that $\alpha + \beta = 1$, one has

$$\Phi(\alpha x + \beta y) \leq \alpha \Phi(x) + \beta \Phi(y).$$

Since

$$\Phi(\alpha x + \beta y) = \psi\big(A(\alpha x + \beta y) + b\big) = \psi(\alpha Ax + \beta Ay + b)$$
$$= \psi\big(\alpha Ax + \alpha b + \beta Ay + \beta b - (\alpha + \beta)b + b\big)$$
$$= \psi\big(\alpha (Ax + b) + \beta (Ay + b)\big)$$
$$\leq \alpha \Phi(x) + \beta \Phi(y),$$

the required inequality follows.   □

Exercise 3.42 Show that $\psi : \mathbb{R}_+^n \to \mathbb{R}$ defined by $\psi(x) = \ln(\frac{1}{x^1} + \cdots + \frac{1}{x^n})$ is a convex function.

Lemma 3.43 *For every $\epsilon > 0$, the function $\psi_\epsilon : \mathbb{R} \to \mathbb{R}_+$ defined by*

$$\psi_\epsilon(x) = \epsilon \ln \left(e^{-\frac{x}{\epsilon}} + e^{\frac{x}{\epsilon}}\right) = \epsilon \ln \left(2 \operatorname{ch} \tfrac{x}{\epsilon}\right) \tag{3.43}$$

is a convex function of class $C^\infty(\mathbb{R})$, and it satisfies

$$0 < \psi_\epsilon(x) - |x| \leq \epsilon \ln 2 \text{ for all } x \in \mathbb{R}, \tag{3.44}$$

$$\psi_\epsilon'(x) = \operatorname{th} \tfrac{x}{\epsilon}, \qquad \psi_\epsilon''(x) = \frac{1}{\epsilon \operatorname{ch}^2 \frac{x}{\epsilon}}, \qquad \underset{x \in \mathbb{R}}{\arg \min} \, \psi_\epsilon(x) = 0, \tag{3.45}$$

and the equality in (3.44) holds true if and only if $x = 0$.

Proof Putting $n = 2$, $x^1 = -\frac{x}{\epsilon}$, and $x^2 = \frac{x}{\epsilon}$, convexity of the function ψ_ϵ follows from Lemma 3.40. In order to prove (3.44), notice that

$$\psi_\epsilon(x) - |x| = \epsilon \ln \left(2 \operatorname{ch} \tfrac{x}{\epsilon}\right) - \epsilon \tfrac{|x|}{\epsilon}$$

$$= \epsilon \left(\ln \left(2 \operatorname{ch} \tfrac{x}{\epsilon}\right) - \ln \exp \left(\tfrac{|x|}{\epsilon}\right) \right) = \epsilon \ln \frac{2 \operatorname{ch} \frac{x}{\epsilon}}{\exp \left(\frac{|x|}{\epsilon}\right)}.$$

Since for every $u \in \mathbb{R}$ (see Exercise 3.44) $1 < \dfrac{2\operatorname{ch} u}{\exp(|u|)} \leq 2$, and since the logarithmic function is monotonous, the previous equality implies

$$\epsilon \ln 1 < \psi_\epsilon(x) - |x| \leq \epsilon \ln 2.$$

Formulas (3.45) follow directly. $\square$

(a) Functions $x \mapsto |x|$ and $x \mapsto \psi_\epsilon(x)$

(b) Function $u \mapsto \dfrac{2\operatorname{ch} u}{e^{|u|}}$

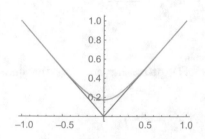
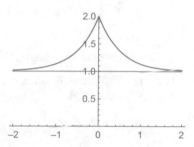

Fig. 3.8 Smooth approximation of the function $x \mapsto |x|$, and the function $u \mapsto \dfrac{2\operatorname{ch} u}{e^{|u|}}$

Exercise 3.44 Prove that for every $u \in \mathbb{R}$ the following holds true (see Fig. 3.8b):

$$1 < \frac{2\operatorname{ch} u}{e^{|u|}} \leq 2.$$

In view of (3.44), note that the function $x \mapsto |x|$, $x \in \mathbb{R}$, can be approximated by the function ψ_ϵ (see Fig. 3.8a).

In general, the non-differentiable function $f \colon \mathbb{R}^p \to \mathbb{R}$, defined as $f(z) = \max_{j=1,\ldots,p} z^j$, can be approximated by the differentiable function

$$\psi_\epsilon(z) = \psi_\epsilon(z^1, \ldots, z^p) = \epsilon \ln \sum_{j=1}^{p} \exp\left(\tfrac{z^j}{\epsilon}\right). \tag{3.46}$$

Namely,

$$\psi_\epsilon(z) - f(z) = \epsilon \ln \sum_{j=1}^{p} \exp\left(\tfrac{z^j}{\epsilon}\right) - \epsilon \frac{\max\limits_{i=1,\ldots,p} z^i}{\epsilon}$$

$$= \epsilon\left(\ln \sum_{j=1}^{p} \exp\left(\frac{z^j}{\epsilon}\right) - \ln \exp\left(\frac{\max z^i}{\epsilon}\right)\right)$$

$$= \epsilon \ln \frac{\sum_{j=1}^{p} \exp\left(\tfrac{z^j}{\epsilon}\right)}{\exp\left(\tfrac{\max z^i}{\epsilon}\right)} = \epsilon \ln \sum_{j=1}^{p} \exp\left(\frac{z^j - \max z^i}{\epsilon}\right)$$

$$\leq \epsilon \ln \sum_{j=1}^{p} e^0 = \epsilon \ln p.$$

Moreover, since $\min_{j=1,\ldots,p} z^j = -\max_{j=1,\ldots,p} (-z^j)$, we can use this result to approximate the function $F(c_1, \ldots, c_k) = \sum_{i=1}^{m} \min_{1 \leq j \leq k} d(c_j, a_i)$ by

$$F_\epsilon(c_1, \ldots, c_k) = -\epsilon \sum_{i=1}^{m} \ln \sum_{j=1}^{k} \exp\left(-\frac{d(c_j, a_i)}{\epsilon}\right). \tag{3.47}$$

We are now ready to prove Theorem 3.39.

Proof of Theorem 3.39 In accordance with (3.47), define the auxiliary function $F_\varepsilon \colon \mathbb{R}^{kn} \to \mathbb{R}_+$ with

$$F_\varepsilon(x) = -\varepsilon \sum_{i=1}^{m} \ln \sum_{j=1}^{k} \exp\left(-\frac{\|x_j - a_i\|^2}{\varepsilon}\right),$$

where $x_j \in \mathbb{R}^n$, $j = 1, \ldots, k$ and $x = (x_1, \ldots, x_k) \in \mathbb{R}^{kn}$. Then, according to [96], the following holds true:

$$0 \leq F(x) - F_\varepsilon(x) \leq \varepsilon m \ln k.$$

Therefore, for $u, v \in \mathbb{R}^{kn}$

$$\begin{aligned}
|F(u) - F(v)| &= |(F(u) - F_\varepsilon(u)) + (F_\varepsilon(v) - F(v)) + (F_\varepsilon(u) - F_\varepsilon(v))| \\
&\leq |F(u) - F_\varepsilon(u)| + |F_\varepsilon(v) - F(v)| + |F_\varepsilon(u) - F_\varepsilon(v)| \\
&\leq 2\varepsilon m \ln k + |F_\varepsilon(u) - F_\varepsilon(v)|. \tag{3.48}
\end{aligned}$$

Since

$$\frac{\partial F_\varepsilon(x)}{\partial x_p} := \left(\frac{\partial F_\varepsilon(x)}{\partial x_p^1}, \ldots, \frac{\partial F_\varepsilon(x)}{\partial x_p^n}\right) = 2 \sum_{i=1}^{m} \frac{(x_p - a_i) \exp\left(-\frac{\|x_p - a_i\|^2}{\varepsilon}\right)}{\sum_{j=1}^{k} \exp\left(-\frac{\|x_j - a_i\|^2}{\varepsilon}\right)},$$

we obtain

$$\left\|\frac{\partial F_\varepsilon(x)}{\partial x_p}\right\| \leq 2 \sum_{i=1}^{m} \|x_p - a_i\| \leq 2 \sum_{i=1}^{m} \max_{j=1,\ldots,m} \|a_i - a_j\|$$

$$\leq 2m \max_{i,j \in \{1,\ldots,m\}} \|a_i - a_j\|, \quad p = 1, \ldots, k,$$

i.e. the gradient $\nabla F_\varepsilon(x)$ is continuous and bounded on Δ^k. Using the Lagrange intermediate value theorem for the function F_ϵ on Δ^k, we conclude that there exists an $L > 0$ (not depending on ε) such that

$$|F_\varepsilon(u) - F_\varepsilon(v)| \le L\|u - v\|, \quad u, v \in \Delta^k.$$

Finally, for $\varepsilon \to 0^+$, (3.48) implies that $|F(u) - F(v)| \le L\|u - v\|$. $\square$

The following lemma and theorem show the connection between the objective function $\mathcal{F}$ defined by (3.5) and the objective function F defined by (3.40).

Lemma 3.45 *Let $\mathcal{A} = \{a_i \in \mathbb{R}^n : i = 1, \ldots, m\}$ be a finite set in $\mathbb{R}^n$, $z_1, \ldots, z_k \in \mathbb{R}^n$ mutually distinct points, and $d \colon \mathbb{R}^n \times \mathbb{R}^n \to \mathbb{R}_+$ a distance-like function. In addition, let $\Pi = \{\pi_1(z_1), \ldots, \pi_k(z_k)\}$ be the partition whose clusters were obtained by minimal distance principle and let $c_j \in \arg\min_{x \in \mathbb{R}^n} \sum_{a \in \pi_j} d(x, a)$, $j = 1, \ldots, k$, be their centers. Then*

$$F(z_1, \ldots, z_k) \overset{(\star)}{\ge} \mathcal{F}(\Pi) \overset{(\star\star)}{\ge} F(c_1, \ldots, c_k), \tag{3.49}$$

while inequalities $(\star)$ and $(\star\star)$ turn to equalities if and only if $z_j = c_j$ for every $j = 1, \ldots, k$.

Proof In order to prove the inequality $(\star)$ we split $\sum_{i=1}^{m}$ into k sums $\sum_{j=1}^{k} \sum_{a \in \pi_j}$.

$$
\begin{aligned}
F(z_1, \ldots, z_k) &= \sum_{i=1}^{m} \min\{d(z_1, a_i), \ldots, d(z_k, a_i)\} \\
&= \sum_{j=1}^{k} \sum_{a_i \in \pi_j} \min\{d(z_1, a_i), \ldots, d(z_k, a_i)\} \\
&= \sum_{j=1}^{k} \sum_{a_i \in \pi_j} d(z_j, a_i) \\
&\ge \sum_{j=1}^{k} \sum_{a_i \in \pi_j} d(c_j, a_i) = \mathcal{F}(\{\pi_1, \ldots, \pi_k\}).
\end{aligned}
$$

To prove $(\star\star)$, first notice that for every $a \in \pi_j$ one has

$$d(c_j, a) \ge \min\{d(c_1, a), \ldots, d(c_k, a)\}.$$

Therefore,

$$\mathcal{F}(\{\pi_1, \ldots, \pi_k\}) = \sum_{j=1}^{k} \sum_{a_i \in \pi_j} d(c_j, a_i)$$

$$\geq \sum_{j=1}^{k} \sum_{a_i \in \pi_j} \min\{d(c_1, a_i), \ldots, d(c_k, a_i)\}$$

$$= \sum_{i=1}^{m} \min\{d(c_1, a_i), \ldots, d(c_k, a_i)\} = F(c_1, \ldots, c_k). \qquad \square$$

Theorem 3.46 *Let $\mathcal{A} = \{a_i \in \mathbb{R}^n : i = 1, \ldots, m\}$. Then:*

(*i*) $c^\star = (c_1^\star, \ldots, c_k^\star) \in \underset{c_1,\ldots,c_k \in \mathbb{R}^n}{arg\,min} \ F(c_1, \ldots, c_k)$ *if and only if*
$$\Pi^\star = \{\pi_1^\star(c_1^\star), \ldots, \pi_k^\star(c_k^\star)\} \in \underset{\Pi \in \mathcal{P}(\mathcal{A};k)}{arg\,min} \ \mathcal{F}(\Pi),$$

(*ii*) $\underset{c_1,\ldots,c_k \in \mathbb{R}^n}{min} F(c_1, \ldots, c_k) = \underset{\Pi \in \mathcal{P}(\mathcal{A};k)}{min} \mathcal{F}(\Pi).$

Proof

(*i*) Let $c^\star = (c_1^\star, \ldots, c_k^\star) \in \underset{c_1,\ldots,c_k \in \mathbb{R}^n}{arg\,min} F(c_1, \ldots, c_k)$. Denote by $\pi_j^\star$ the

corresponding clusters obtained by minimal distance principle, and let $\Pi^\star = \{\pi_1^\star, \ldots, \pi_k^\star\}$. According to Lemma 3.45

$$F(c^\star) = \mathcal{F}(\Pi^\star). \tag{3.50}$$

We claim that

$$\Pi^\star \in \underset{\Pi \in \mathcal{P}(\mathcal{A};k)}{arg\,min} \ \mathcal{F}(\Pi). \tag{3.51}$$

Namely, if there existed a partition $\mathcal{N}^\star = \{v_1^\star, \ldots, v_k^\star\} \in \mathcal{P}(\mathcal{A}; k)$ with cluster centers $\zeta^\star = (\zeta_1^\star, \ldots, \zeta_k^\star)$ such that $\mathcal{F}(\mathcal{N}^\star) < \mathcal{F}(\Pi^\star)$, we would have

$$F(\zeta^\star) \overset{\text{Lemma } 3.45}{=} \mathcal{F}(\mathcal{N}^\star) < \mathcal{F}(\Pi^\star) \overset{\text{Lemma } 3.45}{=} F(c^\star),$$

which is not possible since $c^\star \in \arg\min_{c \in \mathbb{R}^{n \times k}} F(c)$.

(*ii*) Let $\Pi^\star = \{\pi_1^\star, \ldots, \pi_k^\star\} \in \underset{\Pi \in \mathcal{P}(\mathcal{A};k)}{arg\,min} \ \mathcal{F}(\Pi)$. Denote by $c^\star = (c_1^\star, \ldots, c_k^\star)$ where

$c_1^\star, \ldots, c_k^\star$ are cluster centers of partition $\Pi^\star$. According to Lemma 3.45

$$F(c^\star) = \mathcal{F}(\Pi^\star). \tag{3.52}$$

We claim that

$$c^\star \in \arg\min_{c \in \mathbb{R}^{n \times k}} F(c). \tag{3.53}$$

Namely, if there existed a $\zeta^\star = (\zeta_1^\star, \ldots, \zeta_k^\star)$ such that $F(\zeta^\star) < F(c^\star)$, then the partition $\mathcal{N}^\star(\zeta^\star)$ would satisfy

$$\mathcal{F}(\Pi^\star) \overset{\text{Lemma 3.45}}{=} F(c^\star) > F(\zeta^\star) \overset{\text{Lemma 3.45}}{=} \mathcal{F}(\mathcal{N}^\star),$$

which is not possible since $\Pi^\star \in \arg\min\limits_{\Pi \in \mathcal{P}(\mathcal{A};k)} \mathcal{F}(\Pi)$. $\square$

Example 3.47 Let $\mathcal{A} = \{1, 3, 4, 8\}$ be a set with $m = 4$ data. Table 3.8 lists some values of objective functions $\mathcal{F}_{LS}$ and F_{LS} supporting claims of Lemma 3.45 and Theorem 3.46. For the optimal partition the inequality $(\star)$ becomes equality, while z_1, z_2 coincide with cluster centers (the fourth row).

Table 3.8 Comparing values of objective functions $\mathcal{F}_{LS}$ and F_{LS} for $\mathcal{A} = \{1, 3, 4, 8\}$

	z_1	z_2	$\mathcal{F}_{LS}(z_1, z_2)$	π_1	π_2	c_1	c_2	$\mathcal{F}_{LS}(\Pi)$	$F_{LS}(c_1, c_2)$
1.	1	4	17	$\{1\}$	$\{3,4,8\}$	1	5	14	14
2.	1	5	14	$\{1,3\}$	$\{4,8\}$	2	6	10	10
3.	3	7	6	$\{1,3,4\}$	$\{8\}$	$\frac{8}{3}$	8	$\frac{14}{3}$	$\frac{14}{3}$
4.	$\frac{8}{3}$	8	$\frac{14}{3}$	$\{1,3,4\}$	$\{8\}$	$\frac{8}{3}$	8	$\frac{14}{3}$	$\frac{14}{3}$

Example 3.48 Let $\mathcal{A} = \{16, 11, 2, 9, 2, 8, 15, 19, 8, 17\}$ be a set with $m = 10$ data. Table 3.9 lists some values of objective functions $\mathcal{F}_1$ and F_1 supporting claims of Lemma 3.45 and Theorem 3.46. In particular, pay attention to the third row showing sharp inequality $(\star\star)$.

Table 3.9 Comparing values of objective functions $\mathcal{F}_1$ and F_1

	z_1	z_2	$F_1(z_1, z_2)$	π_1	π_2	(c_1, c_2)	$\mathcal{F}_1(\Pi)$	$F_1(c_1, c_2)$
1.	2	6	55	$\{2,2\}$	$\{8,8,9,11,15,16,17,19\}$	$(2,13)$	31	31
2.	2	13	31	$\{2,2\}$	$\{8,8,9,11,15,16,17,19\}$	$(2,13)$	31	31
3.	3	15	29	$\{2,2,8,8,9\}$	$\{11,15,16,17,19\}$	$(8,16)$	23	21
4.	6	16	25	$\{2,2,8,8,9,11\}$	$\{15,16,17,19\}$	$(8, \frac{33}{2})$	21	21
5.	8	16	21	$\{2,2,8,8,9,11\}$	$\{15,16,17,19\}$	$(8, \frac{33}{2})$	21	21

Exercise 3.49 Carry out a similar verification as in Example 3.47 using the ℓ_1 metric function and also a similar verification as in Example 3.48 using the LS-distance-like function.

Using Theorem 3.46 we are now ready to prove Theorem 3.7, stating that increasing the number of clusters in optimal partition does not increase the value of the objective function $\mathcal{F}$.

Proof of Theorem 3.7 Let $\hat{c} = (\hat{c}_1, \ldots, \hat{c}_{k-1})$, where $\hat{c}_j$ are cluster centers of the optimal $(k-1)$-partition $\Pi^{(k-1)}$, and let $c^\star = (c_1^\star, \ldots, c_k^\star)$ where $c_j^\star$ are cluster centers of the optimal k-partition $\Pi^{(k)}$. Take a $\zeta \in \mathbb{R}^n \setminus \{\hat{c}_1, \ldots, \hat{c}_{k-1}\}$ and let

$$\delta_{k-1}^i := \min_{1 \le s \le k-1} d(\hat{c}_s, a_i), \quad i = 1, \ldots, m.$$

Then

$$\mathcal{F}(\Pi^{(k-1)}) \overset{\text{Thm}\,3.46}{=} F(\hat{c}) = \sum_{i=1}^m \min\{d(\hat{c}_1, a_i), \ldots, d(\hat{c}_{k-1}, a_i)\} = \sum_{i=1}^m \delta_{k-1}^i$$

$$\ge \sum_{i=1}^m \min\{\delta_{k-1}^i, d(\zeta, a_i)\} \qquad [\Pi^{(k)} \text{ being optimal } k\text{-partition}]$$

$$\ge \sum_{i=1}^m \min\{d(c_1^\star, a_i), \ldots, d(c_k^\star, a_i)\}$$

$$= F(c^\star) \overset{\text{Thm}\,3.46}{=} \mathcal{F}(\Pi^{(k)}),$$

asserting that increasing the number of clusters in the optimal partition does not increase the value of the objective function.   □

Remark 3.50 The above proof of Theorem 3.7 implicitly shows that $\mathcal{F}$ is a monotonous function.

Lemma 3.45 and Theorem 3.46 motivate the following definition.

Definition 3.51 Let $\mathcal{A} = \{a_i \in \mathbb{R}^n : i = 1, \ldots, m\}$ be a finite set in $\mathbb{R}^n$, $d \colon \mathbb{R}^n \times \mathbb{R}^n \to \mathbb{R}_+$ a distance-like function, and $\hat{\Pi} = \{\hat{\pi}_1, \ldots, \hat{\pi}_k\}$ a partition whose cluster centers $\hat{c}_1, \ldots, \hat{c}_k$ are such that the function F attains a local minimum at $(\hat{c}_1, \ldots, \hat{c}_k)$. The partition $\hat{\Pi}$ is called a *locally optimal k-partition* (LOPart) of the set $\mathcal{A}$ provided that

$$\mathcal{F}(\hat{\Pi}) = F(\hat{c}_1, \ldots, \hat{c}_k). \tag{3.54}$$

Chapter 4
Searching for an Optimal Partition

Let $\mathcal{A} = \{a_i = (a_i^1, \ldots, a_i^n) : i = 1, \ldots, m\} \subset \Delta \subset \mathbb{R}^n$, where $\Delta = [\alpha^1, \beta^1] \times \cdots \times [\alpha^n, \beta^n]$. The k-GOPart is obtained by solving the GOP (3.5), or solving the equivalent problem (3.40). If the components $a_i^1, \ldots, a_i^n$ of data $a_i \in \mathcal{A}$, $i = 1, \ldots, m$, do not belong to similar ranges, i.e. the intervals $[\alpha^j, \beta^j]$ considerably differ in lengths, then one has to normalize these intervals. For instance, the bijective mapping $T : \Delta \to [0, 1]^n$ defined by

$$T(x) = (x - \alpha)D, \quad D = \text{diag}\left(\tfrac{1}{\beta^1 - \alpha^1}, \ldots, \tfrac{1}{\beta^n - \alpha^n}\right), \quad \alpha = (\alpha^1, \ldots, \alpha^n) \quad (4.1)$$

transforms $\mathcal{A}$ onto the set $\mathcal{B} = \{T(a_i) : a_i \in \mathcal{A}\} \subset [0, 1]^n$ (see [69, 165]).

Once the clustering of the set $\mathcal{B}$ is done, one has to pull the result back into the set Δ by applying the inverse mapping $T^{-1} : [0, 1]^n \to \Delta$,

$$T^{-1}(x) = xD^{-1} + \alpha. \quad (4.2)$$

4.1 Solving the GOP (3.40) Directly

There is no method to find k-GOPart reliably and effectively. It turns out that using some of the known global optimization methods in this case is not appropriate (see e.g. [78, 133, 171, 172, 193]).

Let us mention a derivative-free, deterministic sampling method for global optimization of a Lipschitz-continuous function $F : \Delta^k \to \mathbb{R}$ named Dividing Rectangles (DIRECT) which was proposed by Jones et al. [89]. The function F is first transformed into $f : [0, 1]^{kn} \to \mathbb{R}$, $f(x) = (F \circ T^{-1})(x)$, where the mapping $T : \Delta^k \to [0, 1]^{kn}$ is a generalization of (4.1) given by

$$T(x) = (x - u)D,$$

© The Author(s), under exclusive license to Springer Nature Switzerland AG 2021
R. Scitovski et al., *Cluster Analysis and Applications*,
https://doi.org/10.1007/978-3-030-74552-3_4

$$D = \operatorname{diag}\left(\tfrac{1}{\beta^1 - \alpha^1}, \ldots, \tfrac{1}{\beta^n - \alpha^n}, \ldots, \tfrac{1}{\beta^1 - \alpha^1}, \ldots, \tfrac{1}{\beta^n - \alpha^n}\right) \in \mathbb{R}^{(kn) \times (kn)},$$

$$u = (\alpha^1, \ldots, \alpha^n, \ldots, \alpha^1, \ldots, \alpha^n) \in \mathbb{R}^{kn},$$

and the mapping $T^{-1} \colon [0, 1]^{kn} \to \Delta^k$ is given by $T^{-1}(x) = xD^{-1} + u$.

By this transformation GOP (3.40) becomes the following GOP:

$$\operatorname*{arg\,min}_{x \in [0,1]^{kn}} f(x), \qquad f(x) = (F \circ T^{-1})(x). \tag{4.3}$$

Note that, if $\hat{x} \in [0, 1]^{kn}$ is an approximation of the solution to GOP (4.3), then the approximation of the solution to GOP (3.40) becomes $\hat{c} = T^{-1}(\hat{x})$, where $F(\hat{c}) = F(T^{-1}(\hat{x})) = f(\hat{x})$.

Furthermore, by means of a standard strategy (see e.g. [55, 56, 64, 65, 69, 88, 89, 131, 154, 170]), the unit hypercube $[0, 1]^{kn}$ is divided into smaller hyperrectangles, among which the so-called potentially optimal ones are first searched for and then further divided. It should be noted that this procedure does not assume knowing the Lipschitz constant $L > 0$.

Remark 4.1 The corresponding *Mathematica*-module DIRECT[] is described in Sect. 9.2, and the link to appropriate *Mathematica*-code is supplied. The *Mathematica*-module is constructed in such a way that one inputs the minimizing function $F \colon \Delta^k \to \mathbb{R}$, $\Delta \subseteq \mathbb{R}^n$. The module transforms the domain Δ onto the hypercube $[0, 1]^n$ and performs the minimization. Finally, the result is transformed back to the domain Δ.

Using the global optimization algorithm DIRECT to solve the problem (3.40) would be numerically very inefficient since the minimizing function has usually a large number of independent variables assorted into k vectors in $\mathbb{R}^n$, and, because the objective function is symmetric in these k vectors, the DIRECT algorithm would search for all $k!$ solutions. Besides, the DIRECT algorithm quickly arrives close to a point of global minimum, but it can be very slow when it needs to attain high accuracy.

This explains why the DIRECT algorithm can generally be used only to find a good initial approximation, and then one has to use some of the known locally optimization methods as in [161, 162]. The best known locally optimization method to do this is the k-means algorithm and its numerous modifications. But, this will result only in some LOPart.

In the case of data with one feature, there are a few modifications and adaptations of the DIRECT algorithm which find k-GOPart (see e.g. [69, 130, 131, 133, 154, 171]).

4.2   *k*-means Algorithm II

4.2.1   Denoting the Objective Function $\mathcal{F}$ using the Membership Matrix

Let $d \colon \mathbb{R}^n \times \mathbb{R}^n \to \mathbb{R}_+$ be a distance-like function, $\mathcal{A} \subset \mathbb{R}^n$ a finite set, $|\mathcal{A}| = m$, and $\Pi = \{\pi_1, \dots, \pi_k\}$ some k-partition with cluster centers $c_j \in \arg\min\limits_{c \in \mathbb{R}^n} \sum\limits_{a \in \pi_j} d(c, a)$, $j = 1, \dots, k$. The objective function $\mathcal{F}$ from (3.40) is then defined as

$$\mathcal{F}(\Pi) = \sum_{j=1}^{k} \sum_{a \in \pi_j} d(c_j, a).$$

The function $\mathcal{F}$ can be also written in a different way by introducing the so-called *membership matrix* $U \in \{0, 1\}^{m \times k}$ with elements

$$u_{ij} = \begin{cases} 1, & \text{if } a_i \in \pi_j \\ 0, & \text{if } a_i \notin \pi_j \end{cases}, \qquad i = 1, \dots m, \quad j = 1, \dots, k. \tag{4.4}$$

The definition (4.4) means that each element $a_i \in \mathcal{A}$ has to belong to exactly one cluster, and the function $\mathcal{F}$ can then be written as (see [20, 21, 184])

$$\mathcal{F}(\Pi) = \Phi(c, U) = \sum_{j=1}^{k} \sum_{i=1}^{m} u_{ij}\, d(c_j, a_i), \tag{4.5}$$

where $\Phi \colon \mathbb{R}^{kn} \times \{0, 1\}^{m \times k} \to \mathbb{R}_+$.

Example 4.2 One LS-optimal 3-partition of the set $\mathcal{A} = \{2, 4, 8, 10, 16\}$ is

$$\Pi = \big\{ \{2, 4\}, \{8, 10\}, \{16\} \big\},$$

and the corresponding membership matrix is

$$U = \begin{bmatrix} 1 & 0 & 0 \\ 1 & 0 & 0 \\ 0 & 1 & 0 \\ 0 & 1 & 0 \\ 0 & 0 & 1 \end{bmatrix}.$$

Remark 4.3 Note that elements of the matrix U satisfy

$$\sum_{j=1}^{k} u_{ij} = 1, \quad i = 1, \dots, m \tag{4.6}$$

and

$$1 \le \sum_{i=1}^{m} u_{ij} \le m, \quad j = 1, \ldots, k. \tag{4.7}$$

The equality (4.6) means that each element $a_i \in \mathcal{A}$ has to belong to at least one cluster, and the inequality (4.7) means that each cluster contains at least one, and at most m elements.

Exercise 4.4 How many matrices $U \in \{0, 1\}^{m \times k}$ satisfy the conditions (4.6) and (4.7)? What is the connection between the number of such matrices and the number (3.2) of all k-partitions of the set $\mathcal{A}$ with m elements?

In order to optimize the calculations, the function (4.5) is written as

$$\Phi(c, U) = \sum_{i=1}^{m} \sum_{j=1}^{k} u_{ij} \, d(c_j, a_i), \tag{4.8}$$

and the search for globally optimal k-partition can be written as the following GOP:

$$\operatorname*{arg\,min}_{c \in \mathbb{R}^{kn}, \, U \in \{0,1\}^{m \times k}} \Phi(c, U). \tag{4.9}$$

The optimization function Φ is a non-differentiable non-convex function with a huge number of variables: $n \cdot k + m \cdot k$. For instance, the function Φ in Example 4.2 with only $m = 5$ data has $1 \cdot 3 + 5 \cdot 3 = 18$ independent variables. Only in a very special case when the number of features is $n = 1$ and the number k of clusters is not too large, one can carry out a direct minimization using some of the specialized global optimization methods for symmetric Lipschitz-continuous functions (see [69, 131, 154]), resulting in a sufficiently precise solution.

In the general case there is no optimization method to solve the GOP (4.9) effectively. In the following section we are going to illustrate briefly a class of optimization methods which have been specially investigated in the past 10 years in the context of *Big Data Analysis*, and can be found under the term *Coordinate Minimization algorithms* or *Coordinate Descent Algorithms* (see e.g. [17, 121, 126, 139]).

4.2.2 Coordinate Descent Algorithms

Instead of looking for a global minimum $\operatorname*{arg\,min}_{x \in \mathbb{R}^n} f(x)$ of a function $f : \mathbb{R}^n \to \mathbb{R}$, we will be satisfied by finding a point of local minimum or even by finding a stationary point. Let us represent vectors $x \in \mathbb{R}^n$ in the form $x = (u, v)$, $u \in \mathbb{R}^p$, $v \in \mathbb{R}^q$, $p + q = n$. Choosing some appropriate initial approximation (u_0, v_0), the iterative process can be constructed as follows:

$$f(u_0, v_0) \geq f(u_0, \arg\min_{v \in \mathbb{R}^q} f(u_0, v)) =: f(u_0, v_1)$$

$$\geq f(\arg\min_{u \in \mathbb{R}^p} f(u, v_1), v_1) =: f(u_1, v_1) \geq \cdots$$

or as

$$f(u_0, v_0) \geq f(\arg\min_{u \in \mathbb{R}^p} f(u, v_0), v_0) =: f(u_1, v_0)$$

$$\geq f(u_1, \arg\min_{v \in \mathbb{R}^q} f(u_1, v)) =: f(u_1, v_1) \geq \cdots$$

This will generate a sequence of successive approximations $x^{(0)} = (u_0, v_0)$, $x^{(1)} = (u_1, v_1)$, ..., for which the sequence of function values is non-increasing. For a strictly convex differentiable function f the process will converge to the global minimum (see Example 4.5). As shown by Example 4.6, if the function f is not strictly convex, there is no guarantee that the algorithm will converge. This algorithm can be defined for differentiable but also for non-differentiable functions.

(a) $f(x, y) = 2x^3 + xy^2 + 5x^2 + y^2$ (b) $f(x, y) = |x + y| + 3|x - y|$

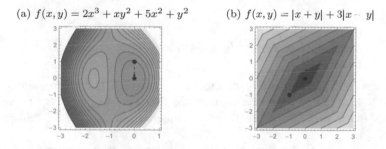

Fig. 4.1 Finding points of global and local minima

Example 4.5 Find the local minimum of the function $f : [-3, 1] \times [-3, 3] \to \mathbb{R}$,

$$f(x, y) = 2x^3 + xy^2 + 5x^2 + y^2, \tag{4.10}$$

whose `ContourPlot` is shown in Fig. 4.1a. Check, using calculus, that f has four stationary points: $S_1 = (-\frac{5}{3}, 0)$, $S_2 = (-1, -2)$, $S_3 = (-1, 2)$, and $S_4 = (0, 0)$, that f attains its local maximum at S_1 and its local minimum at S_4, and that S_2 and S_3 are saddle-points.

Let us find the local minimum in the following way:

1. Choose $y_0 = 1$ and define the function $f_x(x) := f(x, y_0) = 1 + x + 5x^2 + 2x^3$. The function f_x attains its local minimum at $x_0 = -0.10685$;
2. Define the function $f_y(y) := f(x_0, y) = 0.054645 + 0.89315\, y^2$ which attains its local minimum at $y_1 = 0$;

3. Define the function $f_x(x) := f(x, y_1) = 5x^2 + 2x^3$ which attains its local minimum at $x_1 = 0$;
4. Define the function $f_y(y) := f(x_1, y) = y^2$ which attains its local minimum at $y_2 = 0$.

Here we stop the iterative process since the new values are going to repeat the ones we have already obtained. Thus $(0, 0)$ is the point of local minimum of the function f (see the red dots in Fig. 4.1a).

However, as the next example shows, the described procedure does not always perform nicely as in the previous example.

Example 4.6 The function $f : [-3, 3] \times [-3, 3] \to \mathbb{R}$, $f(x, y) = |x+y|+3|x-y|$ attains its global minimum at $(0, 0)$ (its ContourPlot is depicted in Fig. 4.1b). Try to find this minimum in a similar way as previously, starting at $y_0 = -1$.

The function $f_x(x) := f(x, y_0) = |x - 1| + 3|x + 1|$ attains its minimum at $x_0 = -1$. In the next step the function $f_y(y) := f(x_0, y) = |y - 1| + 3|y + 1|$ attains its minimum at $y_1 = -1$. Further procedures repeat themselves, thus not producing the point of global minimum.

4.2.3 Standard k-means Algorithm

Numerous methods described in the literature are devised to find a stationary point or, in the best case, a local minimum of the function (4.8). But for the partition obtained in this way, there is usually no information how close, or far, it is from the optimal one.

The most popular algorithm of this type is the well known *k-means algorithm* already mentioned on page 38. We will construct now the k-means algorithm using the notation (4.8) for the objective function Φ, by making use of the membership matrix and the Coordinate Descent Algorithm. Starting at some $c^{(0)} = (c_1^{(0)}, \ldots, c_k^{(0)}) \in \mathbb{R}^{kn}$, $c_j^{(0)} \in \mathbb{R}^n$, successively repeat the following two steps:

Step A: With fixed $c^{(0)} = (c_1^{(0)}, \ldots, c_k^{(0)})$ define clusters $\pi_1, \ldots, \pi_k$ by minimal distance principle (3.39). In this way (4.4) determines the matrix $U^{(1)}$;

Step B: With fixed matrix $U^{(1)}$ find the optimal $c^{(1)} = (c_1^{(1)}, \ldots, c_k^{(1)})$ by solving k optimization problems:

$$c_1^{(1)} \in \arg\min_{x \in \mathbb{R}^n} \sum_{i=1}^{m} u_{i1}^{(1)} d(x, a_i), \ldots, c_k^{(1)} \in \arg\min_{x \in \mathbb{R}^n} \sum_{i=1}^{m} u_{ik}^{(1)} d(x, a_i),$$

and calculate $\Phi(c^{(1)}, U^{(1)})$ using (4.8).

Applying Step A and Step B consecutively will create a sequence $(c^{(1)}, U^{(1)})$, $(c^{(2)}, U^{(2)}), \ldots$. Since there are finitely many choices for the matrix U, at some moment the algorithm will start to repeat itself, thus giving a locally optimal partition. According to the construction, the sequence $\Phi(c^{(1)}, U^{(1)})$, $\Phi(c^{(2)}, U^{(2)})$, ... of respective function values is monotonously decreasing and finite.

Therefore, the *k*-means algorithm is formally written in two steps which alternate consecutively. The algorithm can be stopped when the relative value of objective function drops below a given positive real number $\epsilon_{KM} > 0$

$$\frac{\Phi_r - \Phi_{r+1}}{\Phi_r} < \epsilon_{KM} > 0, \quad \text{where } \Phi_r = \Phi(c^{(r)}, U^{(r)}). \tag{4.11}$$

Algorithm 4.7 (*k*-means Algorithm II)

Step A: (Assignment step). Given a finite subset $\mathcal{A} \subset \mathbb{R}^n$ and k distinct points $z_1, \ldots, z_k \in \mathbb{R}^n$, apply the minimal distance principle to determine clusters π_j, $j = 1, \ldots, k$, i.e. the membership matrix $U \in \{0, 1\}^{m \times k}$, to get the partition $\Pi = \{\pi_1, \ldots, \pi_k\}$.

Step B: (Update step). Given the partition $\Pi = \{\pi_1, \ldots, \pi_k\}$, i.e. the membership matrix $U \in \{0, 1\}^{m \times k}$, determine the corresponding cluster centers $c_1, \ldots, c_k \in \mathbb{R}^n$ and calculate the objective function value $\mathcal{F}(\Pi)$. Set $z_j := c_j$, $j = 1, \ldots, k$.

The following theorem summarizes previous results (c.f. Theorem 3.14).

Theorem 4.8 *The k-means Algorithm 4.7 finds a* LOPart *in finitely many, T, steps, and the resulting sequence of objective function values* $(\mathcal{F}(\Pi^{(t)}))$, $t = 0, 1, \ldots, T$, *is monotonically decreasing.*

Proof We have already concluded that the sequence of objective function values $(\mathcal{F}(\Pi^{(t)}))$ obtained by the *k*-means algorithm is monotonically decreasing, and that it becomes stationary after a finite number of steps. According to Lemma 3.45 and Theorem 3.46, $\mathcal{F}(\Pi^{(T)}) = F(c_1, \ldots, c_k)$ for some $T > 0$, where c_j, $j = 1, \ldots, k$, are cluster centers of the partition $\Pi^{(T)}$. Because of this property, in accordance with Definition 3.51, the partition $\Pi^{(T)}$ is a k-LOPart. □

As was already mentioned on page 38, the solution obtained by the *k*-means algorithm highly depends on the choice of initial approximation (initial centers or initial partition). For this reason there are numerous heuristic methods described in the literature considering methods for choosing initial approximation (see e.g. [8, 14, 100, 166]).

Remark 4.9 It is known that while executing the *k*-means algorithm the following may happen:

- The solution gives only a locally optimal partition (see Example 3.15);
- The number of clusters could be reduced (see Example 4.14);
- Some elements $a \in \mathcal{A}$ can occur on the border of two or several clusters (see Example 4.15, and more detailed in [160]).

(a) $\Pi^{(1)}$: ●●●   ●●●   ✳   ● $\mathcal{F}_{LS}(\Pi^{(1)}) = 196$
                  1 2 3   8 9 10   25

(b) $\Pi^{\star}$: ●●● ✳ ●●●   ● $\mathcal{F}_{LS}(\Pi^{\star}) = 77.5$
                     1 2 3   8 9 10   25

Fig. 4.2 Locally and globally optimal 2-partitions of the set $\mathcal{A} = \{1, 2, 3, 8, 9, 10, 25\}$

Example 4.10 Let $\mathcal{A} = \{1, 2, 3, 8, 9, 10, 25\}$. Applying the LS distance-like function, let us find an optimal 2-partition. Using the initial partition $\Pi^{(1)} = \{\{1, 2, 3\}, \{8, 9, 10, 25\}\}$, the k-means algorithm recognizes precisely $\Pi^{(1)}$ as an optimal 2-partition (see Fig. 4.2a).

Taking $\Pi^{(1)} = \{\{1, 2, 3, 8\}, \{9, 10, 25\}\}$ for the initial partition, the k-means algorithm comes up with optimal 2-partition $\Pi^{\star}$ shown in Fig. 4.2b.

The *Mathematica*-program KMeansPart[] (see Sect. 9.2) can easily check that $\Pi^{\star}$ is the globally optimal 2-partition and that $\Pi^{(1)}$ is only locally optimal 2-partition.

In addition to the values of objective functions, the following table lists also the respective values of CH and DB indexes (see Chap. 5). Notice that the values of objective functions F_{LS} and $\mathcal{F}_{LS}$ coincide on optimal partitions, and the value of CH (resp. DB) is larger (resp. smaller) for the globally optimal partition.

Partitions	$\mathcal{G}$	F_{LS}	$\mathcal{F}_{LS}$	CH	DB
$\Pi^{(1)}$	207.429	196	196	5.29	0.71
$\Pi^{\star}$	325.929	77.5	77.5	21.03	0.18

Exercise 4.11 Using the same *Mathematica*-program KMeansPart[], find the optimal 2-partitions for the set in Example 4.10 but applying the ℓ_1 metric function. Do there exist again different locally and globally optimal partitions?

Example 4.12 Let $\mathcal{A} = \{a_i = (x_i, y_i) : i = 1, \ldots, 8\} \subset \mathbb{R}^2$ be the set, where

i	1	2	3	4	5	6	7	8
x_i	2	3	5	6	7	8	9	10
y_i	3	6	8	5	7	1	5	3

Let us find optimal 2-partition using the LS distance-like function. The k-means algorithm recognizes the initial partition $\Pi^{(1)}$ (shown in Fig. 4.3a) as the optimal 2-partition.

Starting with $\Pi^{(0)}$ (Fig. 4.3b), the k-means algorithm finds $\Pi^{\star}$ (Fig. 4.3c) as the optimal 2-partition.

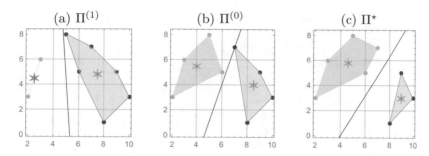

Fig. 4.3 Locally, $\Pi^{(1)}$, and globally, $\Pi^\star$, optimal 2-partitions of the set $\mathcal{A}$

The *Mathematica*-program KMeansCen (see Sect. 9.2) can easily check that $\Pi^\star$ is the globally optimal 2-partition and that $\Pi^{(1)}$ is only a locally optimal 2-partition.

In addition to the values of objective functions, the following table lists also respective values of CH and DB indexes. Notice that the values of objective functions F_{LS} and $\mathcal{F}_{LS}$ coincide on optimal partitions, and the value of CH (resp. DB) index is larger (resp. smaller) for the globally optimal partition.

Partitions	$\mathcal{G}$	$\mathcal{F}_{LS}$	F_{LS}	CH	DB
$\Pi^{(1)}$	37.67	55.33	55.33	4.08	.89
$\Pi^\star$	51	42	42	7.28	.83

Exercise 4.13 Using the same *Mathematica*-program, find the optimal 2-partitions for the set in Example 4.12 but applying the ℓ_1 metric function. Do there exist again different locally and globally optimal partitions?

The following example illustrates the fact that the k-means algorithm can reduce the number of clusters.

Example 4.14 We will apply the k-means algorithm to the data set $\mathcal{A} = \{(0,0), (3,2), (3,3), (3,4), (4,4), (9,6), (10,5)\}$ using the LS distance-like function and initial centers $(0,0)$, $(9,6)$, and $(10,5)$.

Step A produces the partition with three clusters $\Pi^{(1)} = \{\pi_1, \pi_2, \pi_3\}$, where $\pi_1 = \{(0,0), (3,2), (3,3), (3,4)\}$, $\pi_2 = \{(4,4), (9,6)\}$, and $\pi_3 = \{(10,5)\}$. Their centroids are $c_1 = (2.25, 2.25)$, $c_2 = (6.5, 5)$, and $c_3 = (10, 5)$, and the objective function value is $\mathcal{F}_{LS}(\Pi^{(1)}) = 30$ (see Fig. 4.4a). In the next, at the same time the final step, we obtain the locally optimal 2-partition $\Pi^{(2)} = \{\pi_1, \pi_2\}$ with clusters $\pi_1 = \{(0,0), (3,2), (3,3), (3,4), (4,4)\}$ and $\pi_2 = \{(9,6), (10,5)\}$. Centroids of these clusters are $c_1 = (2.6, 2.6)$ and $c_2 = (9.5, 5, 5)$, and the objective function value is $\mathcal{F}_{LS}(\Pi^{(2)}) = 21.4$ (see Fig. 4.4b).

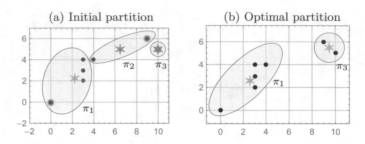

Fig. 4.4 Application of k-means algorithm eliminates the cluster π_2

The next example shows how a single point may result in different locally optimal k-partitions.

Example 4.15 ([160]) Define the data set as follows: choose a datum point $a_0 \in \mathbb{R}^2$ and randomly choose five assignment points $z_1, \ldots, z_5 \in \mathbb{R}^2$ on a circle centered at a_0. In the neighborhood of each point z_j randomly generate points by contaminating the coordinates of z_j by binormal random additive errors with mean vector $(0, 0) \in \mathbb{R}^2$ and the identity covariance matrix.

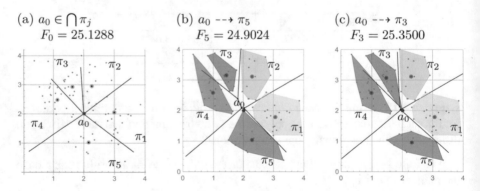

Fig. 4.5 Looking for optimal position of the point a_0

According to the minimal distance principle, clusters $\pi_j = \pi(z_j)$, $j = 1, \ldots, 5$, are defined by assignment points $z_1, \ldots, z_5$. Thereby, the point a_0 lies on the common border of all five clusters (Fig. 4.5a).

Figure 4.5b shows a locally optimal partition where the point a_0 is attached to the cluster π_5. Similarly, Fig. 4.5c shows another locally optimal partition, where the point a_0 is attached to the cluster π_3.

Remark 4.16 The corresponding *Mathematica*-modules KMeansPart[] and KMeansCen[] are described in Sect. 9.2 and links to appropriate *Mathematica*-codes are supplied. These modules perform the k-means algorithm for data sets $\mathcal{A} \subset \mathbb{R}$ or $\mathcal{A} \subset \mathbb{R}^2$, using the initial partition or the initial centers respectively, allowing the choice of the p-distance like function for $p \in \{1, 2\}$. Available are also options for graphical presentation and for printout of intermediate results, allowing the results to be printed as fractions or as decimal numbers.

The standard k-means algorithm for a data set $\mathcal{A} \subset \mathbb{R}^n$ with weights can be performed by the *Mathematica*-module WKMeans[] which is described in Sect. 9.2, where the link to appropriate *Mathematica*-code is supplied.

4.2.4 k-means Algorithm with Multiple Activations

While searching for an optimal k-partition, one can try to circumvent the problem of choosing a good initial approximation by carrying out the standard k-means algorithm several times, each time with new randomly chosen initial centers, and keeping the currently best partition (see [10, 100]). This approach is described in the following algorithm:

Algorithm 1 (k-means algorithm with multiple activations)

Input: $\mathcal{A} \subset \Delta \subset \mathbb{R}^n$ {Data set}; $k \geq 2$, $It > 1$;
1: Set $c^{(0)} \in \Delta^k$ randomly;
2: Apply the k-means algorithm to the set $\mathcal{A}$ with initial center $c^{(0)}$, denote the solution by $\hat{c} = \hat{c}^{(0)}$ and set $F_0 = F(\hat{c})$;
3: **for** $i = 1$ to It **do**
4:     Set $c^{(i)} \subset \Delta^k$ randomly;
5:     Apply the k-means algorithm to the set $\mathcal{A}$ with initial center $c^{(i)}$, denote the solution by $\hat{c}^{(i)}$ and set $F_1 = F(\hat{c}^{(i)})$;
6:     **if** $F_1 \leq F_0$ **then**
7:         Set $\hat{c} = \hat{c}^{(i)}$ and set $F_0 = F_1$;
8:     **end if**
9: **end for**
Output: $\{\hat{c}, F(\hat{c})\}$.

Remark 4.17 Notice that line 5 covers the case when the k-means algorithm reduces the number of clusters. If this happens, the value of function F increases, hence such a partition ceases to be a candidate for optimal k-partition.

4.3 Incremental Algorithm

Let $\mathcal{A} = \{a_i \in \mathbb{R}^n : i = 1, \ldots, m\}$ be a data set and let d be some distance-like function. Suppose that the number of clusters into which the set $\mathcal{A}$ should be divided is not known in advance.

In Sect. 3.4 we have defined the objective function $F : \mathbb{R}^{kn} \to \mathbb{R}_+$

$$F(z_1, \ldots, z_k) = \sum_{i=1}^{m} \min\{d(z_1, a_i), \ldots, d(z_k, a_i)\}, \tag{4.12}$$

which, according to Lemma 3.45, satisfies

$$F(z_1, \ldots, z_k) \geq \mathcal{F}(\Pi) \geq F(c_1, \ldots, c_k),$$

where $\Pi = \{\pi_1(c_1), \ldots, \pi_1(c_k)\}$ is the partition of $\mathcal{A}$ with cluster centers $c_1, \ldots, c_k$. According to Theorem 3.46 the values of objective functions $\mathcal{F}$ and F coincide for k-LOPart $\hat{\Pi} = \{\hat{\pi}_1(\hat{c}_1), \ldots, \hat{\pi}_1(\hat{c}_k)\}$ with cluster centers $\hat{c}_1, \ldots, \hat{c}_k$. As already mentioned at the beginning of this chapter, there is no numerically efficient method to minimize the function (4.12).

However, based on the methods developed in [8, 9, 11, 13–15, 116, 166], we are going to construct an incremental algorithm which will consecutively search for optimal partitions with 2, 3, ... clusters. In order to decide which of these partitions is the most appropriate one, we shall discuss various indexes in Chap. 5.

Let us describe the general incremental algorithm. One can show that it is enough to carry out the search for cluster centers of optimal partitions on a hyperrectangle which contains the set $\mathcal{A}$ (see [154]). Let $\Delta \subset \mathbb{R}^n$ be such a hyperrectangle. For simplicity, start the procedure at some initial point $z_1 \in \Delta$ (e.g. the centroid of the set $\mathcal{A}$). To find the next point $z_2 \in \Delta$, we are going to solve the optimization problem

$$z_2 \in \underset{x \in \Delta}{\arg\min} \, \Phi(z_1, x), \qquad \Phi(z_1, x) = \sum_{i=1}^{m} \min\{d(z_1, a_i), d(x, a_i)\}. \tag{4.13}$$

Notice that the structure of the function Φ corresponds to the structure of the function F in (4.12).

Applying the k-means algorithm with the set $\{z_1, z_2\}$ one obtains the centers $\hat{c}_1$ and $\hat{c}_2$ of a 2-LOPart.

The next point $z_3 \in \mathbb{R}^n$ is found by solving the optimization problem

$$z_3 \in \underset{x \in \Delta}{\arg\min} \, \Phi(\hat{c}_1, \hat{c}_2, x), \qquad \Phi(\hat{c}_1, \hat{c}_2, x) = \sum_{i=1}^{m} \min\{d(\hat{c}_1, a_i), d(\hat{c}_2, a_i), d(x, a_i)\}.$$

Applying the k-means algorithm with the set $\{\hat{c}_1, \hat{c}_2, z_3\}$ one obtains the centers $c_1^\star$, $c_2^\star$, and $c_3^\star$ of a 3-LOPart.

In general, knowing k centers $c_1, \ldots, c_k$, one obtains an approximation of the next center by solving the GOP

$$c_{k+1} \in \underset{x \in \Delta}{\arg\min} \, \Phi(x), \qquad \Phi(x) := \sum_{i=1}^{m} \min\{\delta_k^i, d(x, a_i)\}, \tag{4.14}$$

where $\delta_k^i = \min_{1 \le s \le k} d(c_s, a_i)$. After that, applying the k-means algorithm with centers $c_1, \ldots, c_{k+1}$ one obtains centers $c_1^\star, \ldots, c_k^\star, c_{k+1}^\star$ of an optimal $(k+1)$-partitions $\Pi^{(k+1)}$. The whole procedure is described in Algorithm 2.

Algorithm 2 Incremental algorithm

Input: $\mathcal{A} \subset \mathbb{R}^n$ {data set}; $\epsilon_{\text{INC}} > 0$;

1: Determine a hyperrectangle Δ containing the set $\mathcal{A}$, choose $c_1 \in \Delta$, calculate $F_1 = F(c_1)$, solve GOP (4.13), and denote the solution by c_2;

2: Apply the k-means algorithm with centers c_1, c_2, denote the solution by $c^\star = \{c_1^\star, c_2^\star\}$ and the corresponding objective function value by $F_2 := F(c^\star)$;

3: Set $c_1 = c_1^\star; c_2 = c_2^\star; k = 2; F_k =: F_2$;

4: For centers $c_1, \ldots, c_k$ solve the GOP $\arg\min_{x \in \Delta} \Phi(x)$, where

$$\Phi(x) := \sum_{i=1}^m \min\{\delta_k^i, d(x, a_i)\}, \quad \delta_k^i = \min_{1 \le s \le k} d(c_s, a_i), \qquad (4.15)$$

and denote the solution by c_{k+1};

5: Apply the k-means algorithm with centers $c_1, \ldots, c_k, c_{k+1}$, denote the solution by $c^\star = \{c_1^\star, \ldots, c_k^\star, c_{k+1}^\star\}$ and the corresponding objective function value by $F_{k+1} := F(c^\star)$;

6: **if** $\frac{1}{F_1}(F_k - F_{k+1}) > \epsilon_{\text{INC}}$ **then**

7:    Set $c_1 = c_1^\star; \ldots; c_{k+1} = c_{k+1}^\star; F_k = F_{k+1}; k = k+1$ and go to Step 4;

8: **end if**

Output: $\{c_1^\star, \ldots, c_k^\star, c_{k+1}^\star\}$

Remark 4.18 Let us point out a few important comments regarding effectiveness of Algorithm 2:

- One could have started the algorithm in a similar way by using more than one center.
- For solving the optimization problems (4.13), (4.14), and (4.15) one can use the global optimization algorithm DIRECT (see [55, 64, 69, 89]). One can show that in real-life applications, in order to get a sufficiently good initial approximation for the k-means algorithm which will then give LOPart, it suffices to carry out only several, say 10 iterations of the optimization algorithm DIRECT (see [161, 162]).
- Since the sequence $F_1, F_2, \ldots$ of objective function values is monotonically decreasing, the decision when to stop the iterative process is based on the criterion (see e.g. [10, 14]) $\frac{F_k - F_{k+1}}{F_1} < \epsilon_{\text{INC}}$ for some small $\epsilon_{\text{INC}} > 0$ (say 0.05). In Example 4.20, the third run of the algorithm gives $\frac{F_2 - F_3}{F_1} = 0.04$.

Remark 4.19 The corresponding *Mathematica*-module Inc[] is described in Sect. 9.2, and the link to appropriate *Mathematica*-code is supplied.

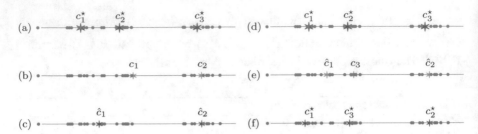

Fig. 4.6 Finding the next center

Example 4.20 We are given three centers $c_1^\star = 0.2$, $c_2^\star = 0.4$, and $c_3^\star = 0.8$, and in the neighborhood of each one 10 random numbers we generated from the normal distribution. This gives the set $\mathcal{A}$ of 30 elements (see Fig. 4.6a or Fig. 4.6d). Applying the previously described Incremental algorithm using the LS distance-like function, we will try to reconstruct centers from which the set $\mathcal{A}$ was set up.

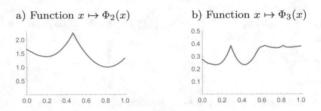

Fig. 4.7 Finding the second and the third center

First choose $c_1 = \text{mean}(\mathcal{A})$, find $F_1 = \sum\limits_{a \in \mathcal{A}} (c_1 - a)^2$, and define the function (see Fig. 4.7a)

$$\Phi_2(x) := \sum_{i=1}^{m} \min\{(c_1 - a_i)^2, (x - a_i)^2\}.$$

Using the *Mathematica*-module `NMinimize[]`, solve the GOP $\underset{x \in [0,1]}{\arg\min} \Phi_2(x)$ and denote the solution by c_2 (Fig. 4.6b).

Next, apply the k-means algorithm to the set $\mathcal{A}$ with initial centers c_1, c_2 to obtain optimal 2-partition of the set $\mathcal{A}$ with centers $\hat{c}_1, \hat{c}_2$ (Fig. 4.6c). Set $F_2 = F_{LS}(\hat{c}_1, \hat{c}_2)$.

In order to find optimal 3-partition of $\mathcal{A}$, define the auxiliary function (see Fig. 4.7b)

$$\Phi_3(x) := \sum_{i=1}^{m} \min\{(\hat{c}_1 - a_i)^2, (\hat{c}_2 - a_i)^2, (x - a_i)^2\}.$$

Using the *Mathematica*-module NMinimize [], solve the GOP $\arg\min \Phi_3(x)$ and denote the solution by c_3 (Fig. 4.6e). $\quad x\in[0,1]$

Next, apply the k-means algorithm to $\mathcal{A}$ with initial centers $\hat{c}_1$, $\hat{c}_2$, c_3, and obtain optimal 3-partition of the set $\mathcal{A}$ with centers $c_1^\star, c_2^\star, c_3^\star$ (which in this case do coincide quite well with the original centers, see Fig. 4.6f). Finally, set $F_3 = F_{LS}(c_1^\star, c_2^\star, c_3^\star)$, and note that $\frac{F_2 - F_3}{F_1} = 0.04$.

Exercise 4.21 Devise and carry out the Incremental algorithm to the set $\mathcal{A}$ from Example 4.20 using the ℓ_1 metric function.

Example 4.22 We are going to apply the Incremental algorithm to the set $\mathcal{A} = \{2, 4, 5, 8, 9, 10\}$ using the ℓ_1 metric function.

Notice that, in this particular case, $\Delta = [2, 10]$. For the initial center c_1 let us choose $\text{med}(\mathcal{A}) = 6.5$ for which the objective function value is $F_1(c_1) = 16$. Choosing the next center arbitrarily as $c_2 = 2$ and applying the k-means algorithm with centers c_1 and c_2, we get, after two iterations, 2-LOPart $\hat{\Pi} = \{\{2, 4, 5\}, \{8, 9, 10\}\}$ with cluster centers $\hat{c}_1 = 4$, $\hat{c}_2 = 9$, and objective function value $F_1(\hat{c}_1, \hat{c}_2) = 5$.

Choosing the next center by solving the optimization problem

$$\arg\min_{x\in\Delta} \Phi(x), \qquad \Phi(x) = \sum_{i=1}^{6} \min\{|c_1 - a_i|, |x - a_i|\},$$

(sketch the graph of this function), we obtain $c_2 = 9$. Applying the k-means algorithm to centers c_1, c_2 we obtain, already in the first iteration, the same solution as above. Check the result using *Mathematica*-module Inc [] (see Sect. 9.2).

Example 4.23 We are going to apply the Incremental algorithm to the set $\mathcal{A} = \{(5, 2), (3, 8), (4, 4), (2, 2), (9, 7), (3, 9), (6, 4)\}$ (see Fig. 4.8) using the ℓ_1 metric function.

a) Arbitrary choice

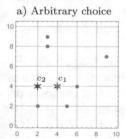

b) Optimized choice

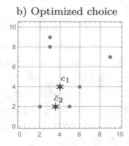

c) ℓ_1-optimal 2-partition

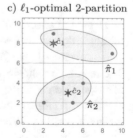

Fig. 4.8 Incremental algorithm: searching for ℓ_1-optimal 2-partition using an arbitrary, and then the optimized choice for the second center

Obviously, $\mathcal{A} \subset [0, 10] \times [0, 10] =: \Delta$. For the initial center c_1 let us choose $c_1 = \text{med}(\mathcal{A}) = (4, 4)$ for which $F_1(c_1) = 28$. Choosing arbitrarily the next center to be $c_2 = (2, 4)$ (see Fig. 4.8a), and applying the k-means algorithm to centers c_1, c_2, after two iterations we obtain the 2-LOPart $\hat{\Pi} = \{\hat{\pi}_1, \hat{\pi}_2\}$, $\hat{\pi}_1 = \{(3, 8), (9, 7), (3, 9)\}$, $\hat{\pi}_2 = \{(5, 2), (4, 4), (2, 2), (6, 4)\}$, with cluster centers $\hat{c}_1 = (3, 8)$, $\hat{c}_2 = (4.5, 3)$, and the value of objective function $F_1(\hat{c}_1, \hat{c}_2) = 17$ (see Fig. 4.8c).

Choosing the next center by solving the optimization problem

$$\arg\min_{x \in \Delta} \Phi(x), \qquad \Phi(x) = \sum_{i=1}^{7} \min\{\|c_1 - a_i\|_1, \|x - a_i\|_1\},$$

we obtain $\tilde{c}_2 = (3.57864, 2)$ (see Fig. 4.8b) (draw the `ContourPlot` of this function). Applying the k-means algorithm to centers c_1, $\tilde{c}_2$, we obtain, already in the first iteration, the earlier solution (see Fig. 4.8c). Check the result using *Mathematica*-module `Inc[]` (see Sect. 9.2).

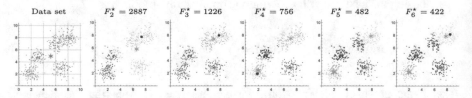

Fig. 4.9 Incremental algorithm

Example 4.24 Take five points in the plane: $C_1 = (2, 2)$, $C_2 = (3, 5)$, $C_3 = (6, 7)$, $C_4 = (7, 3)$, and $C_5 = (8, 8)$. In the neighborhood of each one generate 200 random points from the bivariate normal distribution $\mathcal{N}(C_j, 0.5\,I)$ with expectation $C_j \in \mathbb{R}^2$ and the covariance matrix $\sigma^2 I$, where $\sigma^2 = 0.5$. The corresponding data set $\mathcal{A}$ consists of $m = 1000$ data, depicted in Fig. 4.9.

Using the *Mathematica*-module `Inc[]` (see Sect. 9.2), we applied the incremental algorithm to this set using the LS distance-like function and the initial center $c_1 = \text{mean}(\mathcal{A}) \approx (5.21, 5.00)$, to determine LOParts with $2, \dots, 6$ clusters. The results are shown in Fig. 4.9. The captions above each figure show the values of respective objective functions. The red points denote centroids obtained by the algorithm DIRECT, and the gray stars denote the corrections obtained by the k-means algorithm. What remained open was the question which of the obtained partitions had the most adequate number of clusters.

Example 4.25 Define the data set $\mathcal{A} \subset \mathbb{R}^2$ as follows:

```
In[1]:= SeedRandom[1213];
        c={{-8,-6},{-3,-8},{-1,-1},{8,4},{6,9},{-1,8}}; m=200;
        kov = 1.5 {{1,0},{0,1}};
```

```
data = Table[RandomVariate[
MultinormalDistribution[c[[i]], kov], m], {i,Length[c]}
            ]];
A = Flatten[data, 1];
```

The set $\mathcal{A}$ consists of $m = 1200$ data shown in Fig. 4.10a. We applied the incremental algorithm to this set using the LS distance-like function and determined LOParts with $2, \ldots, 8$ clusters. The results are shown in Fig. 4.10. Captions above each figure show the values of respective objective functions. What is left open is the question which of the obtained partitions has the most adequate number of clusters.

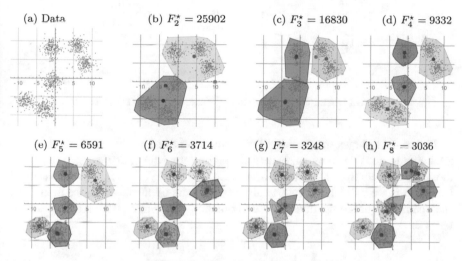

(a) Data   (b) $F_2^\star = 25902$ (c) $F_3^\star = 16830$ (d) $F_4^\star = 9332$

(e) $F_5^\star = 6591$ (f) $F_6^\star = 3714$ (g) $F_7^\star = 3248$ (h) $F_8^\star = 3036$

Fig. 4.10 Incremental algorithm. Red points denote centroids obtained by the algorithm DIRECT, and blue points are corrections obtained by the k-means algorithm

4.4 Hierarchical Algorithms

One possibility to search for optimal partition is to use some of the so-called hierarchical algorithms. These algorithms are used mostly in humanities, biology, medicine, archeology, but also in computer sciences [94, 115, 180, 184].

4.4.1 Introduction and Motivation

The basic idea of hierarchical algorithms is, starting from a known partition $\Pi^{(k)} = \{\pi_1, \ldots, \pi_k\}$ of a set $\mathcal{A} = \{a_i \in \mathbb{R}^n : i = 1, \ldots, m\}$ consisting of $1 < k \leq m$ clusters, to construct a new partition $\Pi^{(r)}$ consisting of r clusters, in such a way that either at least two clusters are merged into one ($r < k$), or that one cluster is split into at least two clusters ($r > k$). In the former case we are talking about *agglomerative algorithms* and in the latter case about *divisive algorithms*.

But let us first introduce the notion of nested partitions.

Definition 4.26 A partition $\Pi^{(k)}$ is said to be ***nested*** in partition $\Pi^{(r)}$, denoted as $\Pi^{(k)} \sqsubset \Pi^{(r)}$, if:

  (i) $r < k$, and
 (ii) each cluster in $\Pi^{(k)}$ is a subset of some cluster in $\Pi^{(r)}$.

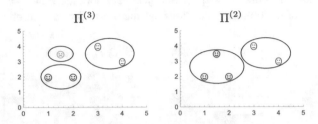

Fig. 4.11 Partition $\Pi^{(3)}$ is nested in the partition $\Pi^{(2)}$ ($\Pi^{(3)} \sqsubset \Pi^{(2)}$)

Figure 4.11 shows the partition $\Pi^{(3)}$ which is nested in the partition $\Pi^{(2)}$. We are going to consider only agglomerative algorithms which, at each step, merge at most two clusters from the current k-partition $\Pi^{(k)} = \{\pi_1, \ldots, \pi_k\}$. To choose these two clusters we will have to look at all possible pairs of clusters, and the number of such pairs is equal to the number of combinations without repetition of k elements taken 2:

$$\binom{k}{2} = \frac{k!}{2!(k-2)!} = \frac{k(k-1)}{2}.$$

Given a distance-like function $d \colon \mathbb{R}^n \times \mathbb{R}^n \to \mathbb{R}_+$ (as in Chap. 2), one can speak about *measure of similarity* or *dissimilarity* between two clusters by considering various measures of distance between clusters [96, 182, 184]. In general, there are several ways to define the distance between two finite sets:

$$D_c(A, B) = d(c_A, c_B) \qquad\qquad \text{[set-centers distance]} \qquad (4.16)$$

$$D_{\min}(A, B) = \min_{a \in A,\, b \in B} d(a, b) \qquad\qquad \text{[minimal distance]} \qquad (4.17)$$

$$D_{\max}(A, B) = \max_{a \in A,\, b \in B} d(a, b) \qquad\qquad \text{[maximal distance]} \qquad (4.18)$$

$$D_{\mathrm{avg}}(A, B) = \frac{1}{|A|\,|B|} \sum_{a \in A} \sum_{b \in B} d(a, b) \qquad\qquad \text{[average distance]} \qquad (4.19)$$

$$D_H(A, B) = \max\{\max_{a \in A} \min_{b \in B} d(a, b),\ \max_{b \in B} \min_{a \in A} d(a, b)\}$$

$$\text{[Hausdorff distance]} \qquad (4.20)$$

Exercise 4.27 What is the Hausdorff distance between the sets A and B in Fig. 4.12?

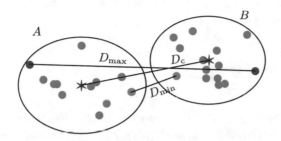

Fig. 4.12 Various ways to measure the distance between sets A and B

Example 4.28 Figure 4.13 shows the sets A (red dots) and B (blue stars) in three different relative positions, and Table 4.1 lists their D_c, D_{min}, and Hausdorff distances for LS distance-like function and for ℓ_1 metric function.

We will consider in more detail the use of similarity measure between clusters defined by the set-centers distance (4.16) for hitherto discussed distance-like functions.

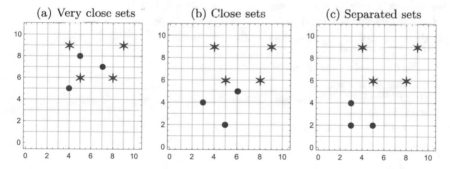

Fig. 4.13 Distances between the sets A (red dots) and B (blue stars) in various relative positions

Let $\mathcal{A} = \{a_i \in \mathbb{R}^n : i = 1, \ldots, m\}$ be a given set. Agglomerative algorithm starts from some partition $\Pi^{(\mu)}$ consisting of μ clusters and winds up with a partition $\Pi^{(k)}$ consisting of k clusters, $1 \leq k < \mu \leq m$. In most cases we will start the algorithm with the partition having m clusters

$$\Pi^{(m)} = \{\{a_1\}, \{a_2\}, \ldots, \{a_m\}\},$$

i.e. with the partition whose all clusters are singletons. The distance D between clusters will be defined as the distance between their centers using the LS distance-like or ℓ_1 metric function.

Table 4.1 Various measures of distances between sets A and B

	Very close sets	Close sets	Separated sets
D_c for ℓ_1 metric function	2	5	9
D_c for LS distance-like function	2.05	18	31.4
$D_{\min}$ for ℓ_1 metric function	2	2	4
$D_{\min}$ for LS distance-like function	2	2	8
D_H for ℓ_1 metric function	4	7	11
D_H for LS distance-like function	8	25	61

Notice that for the partition $\Pi^{(m)}$ the value of objective function $\mathcal{F}$ equals zero. In the first step we choose two most similar clusters, i.e. two closest elements of the set $\mathcal{A}$, and merge them to form a single cluster. According to Theorem 3.7 this will increase the value of objective function $\mathcal{F}$.

The algorithm can be stopped on partition with a specified number of clusters, and it is possible to consider the problem of finding a partition with the most appropriate number of clusters [94, 116, 166, 188]. We are going to discuss this problem in more detail in Chap. 5.

Algorithm 3 (Agglomerative Nesting (AGNES))

Input: $\mathcal{A} = \{a_i \in \mathbb{R}^n : i = 1, \ldots, m\}, \quad 1 < k < m, \quad \mu = 0$;

1: Define the initial partition $\Pi^{(m)} = \{\pi_1, \ldots, \pi_m\}, \pi_j = \{a_j\}$;

2: Construct the similarity matrix for partition $\Pi^{(m-\mu)}$
    $R_{m-\mu} \in \mathbb{R}^{(m-\mu) \times (m-\mu)}, r_{ij} = D(\pi_i, \pi_j)$;

3: Solve the optimization problem $\{i_0, j_0\} \subseteq \underset{1 < i < j \leq m}{\arg\min}\, r_{ij}$;

4: Construct the new partition
    $\Pi^{(m-\mu-1)} = \left(\Pi^{(m-\mu)} \setminus \{\pi_{i_0}, \pi_{j_0}\}\right) \bigcup \{\pi_{i_0} \cup \pi_{j_0}\}$;

5: **if** $\mu < m - k$, **then**

6:    Set $\mu := \mu + 1$ and go to Step 2;

7: **else**

8:    STOP;

9: **end if**

Output: $\{\Pi^{(k)}\}$.

In Step 3 the algorithm looks for the position (i_0, j_0) of the smallest element in the upper triangle of the similarity matrix $R_{m-\mu}$. In Step 4 the clusters π_{i_0} and π_{j_0} are merged into a single cluster, thus forming the new partition. Step 5 checks the stopping criterion of the algorithm.

A useful and illustrative demonstration of Algorithm 3 can be done using *dendrogram*, which shows every step of the algorithm and gives the similarity level. Let us illustrate the algorithm by the following simple example.

Example 4.29 We will start Algorithm AGNES on the set $\mathcal{A} = \{a_i = (x_i, y_i) : i = 1, \ldots, 8\} \subset \mathbb{R}^2$ given by

i	1	2	3	4	5	6	7	8
x_i	2	3	4	6	7	8	9	9
y_i	7	9	8	6	5	2	1	3

(see Fig. 4.14a), starting with the partition $\Pi^{(m)} = \{\{a_1\}, \ldots, \{a_m\}\}$, using similarity measure (4.16), and applying ℓ_1 metric function (see Fig. 4.12):

$$D_1(A, B) = \|c_A - c_B\|_1, \quad c_A = \underset{a \in A}{\mathrm{med}}\, a, \quad c_B = \underset{b \in B}{\mathrm{med}}\, b.$$

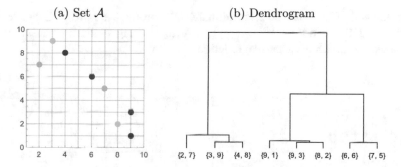

(a) Set $\mathcal{A}$ (b) Dendrogram

Fig. 4.14 Set $\mathcal{A}$ and the corresponding dendrogram using the ℓ_1 metric function

All partitions of the set $\mathcal{A}$ can be found using *Mathematica* computation system (see Fig. 4.14b):

```
In[1]:= Needs["HierarchicalClustering`"]
In[2]:= DendrogramPlot[A, Linkage->"Median", HighlightLevel->2, LeafLabels->(# &)]
```

Let us show how to obtain the same thing using Algorithm AGNES. In the first pass through the Algorithm AGNES we start from the 8-partition $\Pi^{(8)}$, find the cluster centers (in this case these are elements of $\mathcal{A}$ themselves), and create the upper triangle of the similarity matrix R_8. Here we look for the smallest element (if there are several such—take any), which specifies the two most similar clusters. In our case we choose singleton clusters $\{(3, 9)\}$ and $\{(4, 8)\}$. They are removed from the partition and the new two-element cluster $\{(3, 9), (4, 8)\}$ is added. This results in the optimal 7-partition $\Pi^{(7)}$ shown in Fig. 4.15.

By repeating the procedure we obtain the remaining optimal partitions $\Pi^{(6)}$, $\Pi^{(5)}$, $\Pi^{(4)}$, $\Pi^{(3)}$, $\Pi^{(2)}$, $\Pi^{(1)}$, in turn.

$$R_8 = \begin{bmatrix} 0 & 3 & 3 & 5 & 7 & 11 & 13 & 11 \\ & 0 & 2 & 6 & 8 & 12 & 14 & 12 \\ & & 0 & 4 & 6 & 10 & 12 & 10 \\ & & & 0 & 2 & 6 & 8 & 6 \\ & & & & 0 & 4 & 6 & 4 \\ & & & & & 0 & 2 & 2 \\ & & & & & & 0 & 2 \\ & & & & & & & 0 \end{bmatrix}; \quad R_7 = \begin{bmatrix} 0 & 5 & 7 & 11 & 13 & 11 & 3 \\ & 0 & 2 & 6 & 8 & 6 & 5 \\ & & 0 & 4 & 6 & 4 & 7 \\ & & & 0 & 2 & 2 & 11 \\ & & & & 0 & 2 & 13 \\ & & & & & 0 & 11 \\ & & & & & & 0 \end{bmatrix}; \quad R_6 = \begin{bmatrix} 0 & 11 & 13 & 11 & 3 & 6 \\ & 0 & 2 & 2 & 11 & 5 \\ & & 0 & 2 & 13 & 7 \\ & & & 0 & 11 & 5 \\ & & & & 0 & 6 \\ & & & & & 0 \end{bmatrix};$$

$$R_5 = \begin{bmatrix} 0 & 11 & 3 & 6 & 12 \\ & 0 & 11 & 5 & 2 \\ & & 0 & 6 & 12 \\ & & & 0 & 6 \\ & & & & 0 \end{bmatrix}; \quad R_4 = \begin{bmatrix} 0 & 3 & 6 & 12 \\ & 0 & 6 & 12 \\ & & 0 & 6 \\ & & & 0 \end{bmatrix}; \quad R_3 = \begin{bmatrix} 0 & 6 & 6 \\ & 0 & 12 \\ & & 0 \end{bmatrix}.$$

Exercise 4.30 Carry out the Algorithm AGNES on the set $\mathcal{A}$ from Exercise 4.29 starting with the partition $\Pi^{(m)} = \{\{a_1\}, \ldots, \{a_m\}\}$ using similarity measure (4.16) and applying the LS distance-like function.

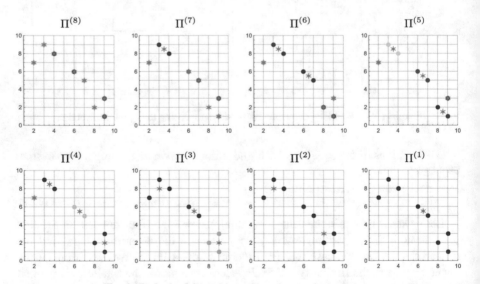

Fig. 4.15 Optimal k-partitions. Stars denote cluster centers

Exercise 4.31 Starting with the partition $\Pi^{(m)} = \{\{a_1\}, \ldots, \{a_m\}\}$, carry out the Algorithm AGNES on the set $\mathcal{A}$ from Example 4.29 using similarity measure (4.17) and applying the ℓ_1 distance function.

Exercise 4.32 Starting with the partition $\Pi^{(m)} = \{\{a_1\}, \ldots, \{a_m\}\}$, carry out the Algorithm AGNES on the set $\mathcal{A}$ from Example 4.29 using similarity measure (4.17) and applying the LS distance-like function.

Remark 4.33 The corresponding *Mathematica*-module `AgglNest [ ]` is described in Sect. 9.2, where the link to appropriate *Mathematica*-code is also supplied.

4.4.2 Applying the Least Squares Principle

Let $\Pi^{(k)} = \{\pi_1, \ldots, \pi_k\}$ be a k-partition of a finite set $\mathcal{A} \subset \mathbb{R}^n$. Using the LS distance-like function, we define the *similarity* (distance) *between clusters* π_r and $\pi_s \in \Pi^{(k)}$ as the distance between their centroids:

$$D_{LS}(A, B) = \|c_r - c_s\|^2, \qquad c_r = \text{mean}(\pi_r), \quad c_s = \text{mean}(\pi_s). \qquad (4.21)$$

The assertion of the following exercise will help us with the proof of Theorem 4.35 providing an explicit formula for the centroid of the union of two clusters and for the value of objective function on this union.

Exercise 4.34 Prove, similarly as in the proof of Lemma 3.20, that for a set $A = \{a_i \in \mathbb{R}^n : i = 1, \ldots, p\}$ and its centroid $c_A = \text{mean } A$, the following holds true:

$$\sum_{i=1}^{p}(a_i - c_A) = 0, \qquad (4.22)$$

$$\sum_{i=1}^{p} \|a_i - x\|^2 = \sum_{i=1}^{p} \|a_i - c_A\|^2 + p \|x - c_A\|^2, \quad x \in \mathbb{R}^n. \qquad (4.23)$$

Theorem 4.35 *Let the set* $\mathcal{A} = \{a_i \in \mathbb{R}^n : i = 1, \ldots, m\}$ *be the disjoint union of (i.e. union of disjoint) nonempty sets A and B, $\mathcal{A} = A \sqcup B$:*

$$A = \{a_1, \ldots, a_p\}, \quad |A| = p, \quad c_A = \frac{1}{p}\sum_{i=1}^{p} a_i$$

$$B = \{b_1, \ldots, b_q\}, \quad |B| = q, \quad c_B = \frac{1}{q}\sum_{j=1}^{q} b_j.$$

Then the centroid of the set $\mathcal{A} = A \sqcup B$ *is*

$$c = c(\mathcal{A}) = \frac{p}{p+q} c_A + \frac{q}{p+q} c_B, \qquad (4.24)$$

and

$$\mathcal{F}_{LS}(A \sqcup B) = \mathcal{F}_{LS}(A) + \mathcal{F}_{LS}(B) + p \|c_A - c\|^2 + q \|c_B - c\|^2, \qquad (4.25)$$

where $\mathcal{F}_{LS}$ *is the LS objective function (see Sect. 3.1).*

Proof (4.24) follows directly from

$$c = c(A \sqcup B) = \frac{1}{p+q}\left(\sum_{i=1}^{p} a_i + \sum_{j=1}^{q} b_j\right) = \frac{p}{p+q}\frac{1}{p}\sum_{i=1}^{p} a_i + \frac{q}{p+q}\frac{1}{q}\sum_{j=1}^{q} b_j.$$

Using Exercise 4.34 we obtain

$$\mathcal{F}_{LS}(A \sqcup B) = \sum_{i=1}^{p}\|a_i - c\|^2 + \sum_{j=1}^{q}\|b_j - c\|^2$$

$$= \sum_{i=1}^{p}\|a_i - c_A\|^2 + p\,\|c - c_A\|^2 + \sum_{j=1}^{q}\|b_j - c_B\|^2 + q\,\|c - c_B\|^2$$

$$= \mathcal{F}_{LS}(A) + \mathcal{F}_{LS}(B) + p\,\|c_A - c\|^2 + q\,\|c_B - c\|^2. \qquad \square$$

The expression $\Delta := p\,\|c_A - c\|^2 + q\,\|c_B - c\|^2$ in (4.25) can be simplified as

Corollary 4.36 *If the set $A = \{a_i \in \mathbb{R}^n : i = 1, \ldots, m\}$ with centroid $c = \frac{1}{m}\sum_{i=1}^{m} a_i$ is the disjoint union of two sets, $A = A \sqcup B$,*

$$A = \{a_1, \ldots, a_p\}, \quad |A| = p, \quad c_A = \frac{1}{p}\sum_{i=1}^{p} a_i$$

$$B = \{b_1, \ldots, b_q\}, \quad |B| = q, \quad c_B = \frac{1}{q}\sum_{j=1}^{q} b_j$$

then

$$\Delta = p\,\|c_A - c\|^2 + q\,\|c_B - c\|^2 = \frac{pq}{p+q}\,\|c_A - c_B\|^2. \qquad (4.26)$$

Proof By (4.24) we have

$$p\,\|c_A - c\|^2 = p\,\left\|\frac{p}{p+q} c_A + \frac{q}{p+q} c_B - c_A\right\|^2 = \frac{pq^2}{(p+q)^2}\,\|c_A - c_B\|^2,$$

$$q\,\|c_B - c\|^2 = q\,\left\|\frac{p}{p+q} c_A + \frac{q}{p+q} c_B - c_B\right\|^2 = \frac{p^2 q}{(p+q)^2}\,\|c_A - c_B\|^2,$$

and therefore

$$\Delta = \frac{pq^2}{(p+q)^2}\,\|c_A - c_B\|^2 + \frac{p^2 q}{(p+q)^2}\,\|c_A - c_B\|^2,$$

directly implying (4.26). $\qquad \square$

Therefore, as a measure of similarity between clusters A and B, instead of (4.21) one can use the **Ward's distance**

$$D_W(A, B) := \frac{|A||B|}{|A|+|B|} \|c_A - c_B\|^2, \tag{4.27}$$

As Theorem 4.35 and Corollary 4.36 show, if we use the Ward's distance (4.27) as the measure of similarity between two clusters, then merging clusters A and B will increase the value of objective function precisely by $D_W(A, B)$. One can prove [184] that this choice ensures the least possible increase of the objective function $\mathcal{F}_{LS}$.

4.5 DBSCAN Method

So far we have been considering methods for finding optimal partitions using the LS or ℓ_1 distance-like function. Optimal partitions obtained in this way tend to be of spherical shape. Such approach is therefore often called *spherical clustering* (see e.g. [21, 96, 184]).

Now we want to consider data which densely populate arbitrary shapes in $\mathbb{R}^n$. Several such objects in the plane are depicted in Fig. 4.17a–c and in Fig. 4.19. First let us show how to construct such artificial data sets.

Example 4.37 We will construct a data set $\mathcal{A}$ consisting of points derived from three discs: $K_1((2, 3), 1.5)$, $K_2((5, 7), 2)$, and $K_3((8, 4), 1)$.

(a) Uniform distributed points   (b) Points inside the disc   (c) Data set $\mathcal{A}$

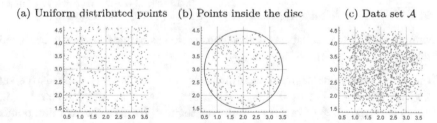

Fig. 4.16 Construction of a data set derived from the disc $K_1((2, 3), 1.5)$

For example, let us show how to construct a data set derived from the disc K_1. Start by populating the rectangle $[0.35, 3.65] \times [1.35, 4.65]$ containing the disc K_1 by *uniformly distributed* points such that there are $n_p = 50$ points per unit square (see Fig 4.16a), and select only those points lying inside the disc K_1 (see Fig 4.16b). Then, to each such point $a \in K_1$ add $n_r = 3$ randomly generated points from *bivariate normal distribution* with expectation $a \in \mathbb{R}^2$ and the covariance matrix $\sigma^2 I$, where $\sigma = 0.025$. This produces 1456 data points derived from disc K_1 with indistinct border shown in Fig 4.16c.

We proceed in a similar way with the other discs. To the data set obtained in the described way, which is now contained in the square $\Delta = [0, 10] \times [0, 10]$, we add additional 2‰ uniformly distributed data points. This produces the data set $\mathcal{A}$ with $|\mathcal{A}| = 4610$ elements shown in Fig 4.17a, with several outliers (red points).

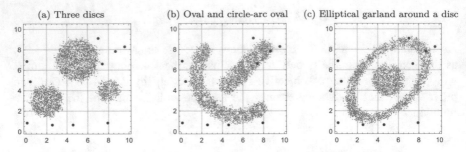

Fig. 4.17 Data sets derived from different geometric objects, with outliers

Remark 4.38 The data set $\mathcal{A}$ can be produced using the `DataDisc[]` module which is described in Sect. 9.3, where the link to appropriate *Mathematica*-code is supplied. The corresponding data set `DataDiscs.txt` can be downloaded from the link posted in Sect. 9.5.

```
In[1]:= Cen={2,3}; r = 1.5; np = 50; sigma = .025; nr = 3;
        data1 = DataDisc[Cen, r, np, sigma, nr, 1];
        Cen={5,7}; r = 2.;
        data2 = DataDisc[Cen, r, np, sigma, nr, 1];
        Cen={8,4}; r = 1;  AG={{0,10},{0,10}};
        data3 = DataDisc[Cen, r, np, sigma, nr, 1];
        app = RandomVariate[UniformDistribution[AG],
                                 Ceiling[2 Length[data]/1000]];
        data = Union[data1[[1]], data2[[1]], data3[[1]],app]
```

Example 4.39 The next data set will be constructed using the so-called generalized circles (see [164, 185]).

Oval The straight line segment determined by two points $\mu, \nu \in \mathbb{R}^2$ is the set $[\mu, \nu] := \{s\mu + (1 - s)\nu : 0 \le s \le 1\} \subset \mathbb{R}^2$. Given a positive number $r > 0$, the *oval* $\mathsf{Oval}([\mu, \nu], r)$ of radius r around the segment $[\mu, \nu]$ is the set of all points in $\mathbb{R}^2$ at distance r from the segment $[\mu, \nu]$, i.e.

$$\mathsf{Oval}([\mu, \nu], r) = \{a \in \mathbb{R}^2 : \mathfrak{D}(a, [\mu, \nu]) = r\}, \tag{4.28}$$

where $\mathfrak{D}(a, [\mu, \nu]) := \min_{t \in [\mu, \nu]} \|a - t\|$ (see Fig. 4.18a), i.e.

$$\mathfrak{D}(a, [\mu, \nu]) = \begin{cases} \|a - \mu\|, & \text{if } \langle (\nu - \mu), (a - \mu) \rangle < 0, \\ \|a - \nu\|, & \text{if } \langle (\mu - \nu), (a - \nu) \rangle < 0, \\ \left\| \mu \frac{\langle (\nu - \mu), (\nu - a) \rangle}{\|\mu - \nu\|^2} + (1 - \frac{\langle (\nu - \mu), (\nu - a) \rangle}{\|\mu - \nu\|^2})\nu \right\|, & \text{otherwise.} \end{cases} \tag{4.29}$$

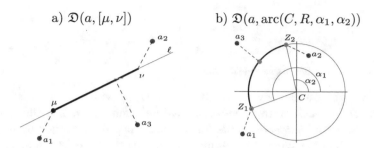

Fig. 4.18 Distances from various points to a segment and to a circle-arc

Similarly as was done in Example 4.37, we generated a data set with indistinct border, consisting of 1960 points in the rectangle $[3.54, 8.96] \times [3.54, 8.96]$, which contains the oval $\mathsf{Oval}([(4.5, 4.5), (8, 8)], 0.8)$ (see Fig. 4.17b).

Circle-arc oval The circle-arc $\mathrm{arc}(C, R, \alpha_1, \alpha_2)$ is part of the circle $K(C, R)$, centered at C with radius R, from α_1 to α_2, i.e. lying between points $Z_1 = C + (R\cos\alpha_1, R\sin\alpha_1)$ and $Z_2 = C + (R\cos\alpha_2, R\sin\alpha_2)$ (see Fig. 4.18b). Given a positive number $r > 0$, the *circle-arc oval*, denoted $\mathsf{CAoval}(\mathrm{arc}(C, R, \alpha_1, \alpha_2), r)$, is the set of all points in the plane at distance r from the circle-arc $\mathrm{arc}(C, R, \alpha_1, \alpha_2)$, i.e.

$$\mathsf{CAoval}(\mathrm{arc}(C, R, \alpha_1, \alpha_2), r) = \{a \in \mathbb{R}^2 : \mathfrak{D}(a, \mathrm{arc}(C, R, \alpha_1, \alpha_2)) = r\}, \qquad (4.30)$$

where $\mathfrak{D}(a, \mathrm{arc}(C, R, \alpha_1, \alpha_2)) = \min_{t \in \mathrm{arc}(C, R, \alpha_1, \alpha_2)} \|a - t\|$, i.e.

$$\mathfrak{D}(a, \mathrm{arc}(C, R, \alpha_1, \alpha_2)) = \begin{cases} \big|\|a - C\| - r\big|, & \varphi_1 + \varphi_2 = \alpha_2 - \alpha_1 \\ \min\{\|a - Z_1\|, \|a - Z_2\|\}, & \text{otherwise} \end{cases}, \qquad (4.31)$$

with φ_i being the angle between vectors $\overrightarrow{Ca}$ and $\overrightarrow{CZ_i}$ (see Fig. 4.18b).

Similarly as we have done in Example 4.37, we generated a data set with indistinct border which consists of 2844 points contained in the rectangle $[0.78, 10.22] \times [0.78, 10.22]$, and which contains the circle-arc oval $\mathsf{CAoval}(\mathrm{arc}((5.5, 5.5), 4, -1, 2.5), 0.6)$ (see Fig. 4.17b).

To the data set obtained in the described way, which is now contained in the square $\Delta = [0, 10] \times [0, 10]$, we added additional 2‰ uniformly distributed data points. This produced the data set $\mathcal{A}$ with $|\mathcal{A}| = 4814$ points shown in Fig. 4.17b, with several outliers (red points).

Remark 4.40 The data set $\mathcal{A}$ can be produced using the `DataOval[]` and `DataCAoval[]` modules which are described in Sect. 9.3, where links to appropriate *Mathematica*-codes are supplied. The corresponding data set `DataGenCirc.txt` can be downloaded from the link posted in Sect. 9.5.

```
In[1]:= Cen={5.5,5.5}; R=4.; al1=-1; al2=2.5; r=.6; np=50; nr=3; sigma=0.025;
```

```
    data1 = DataCAoval[Cen, R, al1, al2, r, np, sigma, nr, 1];
a={4.5, 4.5}; b={8, 8}; r=.8; AG={{0,10},{0,10}};
    data2 = DataOval[a, b, r, np, sigma, nr, 1];
    app = RandomVariate[UniformDistribution[AG],
                        Ceiling[2 Length[data]/1000]];
data = Union[data1[[1]], data2[[1]],app];
```

Example 4.41 In a similar way as in Example 4.37, we will define a data set derived from the region between two non-parallel ellipses, and a disc. The ellipse will be regarded as the *Mahalanobis circle* (6.14) (see Chap. 6.2), defined by its center $C = (5, 5)$ and the symmetric positive definite matrix $\Sigma = U \operatorname{diag}(a^2, b^2) U^T$, where $U = \begin{bmatrix} \cos \vartheta & -\sin \vartheta \\ \sin \vartheta & \cos \vartheta \end{bmatrix}$, $a = 5$, $b = 3$, $\vartheta = \frac{\pi}{4}$. The region between two ellipses that we are interested in is the set $\{x \in \mathbb{R}^2 : 0.8 \le d_m(C, x, \Sigma) \le 1\}$, where d_m is the Mahalanobis distance-like function determined by Σ (see Chap. 6.2).

Similarly as in Example 4.37, we generated 3 536 points (see Fig 4.17c), to which we also added 1 392 points derived from the disc $K((5, 5), 1.5)$. To the data set obtained in this way, which is contained in the square $\Delta = [0, 10] \times [0, 10]$, we added 2‰ uniformly distributed data points, resulting in the data set $\mathcal{A}$ with $|\mathcal{A}| = 4 938$ points shown in Fig. 4.17c, with several outliers (red points).

Remark 4.42 The data set $\mathcal{A}$ can be produced using `DataEllipseGarland[]` and `DataCircle[]` modules described in Sect. 9.3, where the link to appropriate *Mathematica*-code is also supplied. The corresponding data set `DataEllipseGarland.txt` can be downloaded from the link posted in Sect. 9.5.

```
In[1]:= Cen={5.,5.}; a=5; b=3; th=Pi/4; np=50; nr=3; sigma=0.025;r = 1.5;
        data1 = DataEllipseGarland[Cen, a, b, al, np, sigma, nr, 1];
        data2 = DataCircle[Cen, r, np, sigma, nr, 1];
    AG={{0,10},{0,10}}; app = RandomVariate[UniformDistribution[AG],
                            Ceiling[2 Length[data]/1000]];
    data = Union[data1[[1]], data2[[1]],app];
```

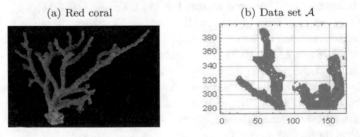

(a) Red coral   (b) Data set $\mathcal{A}$

Fig. 4.19 A data set derived from the photo of a red coral

Example 4.43 The data set $\mathcal{A}$ containing $|\mathcal{A}| = 5 094$ points, shown in Fig. 4.19b, was obtained from a real-world image of a red coral (a detail from the NW corner in

Fig. 4.19a). The corresponding data set `Coral.txt` can be downloaded from the link posted in Sect. 9.5.

In order to search for an optimal partition consisting of such clusters, in the past two decades intensive research has been devoted to the so-called *Density-based clustering methods*. The best known one was proposed in 1996 in [51] by the name DBSCAN (Density-Based Spatial Clustering of Applications with Noise). Density-based clustering techniques are based on the idea that objects which form a dense region should be grouped together into one cluster.

We are going to describe in some more detail the DBSCAN method. This method identifies an optimal partition, its contents, and cluster borders, and in addition, it detects outliers among data. The DBSCAN method has rapidly gained popularity and has been applied in various areas of applications, such as medical images, geography, spam detection, etc. (see e.g. [22, 86, 92]). This algorithm is well described in the following papers: [22, 114, 192, 204], where corresponding pseudo-codes can also be found.

In the following section, we are going to analyze the parameters of the method and then proceed to describe the DBSCAN algorithm itself.

4.5.1 *Parameters* MinPts *and* ϵ

Let $A \subset \mathbb{R}^n$ be a given set of *preponderant homogeneous* data (see Examples 4.37–4.43) and let $d : \mathbb{R}^n \times \mathbb{R}^n \to \mathbb{R}_+$ be some distance-like function. The property of preponderant homogeneity of data means that balls of equal radii contain approximately equal number of data points from the set A, differentiating between points inside the set, called *core points* and points on the border of the set, called *border points* [51]. This assumption can be operationalized in the following way.

Given an $\epsilon > 0$, the (relative) ϵ-*neighborhood* of a point $p \in A$ is defined as

$$N_A(p, \epsilon) = \{q \in A : d(q, p) < \epsilon\} = B(p, \epsilon) \cap A, \tag{4.32}$$

where $B(p, \epsilon) \subset \mathbb{R}^n$ is the open ϵ-ball around p.

If the data set A has the property of preponderant homogeneity, then the ϵ-neighborhood $N_A(p, \epsilon)$ contains approximately equal number MinPts ≥ 2 of data points for almost all core points $p \in A$.

The parameters MinPts and ϵ are interdependent, and in the literature there are several different suggestions for their choice (see e.g. [22, 51, 86, 92]). We will describe the way in which we have chosen the parameters MinPts and ϵ in our papers [146, 163, 164].

First, according to [22], we define MinPts $:= \lfloor \log |\mathcal{A}| \rfloor$,[1] and for each $a \in \mathcal{A}$ we set $\epsilon_a > 0$ to be the radius of the smallest ball centered at a containing at least MinPts elements of the set $\mathcal{A}$. Let $\mathcal{E}(\mathcal{A}) = \{\epsilon_a : a \in \mathcal{A}\}$ be the set of all such radii. We would like to determine the *universal* radius $\epsilon(\mathcal{A})$, such that for each $a \in \mathcal{A}$, the ball centered at a and of radius $\epsilon(\mathcal{A})$ contains at least MinPts elements of $\mathcal{A}$. However, we shall not require this to hold for all points $a \in \mathcal{A}$, i.e. we will not count points which require extremely large radius. Our investigations show that choosing $\epsilon(\mathcal{A})$ to be the 99% quantile of the set $\mathcal{E}(\mathcal{A})$ generates good results. The 99% quantile of the set $\mathcal{E}(\mathcal{A})$ is referred to as **ϵ-density** of the set $\mathcal{A}$ and is denoted by $\epsilon(\mathcal{A})$. Note that for almost all points $a \in \mathcal{A}$, the ball with center a and radius $\epsilon(\mathcal{A})$ contains at least MinPts elements of the set $\mathcal{A}$.

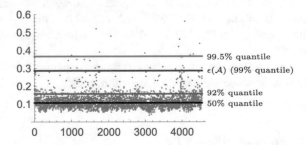

Fig. 4.20 ϵ-density of the set shown in Fig. 4.16a; $\epsilon(\mathcal{A}) = 0.287$

Example 4.44 We will determine the ϵ-density of the set shown in Fig. 4.16a. The parameter MinPts is determined by MinPts $= \lfloor \log |\mathcal{A}| \rfloor = 8$. Figure 4.20 shows the data set $\{(i, \epsilon_{a_i}) : i = 1, \ldots, 4560\}$ and lines that define 50%, 92%, 99%, and 99.5% quantile of the set $\mathcal{E}(\mathcal{A})$. By using 99% quantile of the set $\mathcal{E}(\mathcal{A})$, we obtain $\epsilon(\mathcal{A}) = 0.286558$.

Remark 4.45 ϵ-density of the set $\mathcal{A}$ can be calculated using the *Mathematica*-module EPSILON[] which is described in Sect. 9.2, and the link to appropriate *Mathematica*-code is supplied.

4.5.2   *DBSCAN Algorithm*

Let us mention several basic notions essential for understanding this method and its many modifications.

The point $p \in \mathcal{A}$ is said to be a *core point* if the ball $B(p, \epsilon)$ contains at least MinPts points from $\mathcal{A}$, i.e. $|N_{\mathcal{A}}(p, \epsilon)| \geq$ MinPts. The point $q \in \mathcal{A}$ is said to be

---

[1] $\lfloor x \rfloor$, the *floor of x*, equals x if x in an integer, and $\lfloor x \rfloor$ is the largest integer smaller than x if x is not an integer, cf. the footnote on page 46.

directly density ϵ-reachable from the point p if q is inside the ϵ-neighborhood of p. The point $q \in \mathcal{A}$ is said to be *density ϵ-reachable* from the point p if there is a chain of points $q_1, \ldots, q_r$ in $\mathcal{A}$, such that $q_1 = p$, $q_r = q$, and q_{i+1} is directly density ϵ-reachable from q_i for $1 \le i < r$. Note that all these definitions depend on the values of ϵ and MinPts.

When we talk about the DBSCAN method, cluster, π, will always mean a nonempty subset of $\mathcal{A}$ satisfying the following *maximality* and *connectivity* requirements:

(i) (Maximality) If $p \in \pi$ and q is density ϵ-reachable from p, then $q \in \pi$;
(ii) (Connectivity) If $p, q \in \pi$, then p and q are density ϵ-reachable.

The point $p \subset \mathcal{A}$ is said to be a *border point* if it is not a core point, but it is density ϵ-reachable from some core point. The set of all border points is called the *cluster edge*. *Noise* is the set of points which are neither core points nor border points, implying that noise points do not belong to any cluster.

Selecting the Initial Point

Our algorithm is run by first selecting an initial point c_p which we try to find in the most dense region of the data set. Therefore, we first present the algorithm CorePoint, which, for a given $\epsilon > 0$, in It_0 attempts randomly determines a point $c_p \in \mathcal{A}$ such that its ϵ-neighborhood contains the maximal number of points from $\mathcal{A}$. Because of density of the set $\mathcal{A}$, the parameter It_0 docs not have to be large (in our examples $It_0 = 10$). The output of the CorePoint algorithm is $\{cp, |Cp|\}$, where cp and $|Cp|$ are the selected core point and the number of points in its ϵ-neighborhood, respectively.

Algorithm 4 CorePoint$(\mathcal{A}, \epsilon, It_0)$

Input: $\mathcal{A}$ {Set of data points}; ϵ; It_0; $kk = 1$;
  1: Choose $x \in \mathcal{A}$
  2: Determine $Ne = N_{\mathcal{A}}(x, \epsilon)$, $nr := |Ne|$, $cp := x$, $Cp := Ne$;
  3: Set $\mathcal{A} := \mathcal{A} \setminus \{x\}$;
  4: **while** $kk < It_0$, **do**
  5:     $kk = kk + 1$
  6:     Choose $y \in \mathcal{A}$;
  7:     Set $Ne = N_{\mathcal{A}}(y, \epsilon)$; $nr1 := |N_{\mathcal{A}}(y, \epsilon)|$;
  8:     **if** $nr1 > nr$, **then**
  9:         $nr = nr1$; $cp = y$; $Cp = Ne$
10:     **end if**
11:     $A = \mathcal{A} \setminus \{y\}$
12: **end while**
Output: $\{cp, |Cp|\}$

Remark 4.46 The corresponding *Mathematica*-module CorePoint[] is described in Sect. 9.2, and the link to appropriate *Mathematica*-code is supplied.

Basic DBSCAN Algorithm

The choice of initial point $c_p \in \mathcal{A}$ such that

$$|N_{\mathcal{A}}(c_p, \epsilon)| \geq \text{MinPts}, \tag{4.33}$$

enables us to determine the first cluster π, consisting of all points which are density ϵ-reachable from c_p. At the same time, we determine border points and the noise. The whole procedure is shown in the basic DBSCAN1 algorithm.

Algorithm 5 DBSCAN1($\mathcal{A}$, xx, MinPts, ϵ)

Input: $\mathcal{A}$ {Data set}
Input: ϵ; MinPts; xx;
 1: $B = \mathcal{A}$, $S = \{xx\}$, $SE = \{\}$; $SN = \{\}$; $\mathcal{P} = \{\}$;
    {S: set of visited elements in $\mathcal{A}$; SE: set of edge points SN: set of noise points (outliers)
    in $\mathcal{A}$; B: copy of $\mathcal{A}$; $\mathcal{P}$: partition of the set $\mathcal{A}$}
 2: $Ne = N_{\mathcal{A}}(xx, \epsilon)$;
 3: **if** $1 < |Ne| < \text{MinPts}$, **then**
 4:       put $SE = SE \cup Ne$;
 5: **end if**
 6: **if** $|Ne| < \text{MinPts}$, **then**
 7:       put $SN = SN \cup \{xx\}$
 8: **else**
 9:       put $S = S \cup \{xx\}$, $NeB = B \cap Ne$;
10:       **if** $NeB \neq \emptyset$, **then**
11:            put $\mathcal{P} = \mathcal{P} \cup NeB$, $B = B \setminus NeB$;
12:       **end if**
13:       put $Qu = Ne \setminus S$;
14:       **while** $Qu \neq \emptyset$ **do**
15:            Choose random element $y \in Qu$;
16:            Set $S = S \cup \{y\}$, $Ny = N_{\mathcal{A}}(y, \epsilon)$
17:            **if** $1 < |Ny| < \text{MinPts}$, **then**
18:                 put $SE = SE \cup Ny$
19:            **end if**
20:            **if** $|Ny| \geq \text{MinPts}$, **then**
21:                 put $NeB = B \cap Ny$, $\mathcal{P} = \mathcal{P} \cup NeB$, $B = B \setminus Ny$;
22:                 put $SN = SN \setminus Ny$, $Qu = Qu \cup (Ny \setminus S)$;
23:            **end if**
24:            put $Qu = Qu \setminus \{y\}$;
25:       **end while**
26: **end if**
Output: $\{\mathcal{P}, SE, SN\}$

Remark 4.47 The corresponding *Mathematica*-module DBSCAN1 [] is described in Sect. 9.2, where the link to appropriate *Mathematica*-code is also supplied.

Main DBSCAN Algorithm

Finding an initial point c_p and applying the Algorithm 5, we obtain the first cluster π of the partition Π and its border points. Define $\mathcal{A}_1 = \mathcal{A} \setminus \pi$ and repeat the procedure with the new set $\mathcal{A}_1$. Continue repeating this procedure as long as it is possible to find a point c_p satisfying (4.33). This is described in the Algorithm 6.

Algorithm 6 DBSCAN($\mathcal{A}$, MinPts, ϵ)

Input: $\mathcal{A}$; MinPts; ϵ;
1: Set $\mathcal{B} = \mathcal{A}$; $It_0 = 0.01|\mathcal{A}|$; $\tau = 0.01|\mathcal{A}|$; $\Pi = \emptyset$ (partition of the set $\mathcal{B}$);
    $SSE = \emptyset$ (border points);
2: **while** {cp,|Cp|}=CorePoint[$\mathcal{B}$, ϵ, It_0]; $|Cp| \geq \tau$MinPts, **do**
3:      Call DBSCAN1($\mathcal{B}$, cp, MinPts, ϵ);
        {A cluster π is obtained in which its border points SE are marked}
4:      Add the cluster π to the partition Π,
        add border points SE to SSE and
        set $\mathcal{B} = \mathcal{B} \setminus \pi$;
5: **end while**
6: $Outliers = \mathcal{A} \setminus$ Flatten[Π, 1]
Output: {Π, SSE, $Outliers$}

Remark 4.48 The corresponding *Mathematica*-module DBSCAN[] is described in Sect. 9.2, where the link to appropriate *Mathematica*-code is also supplied.

4.5.3   Numerical Examples

We will illustrate the described DBSCAN algorithm applying it to previously constructed examples using the DBSCAN[] module from the file Methods.nb.

Example 4.49 Consider the data set $\mathcal{A}$, $|\mathcal{A}| = 4560$, from Example 4.37. Following the method described in Sect. 4.5.1 we find MinPts $= 8$ and $\epsilon(\mathcal{A}) = 0.279$ based on the 99% quantile.

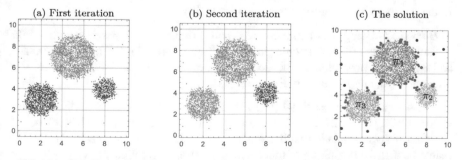

Fig. 4.21 Algorithm 6 produces clusters, their borders (brown points), and noise (red points)

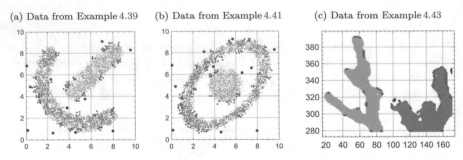

Fig. 4.22 Algorithm 6 produces clusters, their borders (brown points), and noise (red points)

First, the Algorithm 4 produces the initial point (5.528, 5.524) with 74 data points in its ϵ-neighborhood, and then Algorithm 5 finds the first cluster π_1 with $|\pi_1| = 2514$ points, including 72 border points. This cluster is depicted in Fig. 4.21a as the set of gray points.

After that, Algorithm 4 provides the next initial point (0.964, 2.317) with 73 data points in its ϵ-neighborhood, and Algorithm 5 finds the second cluster π_2 with $|\pi_2| = 1456$ points, including 54 border points. This cluster is depicted in Fig. 4.21b as the smaller set of gray points.

Finally, Algorithm 4 produces the next initial point (7.998, 3.305) containing 73 data points in its ϵ-neighborhood, and Algorithm 5 finds the third cluster π_3 with $|\pi_3| = 628$ points, including 14 border points.

In the end we are left with 12 outliers forming the noise of the set $\mathcal{A}$. Figure 4.21c shows the obtained clusters, their borders (brown points), and noise (red points).

Example 4.50 For the data set $\mathcal{A}$, $|\mathcal{A}| = 4814$, from Example 4.39, we find again MinPts $= 8$ and $\epsilon(\mathcal{A}) = 0.298$. The DBSCAN algorithm detects the partition Π with two clusters: π_1 with 2844 points, including 109 border points (oval), and π_2 with 1962 points, including 70 border points (circle-arc oval). In the end there are 8 outliers left (see Fig. 4.22a).

Similarly, for the data set $\mathcal{A}$, $|\mathcal{A}| = 4938$, from Example 4.41 we find again MinPts $= 8$ and $\epsilon(\mathcal{A}) = 0.289$ based on the 99% quantile. The DBSCAN algorithm detects the partition Π with two clusters: π_1 with 3536 points, including 144 border points (ellipse garland), and π_2 with 1391 points, including 33 border points (disc). In the end there are 11 outliers left (see Fig. 4.22b).

Remark 4.51 Using the obtained border points one can get a good estimate of the indistinct borders of such geometric objects (see Chap. 8.1). Examples 8.14, 8.24, and 8.25 show how this can be done for discs in Example 4.49, for ellipses in Example 4.50, and for ovals in Example 8.25.

The data set $\mathcal{A}$, $|\mathcal{A}| = 5094$, from Example 4.43, is contained in the rectangle $\Delta_1 = [19, 171] \times [279, 393]$. It turns out that a better numerical efficiency is achieved if such a data set is first transformed into the unit square $[0, 1]^2$, as was pointed out at the beginning of this chapter. The transformation $\mathsf{T} : \Delta_1 \to [0, 1]^2$ is given by (see (4.1))

$$\mathsf{T}(x) = (x - \alpha)D, \quad D = \mathrm{diag}\left(\tfrac{1}{171-19}, \tfrac{1}{393-279}\right), \quad \alpha = (19, 279).$$

For the transformed set $\mathcal{B} = \{\mathsf{T}(a_i) : a_i \in \mathcal{A}\} \subset [0, 1]^2$ we obtain MinPts $= 8$ and $\epsilon(\mathcal{A}) = 0.0175$. The DBSCAN algorithm finds the partition Π with two clusters: π_1 with 2535 points, including 232 border points, and π_2 with 2583 points, including 122 border points. In the end there are 27 outliers left. The result of using the inverse transformation $\mathsf{T}^{-1} : [0, 1]^2 \to \Delta_1, \mathsf{T}^{-1}(x) = xD^{-1} + \alpha$, to the set $\mathcal{B}$, is shown in the Fig. 4.22c.

Remark 4.52 It makes sense to apply the DBSCAN algorithm if the set $\mathcal{A}$ has a large number of points. The most important advantages of this algorithm are the following: it is possible to recognize non-convex clusters, a partition with the most appropriate number of clusters is obtained automatically, and it is not necessary to use indexes for defining an appropriate number of clusters in a partition.

But, according to [92], the main drawbacks are the following:

- It is rather challenging to determine the proper values of ϵ and MinPts;
- Theoretically, in the worst case, complexity without any special structure is $\mathcal{O}(m^2)$, and therefore the algorithm can take a tremendous amount of CPU-time. But, if a suitable indexing structure is used (e.g. R^*-tree), its performance will be significantly improved and complexity can be reduced to $\mathcal{O}(m \log m)$ (see [71, 192]). In [71], the author introduces a new algorithm for DBSCAN in $\mathbb{R}^2$ which has theoretical running time complexity of $\mathcal{O}(m \log m)$;
- It fails when the border points of two clusters are relatively close;
- DBSCAN needs large volumes of memory support, and it is often faced with the problems of high-dimensional data;
- the DBSCAN algorithm fails to find all clusters of different densities.

In the meantime, various modifications of the DBSCAN method as well as some similar methods came about. One of the most frequently used strategies to improve the DBSCAN algorithm performance is *data indexing*. The use of parallel computing [3] is among the most commonly used indexing techniques and one of the adopted solutions to address this challenge.

An enhanced version of the incremental DBSCAN algorithm for incrementally building and updating arbitrarily shaped clusters in large data sets was introduced in [16]. The main contribution of that paper can be summarized as enhancing the incremental DBSCAN algorithm by limiting the search space to partitions instead of the whole data set, which speeds up the clustering process.

Another popular density-based algorithm called OPTICS (Ordering Points To Identify the Clustering Structure) (see [4]) is based on the concepts of the DBSCAN algorithm, and it identifies nested clusters and the structure of clusters.

Various other modifications of the DBSCAN algorithm can be found in the literature, such as the Incremental DBSCAN algorithm, the partitioning-based DBSCAN algorithm (PDBSCAN) and PACA (the ant clustering algorithm) (see [86]),

SDBDC (Scalable Density-Based Distributed Clustering), and the ST-DBSCAN algorithm (see [22]). The density-based clustering algorithm, ST-DBSCAN (see [22]), has the ability to discover clusters according to non-spatial, spatial, and temporal values of objects. Synchronization with multiple prototypes (SyMP) is a clustering algorithm that can identify clusters of arbitrary shapes in a completely unsupervised fashion (see [61]). In [114] DBSCAN execution time performance for binary data sets and Hamming distances is improved. The hybrid clustering technique called rough-DBSCAN [192] has time complexity of only $\mathcal{O}(m)$. It is shown that for large data sets rough-DBSCAN can find clustering similar to the one found by DBSCAN, but it is consistently faster than DBSCAN. In [98], the authors propose a novel graph-based index structure method *Groups* that accelerates neighbor search operations and is also scalable for high-dimensional data sets. Clustering by fast search and finding density peaks (DPC) has been recently considered in [199] and has become of interest to many machine learning researchers. Recently, particular attention has been paid to methods which can recognize clusters with various data densities (see [4, 38, 50, 163, 204]).

Chapter 5
Indexes

5.1 Choosing a Partition with the Most Appropriate Number of Clusters

One can pose the following question:

> Given a data set $\mathcal{A}$, into how many clusters should one group the elements of $\mathcal{A}$, i.e. how to choose a *partition with the most appropriate number of clusters* (MAPart)?

Answering this question is one of the most complex problems in Cluster Analysis, and there is a vast literature dealing with it [21, 28, 40, 52, 94, 96, 145, 148, 158, 176, 184, 188]. The usual method for attacking the problem is to check various indicators, simply called *indexes*.

In some simple cases, the number of clusters in a partition is determined by the nature of the problem itself.

For instance, it is most natural to group students into $k = 5$ clusters according to their grades,[1] but it is not at all clear into how many clusters should one group all insects, or into how many groups should one divide business firms in a given administrative division.

If the number of clusters into which the given set $\mathcal{A}$ should be divided is not given in advance, it is sensible to choose a partition consisting of clusters which are internally as compact as possible and mutually separated as much as possible. For such a partition we are going to say that it is the partition with the ***most appropriate number of clusters*** — MAPart for short.

We are going to consider only some of the best known indexes which assume using LS distance-like function. Since for the trivial optimal partitions $\Pi^{(1)}$ with one cluster and $\Pi^{(m)}$ with m clusters one has

---

[1] See the footnote on page 12.

© The Author(s), under exclusive license to Springer Nature Switzerland AG 2021
R. Scitovski et al., *Cluster Analysis and Applications*,
https://doi.org/10.1007/978-3-030-74552-3_5

$$k = 1: \qquad \mathcal{F}_{LS}(\Pi^{(1)}) = \sum_{i=1}^{m} \| \operatorname{mean}(\mathcal{A}) - a_i \|^2, \quad \mathcal{F}_{LS}(\Pi^{(m)}) = 0,$$

$$k = m: \qquad \mathcal{G}(\Pi^{(m)}) = \sum_{i=1}^{m} \| a_i - \operatorname{mean}(\mathcal{A}) \|^2, \quad \mathcal{G}(\Pi^{(1)}) = 0,$$

we are going to define these indexes for $1 < k < m$.

5.1.1   Calinski–Harabasz Index

This index was proposed by T. Calinski and J. Harabasz in [28]. Later the Calinski–Harabasz (CH) index underwent numerous improvements and adaptations (see e.g. [21, 158, 188]).

The CH index is defined in such a way that a more internally compact partition whose clusters are better mutually separated has a larger CH value.

Note first (see Theorem 3.7) that the objective function does not increase when the number of clusters in optimal partition increases, i.e. the sequence of objective function values obtained in this way is monotonically decreasing.

When using the LS distance-like function to find an optimal k-partition $\Pi^\star = \{\pi_1^\star, \ldots, \pi_k^\star\}$, the objective function $\mathcal{F}$ can be written as

$$\mathcal{F}_{LS}(\Pi) = \sum_{j=1}^{k} \sum_{a \in \pi_j} \| c_j - a \|^2. \tag{5.1}$$

The value $\mathcal{F}_{LS}(\Pi^\star)$ shows the total dispersion of elements of clusters $\pi_1^\star, \ldots, \pi_k^\star$ from their centroids $c_1^\star, \ldots, c_k^\star$. As we already noticed, the smaller the value of $\mathcal{F}_{LS}$, the smaller the dispersion, meaning that clusters are more inner compact.

Therefore we are going to assume that the CH index of optimal partition $\Pi^\star$ is inversely proportional to the value $\mathcal{F}_{LS}(\Pi^\star)$ of the objective function.

On the other hand, as we noticed earlier in Sect. 3.2.2, p. 43 and p. 50, while searching for an optimal partition, besides minimizing the function $\mathcal{F}_{LS}$, one can maximize the corresponding dual function

$$\mathcal{G}(\Pi) = \sum_{j=1}^{k} |\pi_j| \| c_j - c \|^2, \tag{5.2}$$

where $c = \operatorname*{arg\,min}_{x \in \mathbb{R}^n} \sum_{i=1}^{m} \| x - a_i \|^2 = \frac{1}{m} \sum_{i=1}^{m} a_i$ is the centroid of the whole set $\mathcal{A}$.

The value $\mathcal{G}(\Pi^\star)$ shows the total weighted separation of centroids $c_1^\star, \ldots, c_k^\star$ of clusters $\pi_1^\star, \ldots, \pi_k^\star$. The larger the value of the function $\mathcal{G}$, the larger is the sum of

LS-distances between the centroids c_j^* and the centroid c of the whole set, where the distances are weighted by the number of cluster elements. This means that the centroids c_j^* are mutually maximally separated.

Therefore we are going to assume that the CH index of optimal partition Π^* is proportional to the value $\mathcal{G}(\Pi^*)$ of the dual objective function.

Considering also statistical reasons related to the number m of elements of $\mathcal{A}$ and number k of clusters in the partition Π^*, the measure of inner compactness and external separation of clusters in optimal partition Π^*, when using LS distance-like function, is defined as the number [28, 188]

$$\text{CH}(k) = \frac{\frac{1}{k-1}\mathcal{G}(\Pi^*)}{\frac{1}{m-k}\mathcal{F}_{LS}(\Pi^*)}, \quad 1 < k < m, \tag{5.3}$$

which is called the **CH index** of the partition Π^*. The partition with maximal CH index is deemed MAPart.

Example 5.1 Consider the set $\mathcal{A} = \{2, 4, 8, 10, 16\}$ from Example 3.18, and compare the obtained LS-optimal 3-partition $\Pi_3^* = \{\{2, 4\}, \{8, 10\}, \{16\}\}$ with the LS-optimal 2-partition. Optimal partitions can be determined by using the k-means algorithm with multiple activations (Sect. 4.2.4) or by using the Incremental algorithm (Sect. 4.3).

(a) LS-optimal 2-partition   (b) LS-optimal 3-partition

Fig. 5.1 Choosing partition with more appropriate number of clusters

LS-optimal 2-partition is $\Pi_2^* = \{\{2, 4\}, \{8, 10\}, \{16\}\}$ (Fig. 5.1a) for which

$$\mathcal{F}_{LS}(\Pi_2^*) = 36.67, \quad \mathcal{G}(\Pi_2^*) = 83.33.$$

For LS-optimal 3-partition $\Pi_3^* = \{\{2, 4\}, \{8, 10\}, \{16\}\}$ (Fig. 5.1b) we obtain

$$\mathcal{F}_{LS}(\Pi_3^*) = 4, \quad \mathcal{G}(\Pi_3^*) = 116.$$

The corresponding CH indexes are

$$\text{CH}(2) = \frac{83.33/1}{36.67/3} = 6.82, \quad \text{CH}(3) = \frac{116/2}{4/2} = 29.$$

Since $\text{CH}(3) > \text{CH}(2)$, Π_3^* is the partition with more acceptable (favorable) number of clusters.

The next proposition shows that the CH index of a k-partition attains its largest value at a globally optimal k-partition.

Proposition 5.2 *Let $\mathcal{A} \subset \mathbb{R}^n$ be a finite set, and let Π_1 and Π_2 be two different k-partitions of the set $\mathcal{A}$. Then*

$$\mathrm{CH}(\Pi_1) \geq \mathrm{CH}(\Pi_2) \quad \Leftrightarrow \quad \mathcal{F}_{LS}(\Pi_1) \leq \mathcal{F}_{LS}(\Pi_2). \tag{5.4}$$

Proof Let $m = |\mathcal{A}|$ and $c = \mathrm{mean}(\mathcal{A})$. Denote $\kappa := \sum_{i=1}^{m} \|c - a_i\|^2$. Then by (3.31) we have

$$\mathcal{G}(\Pi_1) = \kappa - \mathcal{F}_{LS}(\Pi_1) \quad \text{and} \quad \mathcal{G}(\Pi_2) = \kappa - \mathcal{F}_{LS}(\Pi_2).$$

Therefore,

$$\mathrm{CH}(\Pi_1) \geq \mathrm{CH}(\Pi_2) \Leftrightarrow \frac{m-k}{k-1} \frac{\mathcal{G}(\Pi_1)}{\mathcal{F}_{LS}(\Pi_1)} \geq \frac{m-k}{k-1} \frac{\mathcal{G}(\Pi_2)}{\mathcal{F}_{LS}(\Pi_2)}$$

$$\Leftrightarrow \frac{\kappa - \mathcal{F}_{LS}(\Pi_1)}{\mathcal{F}_{LS}(\Pi_1)} \geq \frac{\kappa - \mathcal{F}_{LS}(\Pi_2)}{\mathcal{F}_{LS}(\Pi_2)}$$

$$\Leftrightarrow \frac{\kappa}{\mathcal{F}_{LS}(\Pi_1)} - 1 \geq \frac{\kappa}{\mathcal{F}_{LS}(\Pi_2)} - 1$$

$$\Leftrightarrow \mathcal{F}_{LS}(\Pi_1) \leq \mathcal{F}_{LS}(\Pi_2). \qquad \square$$

Exercise 5.3 Let $\mathcal{A} \subset \mathbb{R}^n$ be a finite set and let Π_1 and Π_2 be two different k-partitions of the set $\mathcal{A}$. Prove that

$$\mathrm{CH}(\Pi_1) \geq \mathrm{CH}(\Pi_2) \quad \Leftrightarrow \quad \mathcal{G}(\Pi_1) \geq \mathcal{G}(\Pi_2), \tag{5.5}$$

where $\mathcal{G}$ is the dual function of the objective function $\mathcal{F}_{LS}$.

5.1.2 Davies–Bouldin Index

This index was proposed by D. Davies and D. Bouldin in [40], and it also underwent numerous improvements and adaptations (see e.g. [21, 158, 188, 189, 191]).

The DB index is defined in such a way that a more internally compact partition whose clusters are better mutually separated has a smaller DB value.

The following concept is taken from [189]. Let $c \in \mathbb{R}^2$ be a point in the plane around which there are m random points a_i generated from the bivariate normal distribution $\mathcal{N}(c, \sigma^2 I)$ with the covariance matrix $\sigma^2 I$, where I is the identity 2×2 matrix. This set of points forms a *spherical data set* which will be denoted by $\mathcal{A}$.

From statistical literature [129, 183] it is known that around 68% of all points of the set $\mathcal{A}$ lie inside the circle $K(c, \sigma)$ with center c and radius σ. We call this circle the *main circle* of the data set $\mathcal{A}$.

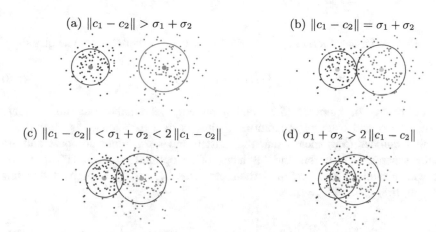

(a) $\|c_1 - c_2\| > \sigma_1 + \sigma_2$

(b) $\|c_1 - c_2\| = \sigma_1 + \sigma_2$

(c) $\|c_1 - c_2\| < \sigma_1 + \sigma_2 < 2\|c_1 - c_2\|$

(d) $\sigma_1 + \sigma_2 > 2\|c_1 - c_2\|$

Fig. 5.2 Different relations between two spherical data sets

Suppose two spherical data sets $\mathcal{A}_1$ and $\mathcal{A}_2$ were constructed in the described way around two different points c_1 and c_2 and different standard deviations σ_1 and σ_2, and let $K_1(c_1, \sigma_1)$ and $K_2(c_2, \sigma_2)$ be the respective main circles.

Figure 5.2 depicts possible relations between sets $\mathcal{A}_1$ and $\mathcal{A}_2$ with respect to the position of their main circles $K_1(c_1, \sigma_1)$ and $K_2(c_2, \sigma_2)$. Figure 5.2a shows the sets $\mathcal{A}_1$ and $\mathcal{A}_2$ whose main circles do not intersect, and for which $\|c_1 - c_2\| > \sigma_1 + \sigma_2$; Fig. 5.2b shows the sets $\mathcal{A}_1$ and $\mathcal{A}_2$ whose main circles touch, and for which $\|c_1 - c_2\| = \sigma_1 + \sigma_2$; etc.

Therefore one can say that the main circles $K_1(c_1, \sigma_1)$ and $K_2(c_2, \sigma_2)$ of the sets $\mathcal{A}_1$ and $\mathcal{A}_2$ intersect if [189]

$$\|c_1 - c_2\| \leq \sigma_1 + \sigma_2, \qquad (5.6)$$

i.e. that the main circles of $\mathcal{A}_1$ and $\mathcal{A}_2$ are separated if

$$\frac{\sigma_1 + \sigma_2}{\|c_1 - c_2\|} < 1,$$

which is also in accordance with the result for the one-dimensional case from [18]: "a mixture of two normal distributions $\mathcal{N}(c_1, \sigma_1^2)$ and $\mathcal{N}(c_2, \sigma_2^2)$ is unimodal if $\|c_1 - c_2\| \leq 2\min\{\sigma_1, \sigma_2\}$," since $2\min\{\sigma_1, \sigma_2\} \leq \sigma_1 + \sigma_2$.

Let $\Pi^\star$ be an optimal partition of the set $\mathcal{A}$ with clusters $\pi_1^\star, \ldots, \pi_k^\star$ and their centroids $c_1^\star, \ldots, c_k^\star$. Fix one of the clusters, say $\pi_j^\star$, and let us look at its relationship to the other clusters. Note that the quantity

$$D_j := \max_{s \neq j} \frac{\sigma_j + \sigma_s}{\|c_j^\star - c_s^\star\|}, \quad \text{where } \sigma_j^2 = \frac{1}{|\pi_j^\star|} \sum_{a \in \pi_j^\star} \|c_j^\star - a\|^2, \tag{5.7}$$

identifies the largest overlap of the cluster $\pi_j^\star$ with any other cluster. The quantity

$$\frac{1}{k}(D_1 + \cdots + D_k) \tag{5.8}$$

is the average of numbers (5.7), and it presents yet another measure of inner compactness and external separation of clusters in a partition. It is clear that a smaller number (5.8) means that the clusters are more inner compact and are better separated. Therefore the DB index of an optimal partition $\Pi^\star$ of the set $\mathcal{A}$ with clusters $\pi_1^\star, \ldots, \pi_k^\star$ and their centroids $c_1^\star, \ldots, c_k^\star$ is defined like this [21, 40, 148, 158, 188, 189]:

$$\mathrm{DB}(k) := \frac{1}{k} \sum_{j=1}^k \max_{s \neq j} \frac{\sigma_j + \sigma_s}{\|c_j^\star - c_s^\star\|}, \quad \sigma_j^2 = \frac{1}{|\pi_j^\star|} \sum_{a \in \pi_j^\star} \|c_j^\star - a\|^2, \ 1 < k < m. \tag{5.9}$$

The partition with the smallest DB index is deemed MAPart.

Example 5.4 Again let us look at the set $\mathcal{A} = \{2, 4, 8, 10, 16\}$ from Example 5.1. We are going to use the DB index to compare optimal partitions $\Pi_2^\star$ and $\Pi_3^\star$.

The LS-optimal 2-partition is $\Pi_2^\star = \{\{2, 4\}, \{8, 10, 16\}\}$. Centers of its clusters are $c_1^\star = 3$ and $c_2^\star = 11.33$, the corresponding σ-values are $\sigma_1 = 1$ and $\sigma_2 = 3.4$, and the DB index is

$$\mathrm{DB}(2) = \frac{1}{2}\left(\frac{\sigma_1 + \sigma_2}{\|c_1^\star - c_2^\star\|} + \frac{\sigma_2 + \sigma_1}{\|c_2^\star - c_1^\star\|}\right) = \frac{\sigma_1 + \sigma_2}{\|c_1^\star - c_2^\star\|} = 0.58788.$$

The LS-optimal 3-partition is $\Pi_3^\star = \{\{2, 4\}, \{8, 10\}, \{16\}\}$. The centers of its clusters are $c_1^\star = 3$, $c_2^\star = 9$, and $c_3^\star = 16$, and the corresponding σ-values are $\sigma_1 = 1$, $\sigma_2 = 1$, and $\sigma_3 = 0$, and the DB index is

$$\mathrm{DB}(3) = \frac{1}{3}\Big(\max\left\{\frac{\sigma_1 + \sigma_2}{\|c_1^\star - c_2^\star\|}, \frac{\sigma_1 + \sigma_3}{\|c_1^\star - c_3^\star\|}\right\} + \max\left\{\frac{\sigma_2 + \sigma_1}{\|c_2^\star - c_1^\star\|}, \frac{\sigma_2 + \sigma_3}{\|c_2^\star - c_3^\star\|}\right\}$$

$$+ \max\left\{\frac{\sigma_3 + \sigma_1}{\|c_3^\star - c_1^\star\|}, \frac{\sigma_3 + \sigma_2}{\|c_3^\star - c_2^\star\|}\right\}\Big) = 0.26984.$$

Since $DB(3) < DB(2)$, the partition $\Pi_3^\star$ is the partition with more acceptable number of clusters, which agrees with the earlier conclusion obtained using the CH index.

Example 5.5 Let us find a partition of the set $\mathcal{A} = \{a_i = (x_i, y_i) : i = 1, \ldots, 12\} \subset \mathbb{R}^2$ with the most acceptable number of clusters, where

i	1	2	3	4	5	6	7	8	9	10	11	12
x_i	1	2	3	2	2	3	4	3	6	8	7	7
y_i	5	4	6	7	1	2	1	1	5	5	4	6

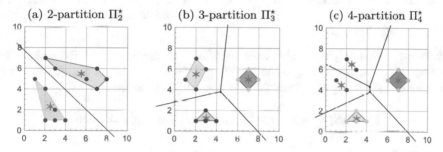

Fig. 5.3   Choosing LS-optimal partition with the most acceptable number of clusters

We will compute the values of objective function $\mathcal{F}_{LS}$, and the values of the CH and DB indexes for optimal LS-partitions $\Pi_2^\star, \ldots, \Pi_6^\star$ with $2, 3, \ldots, 6$ clusters. Figure 5.3 depicts LS-optimal 2-, 3-, and 4-partitions, and Fig. 5.4 shows graphs of the CH and DB indexes for LS-optimal partitions.

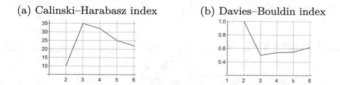

Fig. 5.4   Choosing partition with the most acceptable number of clusters

As one can see, the CH index attains its largest and DB index its smallest value for the partition $\Pi_3^\star$. This means that with a high confidence one can claim that the 3-partition $\Pi_3^\star$ is the MAPart, as was visually expected looking at Fig. 5.3.

Table 5.1 CH and DB indexes for LS-optimal partitions from Example 5.8

Index	$\Pi^{(2)}$	$\Pi^{(3)}$	$\Pi^{(4)}$	$\Pi^{(5)}$	$\Pi^{(6)}$	$\Pi^{(7)}$
CH	17.76	33.65	30.34	26.97	21.78	18.31
DB	0.50	0.42	0.36	0.38	0.25	0.18

Exercise 5.6 Let $\mathcal{A} \subset \mathbb{R}^n$ be a finite set and let Π_1 and Π_2 be two different k-partitions of $\mathcal{A}$. Is it true that

$$\mathrm{DB}(\Pi_1) \leq \mathrm{DB}(\Pi_2) \quad \Leftrightarrow \quad \mathcal{F}_{LS}(\Pi_1) \leq \mathcal{F}_{LS}(\Pi_2) ? \tag{5.10}$$

Remark 5.7 Comparing the values of CH and DB indexes for LOPart and GOPart for the set with one feature in Example 4.10 and for the set with two features in Example 4.12 shows that the value of the CH index (resp. DB index) is larger (resp. smaller) on GOPart.

Example 5.8 Applying the AGNES algorithm to the set $\mathcal{A}$ in Example 4.29 (see Fig. 4.14a) starting from the partition $\Pi^{(m)} = \{\{a_1\}, \dots, \{a_m\}\}$, using the measure of similarity (4.16) and the LS distance-like function

$$D_2(A, B) = \|c_A - c_B\|^2, \quad c_A = \underset{a \in A}{\mathrm{mean}}\, a, \quad c_B = \underset{b \in B}{\mathrm{mean}}\, b,$$

one obtains a similar result as in Example 4.29, but now one can calculate also the values of CH and DB indexes (see Table 5.1).

While the CH index suggests that $\Pi^{(3)}$ is the partition with the most acceptable number of clusters, the DB index suggests that $\Pi^{(7)}$ is the partition with the most acceptable number of clusters.

5.1.3 Silhouette Width Criterion

The **Silhouette Width Criterion** (SWC) is very popular in Cluster Analysis and applications [94, 184, 188]. For a k-LOPart $\Pi^\star = \{\pi_1^\star, \dots, \pi_k^\star\}$ obtained using a distance-like function d, the SWC is defined as follows: for each $a_i \in \pi_r^\star$ one calculates the numbers

$$\alpha_{ir} = \tfrac{1}{|\pi_r^\star|} \sum_{b \in \pi_r^\star} d(a_i, b), \quad \beta_{ir} = \min_{q \neq r} \tfrac{1}{|\pi_q^\star|} \sum_{b \in \pi_q^\star} d(a_i, b), \tag{5.11}$$

and the associated SWC is defined as

$$\text{SWC}(k) = \frac{1}{m} \sum_{i=1}^{m} \frac{\beta_{ir} - \alpha_{ir}}{\max\{\alpha_{ir}, \beta_{ir}\}}, \quad 1 < k < m. \tag{5.12}$$

More inner compact and better separated clusters result in a larger SWC number. The partition with the largest SWC index is deemed MAPart.

Since the numerical procedure to calculate the SWC index is rather demanding, one usually applies the **Simplified Silhouette Width Criterion** (SSWC) which, instead of the average value (5.11), uses the distance between the element $a_i \in \pi_r^\star$ and the centers $c_1^\star, \ldots, c_k^\star$:

$$\text{SSWC}(k) = \frac{1}{m} \sum_{i=1}^{m} \frac{\beta_{ir} - \alpha_{ir}}{\max\{\alpha_{ir}, \beta_{ir}\}}, \quad 1 < k < m, \tag{5.13}$$

where $\alpha_{ir} = d(a_i, c_r^\star)$, $\beta_{ir} = \min_{q \neq r} d(a_i, c_q^\star)$. According to this criterion, partition with the largest SSWC index is deemed MAPart.

5.1.4 Dunn Index

This index was proposed by J. C. Dunn in [48], and it also underwent numerous improvements and adaptations (see e.g. [21, 188]).

The **Dunn index** is defined in such a way that a more internally compact partition with better mutually separated clusters has a smaller Dunn value.

Let $\Pi^\star = \{\pi_1^\star, \ldots, \pi_k^\star\}$ be an optimal k-partition obtained using a distance-like function d. The measure of separation between pairs of clusters in the partition is determined by calculating distances between every pair $\{\pi_i^\star, \pi_j^\star\}$ of clusters

$$D(\pi_i^\star, \pi_j^\star) = \min_{a \in \pi_i^\star, b \in \pi_j^\star} d(a, b), \tag{5.14}$$

and for the measure of inner compactness of clusters $\pi_i^\star \in \Pi^\star$ we take their diameters $\text{diam}\, \pi_i^\star = \max_{a, b \in \pi_i^\star} d(a, b)$.

Therefore the Dunn index of an optimal partition $\Pi^\star$ is proportional to the number $\min_{1 \leq i < j \leq k} D(\pi_i^\star, \pi_j^\star)$ and inversely proportional to the number $\max_{1 \leq s \leq k} \text{diam}\, \pi_s^\star$ and is defined as the quotient of these two values:

$$\text{Dunn}(k) = \frac{\displaystyle\min_{1 \leq i < j \leq k} D(\pi_i^\star, \pi_j^\star)}{\displaystyle\max_{1 \leq s \leq k} \text{diam}\, \pi_s^\star}, \quad 1 < k < m. \tag{5.15}$$

This definition meets the requirement that smaller index value has that partition whose clusters are more inner compact and mutually better separated. According to this criterion, partition with the smallest Dunn index is deemed MAPart.

Although the original Dunn index was defined using the ℓ_2-metric function [48, 188], as for the other aforementioned indexes, we are going to use the LS distance-like function.

Example 5.9 Consider again the set $\mathcal{A}$ from Example 4.25, and let us try to determine the most acceptable number of clusters using CH, DB, SSWC, and Dunn indexes.

Table 5.2 Index values for optimal k-partitions in Example 5.9

	$k = 2$	$k = 3$	$k = 4$	$k = 5$	$k = 6$	$k = 7$	$k = 8$	$k = 9$	$k = 10$
CH	2202	2357	3447	3751	**5554**	5263	4824	4562	4430
DB	0.62	0.67	0.63	0.65	**0.60**	0.80	0.93	1.06	1.14
SSWC	0.84	0.82	0.87	0.88	**0.90**	0.86	0.81	0.78	0.76
Dunn	0.027	0.037	0.027	0.017	0.016	0.015	0.014	**0.008**	0.013

The values of all indexes for optimal partitions $\Pi_2^\star, \ldots, \Pi_{10}^\star$ are given in Table 5.2 (indexes suggesting best acceptable clusters are printed bold), and they are graphically shown in Fig. 5.5. The indexes CH, DB, and SSWC show correctly that $\Pi_6^\star$ is the partition with the most acceptable number of clusters, while in this case the Dunn index suggests falsely that the most acceptable partition is $\Pi_9^\star$.

(a) Calinski–Harabasz   (b) Davies–Bouldin   (c) Simplified Silhouette   (d) Dunn

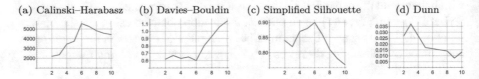

Fig. 5.5 Index values for optimal $2, \ldots, 10$ partitions

It should be said that in general the aforementioned indexes give acceptable conclusions provided the data fit the assumptions based on which the indexes were constructed. As was illustrated by previous examples, the conclusion about the most acceptable number of clusters for sets with relatively small number of data will not be reliable enough. Besides, the conclusion about the most acceptable number of clusters based on the said indexes is not always unambiguous.

Remark 5.10 The CH, DB, SSWC, and Dunn indexes can be calculated using *Mathematica*-modules CH[], DB[], SSWC[], and Dunn[], which are described in Sect. 9.2, where links to appropriate *Mathematica*-codes are also supplied.

5.2 Comparing Two Partitions

The problem of comparing two different partitions $\Pi^{(1)} = \{\pi_1^{(1)}, \ldots, \pi_k^{(1)}\}$ and $\Pi^{(2)} = \{\pi_1^{(2)}, \ldots, \pi_\ell^{(2)}\}$ of a set $\mathcal{A}$ with m elements comes up in various situations, e.g. when one wants to compare the partition obtained by some method, with the original partition on which the method was tested. We are going to show how this can be accomplished using the so-called Rand and Jaccard indexes of these partitions, or by finding the Hausdorff distance between sets of cluster centers of the original and of the obtained partition.

5.2.1 Rand Index of Two Partitions

We will introduce the *Rand index* following [83]. Let $\Pi^{(1)} = \{\pi_1^{(1)}, \ldots, \pi_k^{(1)}\}$ and $\Pi^{(2)} = \{\pi_1^{(2)}, \ldots, \pi_\ell^{(2)}\}$ be two partitions of the set $\mathcal{A}$ with separated clusters.

Definition 5.11 Let $\mathcal{C} = \{(a_i, a_j) \in \mathcal{A} \times \mathcal{A} : 1 \leq i < j \leq m\} \subset \mathbb{R}^{n \times n}$. We say that elements $a_i, a_j \in \mathcal{A}$, $1 \leq i < j \leq m$, are *paired* in $\Pi^{(1)}$, or that the pair $(a_i, a_j) \in \mathcal{C}$ is *paired* in $\Pi^{(1)}$, if a_i and a_j belong to a same cluster in $\Pi^{(1)}$, i.e. if there exists a cluster $\pi_s^{(1)} \in \Pi^{(1)}$ such that $a_i, a_j \in \pi_s^{(1)}$. Analogously one defines "being paired" in the partition $\Pi^{(2)}$.

Note that the set $\mathcal{C}$ has the same number of elements as the set of all combinations without repetitions of $m = |\mathcal{A}|$ elements taken two, hence $|\mathcal{C}| = \binom{m}{2} = \frac{m(m-1)}{2}$.

Definition 5.12 Let us define the following subsets of $\mathcal{C}$:

$\mathcal{C}_1$: all pairs $(a_i, a_j) \in \mathcal{C}$ paired in $\Pi^{(1)}$ and paired in $\Pi^{(2)}$;

$\mathcal{C}_2$: all pairs $(a_i, a_j) \in \mathcal{C}$ paired in $\Pi^{(1)}$ but not paired in $\Pi^{(2)}$;

$\mathcal{C}_3$: all pairs $(a_i, a_j) \in \mathcal{C}$ not paired in $\Pi^{(1)}$ but paired in $\Pi^{(2)}$;

$\mathcal{C}_4$: all pairs $(a_i, a_j) \in \mathcal{C}$ paired neither in $\Pi^{(1)}$ nor in $\Pi^{(2)}$.

Remark 5.13 The set $\{\mathcal{C}_1, \mathcal{C}_2, \mathcal{C}_3, \mathcal{C}_4\}$ is a partition of the set $\mathcal{C}$. Defining $a := |\mathcal{C}_1|$, $b := |\mathcal{C}_2|$, $c := |\mathcal{C}_3|$, and $d := |\mathcal{C}_4|$, the following holds true:

$$a + b + c + d = |\mathcal{C}| = \frac{m(m-1)}{2}.$$

We say that pairs in $\mathcal{C}_1 \cup \mathcal{C}_4$ are *concordant pairs* because they have the same status in both partitions, while pairs in $\mathcal{C}_2 \cup \mathcal{C}_3$ are *discordant pairs* since they do not have the same status in both partitions.

The *Rand index* is defined like this:

$$R(\Pi^{(1)}, \Pi^{(2)}) := \frac{a+d}{a+b+c+d}. \tag{5.16}$$

Obviously $R(\Pi^{(1)}, \Pi^{(2)}) \in [0, 1]$, and two partitions are more similar when the value of Rand index is closer to 1. Note that according to the previous remark, the Rand index (5.16) can be defined as

$$R(\Pi^{(1)}, \Pi^{(2)}) = \frac{2(a+d)}{m(m-1)}.$$

To circumvent symmetry in the Rand index, one defines the *Jaccard index*

$$J(\Pi^{(1)}, \Pi^{(2)}) := \frac{a}{a+b+c}. \tag{5.17}$$

In order to compare partitions $\Pi^{(1)}$ and $\Pi^{(2)}$ one can also define the so-called *confusion matrix* (see [184]), also called the *bonding matrix* in [83]

$$S(\Pi^{(1)}, \Pi^{(2)}) = (s_{ij}), \tag{5.18}$$

where s_{ij} is the number of elements in $\pi_i^{(1)} \cap \pi_j^{(2)}$. This matrix shows a more subtle correspondence between partitions $\Pi^{(1)}$ and $\Pi^{(2)}$.

Exercise 5.14 Let $\mathcal{A} \subset \mathbb{R}^n$ be a finite set. If $U^{(1)}$ is the membership matrix of a k-partition $\Pi^{(1)}$, and $U^{(2)}$ is the membership matrix of an ℓ-partition $\Pi^{(2)}$, show that the confusion matrix can be obtained as

$$S(\Pi^{(1)}, \Pi^{(2)}) = \left(U^{(1)}\right)^T U^{(2)}.$$

Example 5.15 Let $\mathcal{A} = \{i : i = 1, \dots, 10\}$ and consider the two partitions

$$\Pi^{(1)} = \big\{\{1, 2, 3\}, \{5, 7, 9\}, \{4, 6, 8, 10\}\big\},$$

$$\Pi^{(2)} = \big\{\{1, 2, 3, 4\}, \{5, 6, 7, 8, 9, 10\}\big\}.$$

Writing down the corresponding membership matrices, one readily checks that

$$a = |\mathcal{C}_1| = \big|\{(1, 2), (1, 3), (2, 3), (5, 7), (5, 9), (6, 8), (6, 10), (7, 9), (8, 10)\}\big| = 9$$

$$b = |\mathcal{C}_2| = \big|\{(4, 6), (4, 8), (4, 10)\}\big| = 3$$

$$c = |\mathcal{C}_3| = \big|\{(1, 4), (2, 4), (3, 4), (5, 6), (5, 8), (5, 10), (6, 7), (6, 9), (7, 8),$$

$$(7, 10), (8, 9), (9, 10)\}\big| = 12$$

Table 5.3 CPU-times used to calculate Rand and Jaccard indexes and confusion matrix (numbers inside parentheses are times when (5.19)–(5.22) were used)

$|\mathcal{A}_m|$	$\{\Pi_2^{(1)}, \Pi_2^{(2)}\}$	$\{\Pi_2^{(1)}, \Pi_3^{(2)}\}$	$\{\Pi_3^{(1)}, \Pi_4^{(2)}\}$	$\{\Pi_4^{(1)}, \Pi_5^{(2)}\}$
100	1.7 (0)	5.7 (0.02)	7.3 (0)	9.6 (0)
200	12.6 (0)	48.4 (0.02)	53.6 (0.02)	74.9 (0.03)
500	191.3 (0)	779.3 (0.06)	896.2 (0.09)	1279.2 (1.34)
1000	1807.8 (0.02)	7654.2 (1.95)	$\gg$ 1h (8.98)	$\gg$ 1h (20.52)

$$d = |\mathcal{C}_4| = \big|\{(1, 5), (1, 6), (1, 7), (1, 8), (1, 9), (1, 10), (2, 5), (2, 6), (2, 7),$$
$$(2, 8), (2, 9), (2, 10), (3, 5), (3, 6), (3, 7), (3, 8), (3, 9), (3, 10),$$
$$(4, 5), (4, 7), (4, 9)\}\big| = 21$$

and

$$R(\Pi^{(1)}, \Pi^{(2)}) = \tfrac{2}{3} \approx 0.667, \quad J(\Pi^{(1)}, \Pi^{(2)}) = \tfrac{3}{8} = 0.375,$$

$$S(\Pi^{(1)}, \Pi^{(2)}) = \begin{bmatrix} 3 & 0 \\ 0 & 3 \\ 1 & 3 \end{bmatrix}.$$

For large number of data, the CPU time for calculating the Rand and Jaccard indexes according to (5.16) and (5.17), and for the confusion matrix according to (5.18), can be rather long as shown by the following example.

Example 5.16 Let $\mathcal{A}_m = \{i : i = 1, \ldots, m\}$, for $m = 100, 200, 500, 1000$. For each $\mathcal{A}_m$ define pairs of partitions $\{\Pi_k^{(1)} = \{\pi_1^{(1)}, \ldots, \pi_k^{(1)}\}, \Pi_\ell^{(2)} = \{\pi_1^{(2)}, \ldots, \pi_\ell^{(2)}\}\}$ for $k = 2, 3, 4$ and $\ell = 2, 3, 4, 5$.

The CPU-times (in seconds) required for calculating the Rand and Jaccard indexes according to (5.16) and (5.17), and for the confusion matrix according to (5.18) for these partition pairs are listed in Table 5.3 (numbers outside the parentheses).

For that reason far more efficient formulas for calculating the values a, b, c, d appeared in the literature (see e.g. [82, 151]).

For instance, let us see how to find the number $a = |\mathcal{C}_1|$ in a simpler way. Without loss of generality, assume that $\Pi^{(1)} = \{\pi_1^{(1)}, \ldots, \pi_k^{(1)}\}$ and $\Pi^{(2)} = \{\pi_1^{(2)}, \ldots, \pi_\ell^{(2)}\}$ are two partitions of the set $\mathcal{A} = \{i : i = 1, \ldots, m\}$ and let $S \in \mathbb{N}_0^{k \times \ell}$ be the corresponding confusion matrix. Let $\mathcal{C} = \{(i, j) \in \mathcal{A} \times \mathcal{A} : 1 \leq i < j \leq m\}$ and let $\mathcal{C}_1, \ldots, \mathcal{C}_4$ be given by Definition 5.12.

By definition, s_{ij} is the number of elements of the set $\mathcal{A}$ belonging to both clusters $\pi_i^{(1)} \in \Pi^{(1)}$ and $\pi_j^{(2)} \in \Pi^{(2)}$. On the other hand, the set $\mathcal{C}_1$ contains all pairs $(i, j) \in \mathcal{C}$ which are paired in $\Pi^{(1)}$ and paired in $\Pi^{(2)}$. Therefore the number $|\mathcal{C}_1|$ of elements of the set $\mathcal{C}_1$ equals the total number of all two-point subsets of $\pi_i^{(1)} \cap \pi_j^{(2)}$, thus

$$|\mathcal{C}_1| = \sum_{i=1}^{k}\sum_{j=1}^{\ell} \binom{s_{ij}}{2}\binom{s_{ij}}{2} = \tfrac{1}{2}(\kappa - m), \qquad \text{where } \kappa = \sum_{i=1}^{k}\sum_{j=1}^{\ell} s_{ij}^2. \qquad (5.19)$$

Denoting $s_j := \sum_{i=1}^{k} s_{ij}$ and $s_i := \sum_{j=1}^{\ell} s_{ij}$, by similar arguments one can show that

$$|\mathcal{C}_2| = \sum_{i=1}^{k}\binom{s_i}{2} - |\mathcal{C}_1| = \tfrac{1}{2}\Big(\sum_{i=1}^{k} s_i^2 - \kappa\Big), \qquad\qquad\qquad (5.20)$$

$$|\mathcal{C}_3| = \sum_{j=1}^{\ell}\binom{s_j}{2} - |\mathcal{C}_1| = \tfrac{1}{2}\Big(\sum_{j=1}^{\ell} s_j^2 - \kappa\Big), \qquad\qquad\qquad (5.21)$$

$$|\mathcal{C}_4| = \binom{m}{2} - \sum_{i=1}^{k}\binom{s_i}{2} - \sum_{j=1}^{\ell}\binom{s_j}{2} + |\mathcal{C}_1| = \tfrac{1}{2}\Big(\kappa + m^2 - \sum_{i=1}^{k} s_i^2 - \sum_{j=1}^{\ell} s_j^2\Big). \qquad (5.22)$$

Calculating the Rand and Jaccard indexes according to (5.16) and (5.17) and using formulas (5.19)–(5.22) becomes far more efficient (see the corresponding CPU-times in parentheses in Table 5.3).

Example 5.17 Consider the data set $\mathcal{A}$ from Example 4.25. The values of the Rand and Jaccard indexes between the original 6-partition Π and the k-LOParts $\Pi^\star(2), \dots, \Pi^\star(10)$ obtained using the Incremental algorithm and depicted in Fig. 4.10 are shown in Table 5.4 (indexes suggesting best acceptable clusters are printed bold).

Table 5.4 Rand and Jaccard indexes and Hausdorff distances for k-LOpart

Index	$\Pi^\star(2)$	$\Pi^\star(3)$	$\Pi^\star(4)$	$\Pi^\star(5)$	$\Pi^\star(6)$	$\Pi^\star(7)$	$\Pi^\star(8)$	$\Pi^\star(9)$	$\Pi^\star(10)$
Rand	0.903	0.944	0.971	0.982	**0.996**	**0.996**	0.993	0.990	0.986
Jaccard	0.297	0.424	0.588	0.681	**0.899**	0.898	0.831	0.751	0.672
Hausdorff	7.264	5.512	2.784	2.754	**0.534**	1.083	1.848	1.668	1.670

As we expected, the largest values of the Rand and Jaccard indexes are for the 6-LOPart $\Pi^\star(6)$. The confusion matrix (5.18) detects very well the cluster elements

$$S_6 = \begin{bmatrix} 197 & 3 & 0 & 0 & 0 & 0 \\ 3 & 197 & 0 & 0 & 0 & 0 \\ 0 & 0 & 200 & 0 & 0 & 0 \\ 0 & 0 & 0 & 198 & 2 & 0 \\ 0 & 0 & 0 & 0 & 200 & 0 \\ 0 & 0 & 0 & 0 & 0 & 200 \end{bmatrix}.$$

This points to the conclusion that $\Pi^*(6)$ could be deemed MAPart.

Remark 5.18 The Rand index R, Jaccard index J, and the Confusion matrix S can be calculated fast using *Mathematica*-module Rand[]. The sets C_1, C_2, C_3, C_4, the Rand index R, Jaccard index J, and the Confusion matrix S can be calculated according to Definition 5.12 and Remark 5.13 using *Mathematica*-module RandCompare[]. Both modules are described in Sect. 9.2, and the links to appropriate *Mathematica*-codes are supplied.

5.2.2 Application of the Hausdorff Distance

One can also compare two partitions $\Pi^{(1)} = \{\pi_1^{(1)}, \ldots, \pi_\lambda^{(1)}\}$ and $\Pi^{(2)} = \{\pi_1^{(2)}, \ldots, \pi_\ell^{(2)}\}$ by looking at the distance between the set $\{c_1, \ldots, c_k\}$ of cluster centers of $\Pi^{(1)}$ and the set $\{z_1, \ldots, z_\ell\}$ of cluster centers of $\Pi^{(2)}$. To do this, we will use the Hausdorff metric function (4.20) applied to these sets.

The Hausdorff distances between cluster centers of partitions in Example 5.17 and the original 6-partition Π are shown in Table 5.4. It indicates again that the 6-LOPart $\Pi^*(6)$ coincides best with the original 6-partition Π.

Chapter 6
Mahalanobis Data Clustering

In the previous chapter we were dealing with problems of grouping data into spherical clusters. Now we are going to make a step toward real-life problems considering problems of grouping data into ellipsoidal clusters. The nature of the problem itself, or some geometrical reasons, often indicates the necessity of finding partitions with ellipsoidal clusters (see Examples in Sect. 8.1.5).

First we are going to consider the problem in the plane, and later it will be easy to generalize it to $\mathbb{R}^n$.

6.1 Total Least Squares Line in the Plane

Consider the well known problem of finding the best straight line, in the sense of the *Total Least Squares* (TLS), in the plane[1] (see [31, 90, 124]). Given a data set $\mathcal{A} = \{a_i = (x_i, y_i) : i = 1, \ldots, m\} \subset \mathbb{R}^2$, one has to determine parameters a, b, c of the line

$$ax + by + c = 0, \tag{6.1}$$

such that the sum of squares of orthogonal distances from points a_i to the line (6.1) be minimal. In order that (6.1) defines a line, parameters a and b have to satisfy $a^2 + b^2 \neq 0$, ensuring that at least one of them is nonzero. Dividing (6.1) by $\sqrt{a^2 + b^2}$ and denoting $\alpha := \frac{a}{\sqrt{a^2+b^2}}$, $\beta := \frac{b}{\sqrt{a^2+b^2}}$, and $\gamma := \frac{c}{\sqrt{a^2+b^2}}$, the requirement becomes $\alpha^2 + \beta^2 = 1$, and the problem of finding the optimal parameters of the sought TLS-line can be posed as the following GOP in $\mathbb{R}^3$:

---

[1] In the literature, this problem can also be found under the key word *errors-in-variables*.

© The Author(s), under exclusive license to Springer Nature Switzerland AG 2021
R. Scitovski et al., *Cluster Analysis and Applications*,
https://doi.org/10.1007/978-3-030-74552-3_6

117

$$\underset{\alpha,\beta,\gamma \in \mathbb{R}}{\arg\min} \sum_{i=1}^{m}(\alpha x_i + \beta y_i + \gamma)^2, \quad \text{where} \quad \alpha^2 + \beta^2 = 1. \tag{6.2}$$

A line (6.1) satisfying $a^2 + b^2 = 1$ is called a *normalized line*.[2] In the general, weighted case, the following lemma provides means to reduce dimension of this GOP by one.

Lemma 6.1 *Let* $A = \{a_i = (x_i, y_i) : i = 1, \ldots, m\} \subset \mathbb{R}^2$ *be a data set with weights* $w_i > 0$. *The TLS-line passes through the centroid* $(\bar{x}, \bar{y})$ *of* A, *where*

$$\bar{x} = \frac{1}{W}\sum_{i=1}^{m} w_i x_i, \qquad \bar{y} = \frac{1}{W}\sum_{i=1}^{m} w_i y_i, \qquad W = \sum_{i=1}^{m} w_i.$$

Proof Note that if the normalized line $\alpha x + \beta y + \gamma = 0$, $\alpha^2 + \beta^2 = 1$, passes through the centroid $(\bar{x}, \bar{y})$, then

$$\alpha(x - \bar{x}) + \beta(y - \bar{y}) = 0, \quad \alpha^2 + \beta^2 = 1. \tag{6.3}$$

Using (2.7) and linearity of the arithmetic mean,

$$\sum_{i=1}^{m} w_i\big(\alpha x_i + \beta y_i - (-\gamma)\big)^2$$

$$\geq \sum_{i=1}^{m} w_i\big(\alpha x_i + \beta y_i - (\alpha\bar{x} + \beta\bar{y})\big)^2 = \sum_{i=1}^{m} w_i\big(\alpha(x_i - \bar{x}) + \beta(y_i - \bar{y})\big)^2.$$

The equality holds true if and only if $\gamma = -\alpha\bar{x} - \beta\bar{y}$, i.e. among all normalized lines $\alpha x + \beta y + \gamma = 0$, $\alpha^2 + \beta^2 = 1$, the line (6.3) passing through the centroid of data has the least possible weighted sum of squares of orthogonal deviations. $\square$

Exercise 6.2 Let p, given by $\alpha x + \beta y + \gamma = 0$, $\alpha^2 + \beta^2 = 1$, be a normalized line in the plane, and let $T = (x_0, y_0) \in \mathbb{R}^2$ be an arbitrary point. Prove that

(*i*) $n_0 = (\alpha, \beta)$ is the unit normal vector to the line p ;
(*ii*) the unit vector $u_0 = (\beta, -\alpha)$ determines the direction of the line p ;
(*iii*) the Euclidean distance from the point T to the line p is given by

$$d(T, p) = |\alpha x_0 + \beta y_0 + \gamma|, \tag{6.4}$$

in particular, $|\gamma|$ equals the distance from the origin O to the line p ;

---

[2] Line is a geometric object, and (6.1) is one of at least half a dozen type of equations defining a line in the plane. So, to be precise, Eq. (6.1) with $a^2 + b^2 = 1$ is the *normalized equation* of a line, but slightly abusing the terminology, we will say that (6.1) is a *normalized line*.

(*iv*) the orthogonal projection T_p of the point T onto the line p is given by

$$T_p = \langle T, u_0 \rangle u_0 - \gamma \, n_0, \tag{6.5}$$

where $\langle \, , \, \rangle$ denotes the usual scalar product.

Remark 6.3 The distance from a point T_0 to a normalized line, and the orthogonal projection of T_0 to this line, can be calculated using *Mathematica*-module `Proj[]` which is described in Sect. 9.2, and the link to appropriate *Mathematica*-code is supplied.

Exercise 6.4 Let $\mathcal{A} = \{a_i = (x_i, y_i) : i = 1, \dots, m\} \subset \mathbb{R}^2$ be a data set with weights $w_i > 0$. The parameters $\hat{k}$, $\hat{\ell}$ of *the best weighted LS-line* $y = \hat{k}x + \hat{\ell}$ are defined as the solution to the following GOP (see e.g. [31, 144]):

$$\underset{k, \ell \in \mathbb{R}}{\arg\min} \, F_{LS}(k, \ell), \qquad F_{LS}(k, \ell) = \sum_{i=1}^{m} w_i (y_i - k x_i - \ell)^2. \tag{6.6}$$

Does the solution to the problem (6.6) always exist? Does the best LS-line pass through the data's centroid?

Exercise 6.5 Let p be a line in the plane given by $y = kx + \ell$ and let $T = (x_0, y_0)$ be a point. Show that the Euclidean distance from T to the line p, and its orthogonal projection T_p, are given by

$$d(T, p) = \frac{|k x_0 + \ell - y_0|}{\sqrt{k^2 + 1}}, \qquad T_p = \tfrac{1}{k^2+1}(-k\ell + x_0 + k \, y_0, \; \ell + k(x_0 + k y_0)).$$

According to Lemma 6.1, looking for the best TLS-line for a given data set $\mathcal{A} = \{a_i = (x_i, y_i) : i = 1, \dots, m\} \subset \mathbb{R}^2$ in the form of (6.3), instead of solving the GOP (6.2), one can solve the GOP

$$\underset{\alpha, \beta \in \mathbb{R}}{\arg\min} \, F(\alpha, \beta), \quad \text{with} \quad \alpha^2 + \beta^2 = 1, \quad \text{where}$$

$$F(\alpha, \beta) = \sum_{i=1}^{m} w_i \big(\alpha(x_i - \bar{x}) + \beta(y_i - \bar{y}) \big)^2. \tag{6.7}$$

Using

$$B := \begin{bmatrix} x_1 - \bar{x} & y_1 - \bar{y} \\ \vdots & \vdots \\ x_m - \bar{x} & y_m - \bar{y} \end{bmatrix}, \qquad D = \mathrm{diag}(w_1, \dots, w_m), \qquad t = (\alpha, \beta), \tag{6.8}$$

the function F can be written as

$$F(\alpha, \beta) = \|\sqrt{D} \, B \, t^T\|^2, \quad \|t\| = 1. \tag{6.9}$$

Namely, $\|\sqrt{D} \, B \, t^T\|^2 = (\sqrt{D} \, B \, t^T)^T \, (\sqrt{D} \, B \, t^T) = t \, B^T D \, B \, t^T$, and since

$$B^T D B = \begin{bmatrix} x_1 - \bar{x} \cdots x_m - \bar{x} \\ y_1 - \bar{y} \cdots y_m - \bar{y} \end{bmatrix} \cdot \begin{bmatrix} w_1(x_1 - \bar{x}) & w_1(y_1 - \bar{y}) \\ \cdots & \cdots \\ w_m(x_m - \bar{x}) & w_m(y_m - \bar{y}) \end{bmatrix}$$

$$= \begin{bmatrix} \sum_{i=1}^{m} w_i(x_i - \bar{x})^2 & \sum_{i=1}^{m} w_i(x_i - \bar{x})(y_i - \bar{y}) \\ \sum_{i=1}^{m} w_i(x_i - \bar{x})(y_i - \bar{y}) & \sum_{i=1}^{m} w_i(y_i - \bar{y})^2 \end{bmatrix},$$

we get

$$(\alpha, \beta)(B^T D B) (\alpha, \beta)^T = (\alpha, \beta) \begin{bmatrix} \alpha \sum_{i=1}^{m} w_i(x_i - \bar{x})^2 + \beta \sum_{i=1}^{m} w_i(x_i - \bar{x})(y_i - \bar{y}) \\ \alpha \sum_{i=1}^{m} w_i(x_i - \bar{x})(y_i - \bar{y}) + \beta \sum_{i=1}^{m} w_i(y_i - \bar{y})^2 \end{bmatrix}$$

$$= \alpha^2 \sum_{i=1}^{m} w_i(x_i - \bar{x})^2 + 2\alpha\beta \sum_{i=1}^{m} w_i(x_i - \bar{x})(y_i - \bar{y}) + \beta^2 \sum_{i=1}^{m} w_i(y_i - \bar{y})^2$$

$$= \sum_{i=1}^{m} w_i \big(\alpha(x_i - \bar{x}) + \beta(y_i - \bar{y})\big)^2 = F(\alpha, \beta).$$

Exercise 6.6 Show that $B^T D B$ is a positive definite matrix if and only if the points (x_i, y_i), $i = 1, \ldots, m$, are not collinear.

According to [124], the following theorem holds true (for the necessary notions see [19, 31, 107, 179]).

Theorem 6.7 *The function F in (6.7) attains its global minimum at the unit eigenvector $t = (\alpha, \beta)$ corresponding to the smaller eigenvalue of the symmetric positive definite matrix $B^T D B$.*

Proof Notice that $B^T D B \in \mathbb{R}^{2 \times 2}$ is a symmetric positive definite matrix. According to the theorem on diagonalization of such matrices (see e.g. [19, 179]), there exist an orthogonal matrix V and a diagonal matrix $\Delta = \text{diag}(\lambda_1, \lambda_2)$ such that $B^T D B = V \Delta V^T$. The numbers $\lambda_1 \geq \lambda_2 > 0$ are the (positive) eigenvalues, and columns of the matrix V correspond to eigenvectors of the matrix $B^T D B$.

For an arbitrary unit vector $t \in \mathbb{R}^2$ one has

$$\|\sqrt{D} \, B \, t^T\|^2 = (\sqrt{D} \, B \, t^T)^T \, (\sqrt{D} \, B \, t^T) = t \, B^T D \, B \, t^T = t \, V \Delta V^T t^T = s \, \Delta \, s^T,$$

where $s = t \, V$. Since the matrix V is orthogonal, $\|s\| = 1$, and therefore

$$\|\sqrt{D}B\,t^T\|^2 = s\,\Delta\,s^T = \sum_{i=1}^{2}\lambda_i\,s_i^2 \ge \lambda_2.$$

The latter inequality is the consequence of the fact that the minimum of a convex combination of several numbers is attained at the smallest of these numbers. It is not difficult to see that the equality takes place precisely when t is the unit eigenvector corresponding to the smallest (in our two-dimensional case this means the smaller) eigenvalue of the matrix $B^T D B$. ☐

Remark 6.8 The corresponding *Mathematica*-module TLSline[] is described in Sect. 9.2, and the link to appropriate *Mathematica*-code is supplied.

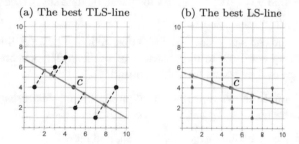

Fig. 6.1 The best TLS and LS lines

Exercise 6.9 Find the explicit formulas for eigenvalues of the matrix $B^T D B$.

Example 6.10 Let $\mathcal{A} = \{(1,4),\ (3,6),\ (4,7),\ (5,2),\ (7,1),\ (9,4)\}$ be a set with weights $w = (1,1,1,1,1,1)$.

Using the module TLSline[] one finds the data's centroid $(\bar{x}, \bar{y}) = (\frac{29}{6}, 4)$ and matrices

$$B = \begin{bmatrix} -23/6 & 0 \\ -11/6 & 2 \\ -5/6 & 3 \\ 1/6 & -2 \\ 13/6 & -3 \\ 25/6 & 0 \end{bmatrix}, \quad D = \operatorname{diag}(1,1,1,1,1,1), \quad B^T D B = \begin{bmatrix} 245/6 & -13 \\ -13 & 26 \end{bmatrix}.$$

Eigenvalues of the matrix $B^T D B$ are $\lambda_1 = 48.38$ and $\lambda_2 = 18.45$, and the unit eigenvector corresponding to the smaller eigenvalue is $t = (0.50,\ 0.86)$. Therefore, the best TLS-line is given by $0.50\,(x-\bar{x})+0.86\,(y-\bar{y}) = 0$, i.e. $y = -0.65\,x+6.85$, shown in Fig. 6.1a.

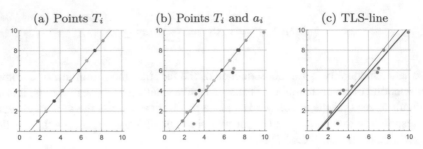

Fig. 6.2 Construction of an artificial data set derived from the line $y = \frac{5}{4}(x - 1)$, and the corresponding TLS-line

The best LS-line $y = -0.32\,x + 5.54$ for the same data set $\mathcal{A}$ (see Exercise 6.4), is shown in Fig. 6.1b.

Remark 6.11 One can show (see e.g. [44]) that both, the best LS-line and the best TLS-line, are very sensitive to the presence of outliers. The sensitivity increases with the distance of data points from the centroid.

The next example shows how one can define artificial data arising from some straight line. Later on, this is going to be used in applications.

Example 6.12 Let p be the line $y = \frac{5}{4}(x - 1)$, and let

$$T_i = \lambda_i A + (1 - \lambda_i)B, \quad \lambda_i = \frac{i}{m}, \quad i = 0, 1, \dots, m,$$

be $m + 1$ uniformly distributed points on the straight line segment between points $A = (1, 0)$ and $B = (9, 10)$. Replacing each point T_i by a random point a_i with added noise from the bivariate normal distribution $\mathcal{N}(0, \sigma^2 I)$ with expectation $0 \in \mathbb{R}^2$ and covariance matrix $\sigma^2 I$, $\sigma = 0.25$, we obtain an artificial data set $\mathcal{A}$. Figure 6.2a shows the line p with points T_i, Fig. 6.2b shows points T_i and the corresponding points a_i, and Fig. 6.2c shows the data set $\mathcal{A}$ and the best TLS-line (red): $-0.756248\,x + 0.654285\,y + 0.825634 = 0$ $[y = 1.15584\,x - 1.26189]$.

Using *Mathematica* computation system, such a data set $\mathcal{A}$ can be obtained in the following way:

```
In[1]:= cA={1,0}; cB={9,10}; m = 10; sigma = .25; A = TT = {}; SeedRandom[5];
        Do[
           lam = s/m;
           T = lam cA + (1 - lam) cB; TT = Append[TT, T];
           A = Append[A, RandomReal[
             MultinormalDistribution[T, sigma*IdentityMatrix[2]],{1}][[1]]];
           , {s, m}];
```

Under some conditions on the data set $\mathcal{A}$ (see Lemma 6.19), the matrix $\Sigma = \frac{1}{W}B^T D B$ is a positive definite matrix. Therefore its eigenvalues are real numbers, and corresponding eigenvectors are mutually orthogonal (see e.g. [19]). The eigenvector corresponding to the larger eigenvalue points in direction of the TLS-line. The unit eigenvector corresponding to the smaller eigenvalue is the unit normal vector $n_0 = (\xi, \eta)$ of the TLS-line, and since the TLS-line passes through the centroid $(\bar{x}, \bar{y})$ of the set $\mathcal{A}$, its equation is

$$\xi (x - \bar{x}) + \eta (y - \bar{y}) = 0.$$

In statistical literature, when the data $a_i = (x_i, y_i)$, $i = 1, \ldots, m$, with weights $w_i = 1$ are interpreted as realization of a random vector $A = (X, Y)$, the matrix $\frac{1}{m}\Sigma$ is called the *sample covariance matrix* (see e.g. [183]), its eigenvectors are called *principal components*, and the eigenvector corresponding to the smaller eigenvalue is called the *second principal component*. We will refer to $\frac{1}{m}\Sigma$ as the covariance matrix. Notice that the first principal component points in the direction of the TLS-line, while the second principal component is perpendicular to that line.

Let us describe yet another interpretation of principal components. Projecting the datum a_i onto some unit vector u defines the number $a_i u^T$ which determines the position of the projection of a_i onto u. Suppose we want to determine the unit vector onto which we project our data, and which we will now denote by u_1, in such a way that the variance of projection data be as large as possible, i.e.

$$u_1 = \underset{u \in \mathbb{R}^2, \|u\|=1}{\arg\max} \frac{1}{m} \sum_{i=1}^{m} (u a_i^T - u \bar{a}^T)^2.$$

Since

$$\frac{1}{m} \sum_{i=1}^{m} (u a_i^T - u \bar{a}^T)^2 = \frac{1}{m} \sum_{i=1}^{m} \left(u (a_i - \bar{a})^T (a_i - \bar{a}) u^T \right) = u \left(\frac{1}{m} \Sigma \right) u^T,$$

we see that the vector u_1 corresponds to the first principal component. Therefore, the first principal component is the vector onto which the data should be projected in order that the variance of projected data be maximal.

Similarly, one can show that the second principal component u_2 is the solution to the following two optimization problems:

$$u_2 = \underset{u \in \mathbb{R}^2, \|u\|=1}{\arg\min} \frac{1}{m} \sum_{i=1}^{m} (u a_i^T - u \bar{a}^T)^2$$

and

$$u_2 = \underset{u \in \mathbb{R}^2, u u_1^T = 0, \|u\|=1}{\arg\max} \frac{1}{m} \sum_{i=1}^{m} (u a_i^T - u \bar{a}^T)^2.$$

6.2 Mahalanobis Distance-Like Function in the Plane

The ellipse $E(O, a, b, \vartheta)$ in the plane, centered at the origin, with semi-axes $a, b >$ 0, and rotated around O by the angle ϑ, has the equation, [111]

$$\frac{(x_1 \cos \vartheta + x_2 \sin \vartheta)^2}{a^2} + \frac{(-x_1 \sin \vartheta + x_2 \cos \vartheta)^2}{b^2} = 1, \tag{6.10}$$

where $x = (x_1, x_2) \in \mathbb{R}^2$.

Denoting

$$\Sigma := U \operatorname{diag}(a^2, b^2) U^T, \qquad U = \begin{bmatrix} \cos \vartheta & -\sin \vartheta \\ \sin \vartheta & \cos \vartheta \end{bmatrix}, \tag{6.11}$$

Equation (6.10) can be written as

$$x \, \Sigma \, x^T = 1, \quad x = (x_1, x_2). \tag{6.12}$$

On the other hand, if $\Sigma \in \mathbb{R}^{2 \times 2}$ is a symmetric positive definite matrix, its eigenvalues are positive real numbers $\lambda_1 \geq \lambda_2 > 0$, and the corresponding eigenvectors e_1, e_2 are orthonormal. Such a matrix Σ defines the ellipse centered at the origin with semi-axes $\sqrt{\lambda_1}, \sqrt{\lambda_2}$ in directions of eigenvectors e_1, e_2 (see e.g. [7, 49, 111, 184]).

Thus we have the following lemma:

Lemma 6.13 *There is a bijection between the set of all ellipses in the plane centered at the origin and the set of all symmetric positive definite 2×2 matrices.*

Example 6.14 Let $K(0, r) \subset \mathbb{R}^2$ be a disc of radius r centered at the origin. Its area equals $r^2 \pi$. Various deformations of this disc, by expanding/contracting in one direction and contracting/expanding in the orthogonal direction, produce ellipses with semi-axes a and b, whose area equals $r^2 \pi$. Therefore $ab = r^2$, regardless of ϑ, (see Fig. 6.3).

Exercise 6.15 For each ellipse in Fig. 6.3 determine the corresponding matrix Σ as in (6.11).

Take an ellipse in Fig. 6.3 and let $\Sigma \in \mathbb{R}^{2 \times 2}$ be the corresponding symmetric positive definite matrix (cf. Lemma 6.13). We want to define a distance-like function $d_m : \mathbb{R}^2 \times \mathbb{R}^2 \to \mathbb{R}_+$ such that all points on this ellipse are at the same distance from the origin, i.e. the ellipse becomes a circle in the space provided with the distance-like function d_m. This will be achieved by considering distances in directions of eigenvectors of the matrix Σ to be inversely proportional to the corresponding eigenvalues. This is precisely what the inverse matrix Σ^{-1} does. Therefore we define the distance-like function $d_m : \mathbb{R}^2 \times \mathbb{R}^2 \to \mathbb{R}_+$ by

$$d_m(x, y; \Sigma) := \| \Sigma^{-1/2}(x - y) \|^2 = (x - y) \Sigma^{-1}(x - y)^T. \tag{6.13}$$

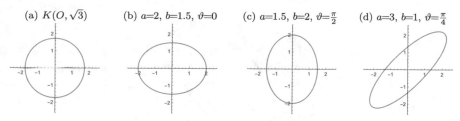

Fig. 6.3 M-circles with radius $r = \sqrt{3}$ derived from the disc $K(O, \sqrt{3})$

The distance-like function d_m is called the **Mahalanobis distance-like function**, and $d_m(x, y; \Sigma)$ is called the **Mahalanobis distance**, or **M-distance** for short, between points $x, y \in \mathbb{R}^2$.

The Example 6.14 motivates the definition of an **M-circle**. Assume that an ellipse, which is determined by a symmetric positive definite matrix Σ, was obtained by deforming a disc with radius $r > 0$ as described in Example 6.14. The area of this ellipse equals the area of that disc. Since the lengths of semi-axes equal square roots of the eigenvalues $\lambda_1 \geq \lambda_2 > 0$, and since $\lambda_1 \cdot \lambda_2 = \det \Sigma$, the area of this ellipse equals $\pi \sqrt{\lambda_1} \sqrt{\lambda_2} = \pi \sqrt{\det \Sigma}$, and it is equal to the area of a disc of radius $r = \sqrt[4]{\det \Sigma}$. Therefore, it does make sense to call this number the **Euclidean radius of the M-circle**.

Definition 6.16 Let Σ be a symmetric positive definite matrix. The **Mahalanobis circle** (**M-circle**) centered at the point $C \in \mathbb{R}^2$, is the set

$$M(C, \Sigma) = \{x \in \mathbb{R}^2 : d_m(x, C; \Sigma) = 1\}. \tag{6.14}$$

With respect to the distance-like function d_m, M-circle $M(C, \Sigma)$ is the circle centered at C with radius 1. But with respect to the Euclidean metric, $M(C, \Sigma)$ is the ellipse centered at C, with semi-axes being $\sqrt{\lambda_1} \geq \sqrt{\lambda_2} > 0$, i.e. the square roots of the eigenvalues of the matrix Σ, and in directions of eigenvectors of the matrix Σ. The Euclidean radius of the M-circle $M(C, \Sigma)$ is defined as

$$r = \sqrt[4]{\det \Sigma}. \tag{6.15}$$

Example 6.17 The matrix $\Sigma = \begin{bmatrix} 2 & 1 \\ 1 & 2 \end{bmatrix}$ defines the M-circle $M(O, \Sigma)$ centered at the origin O and of Euclidean radius $r = \sqrt[4]{3}$. This M-circle originated from the circle $\mathcal{C}(O, \sqrt[4]{3})$ by expanding in the direction of eigenvector $e_1 = \frac{1}{\sqrt{2}}(1, 1)$, and shrinking in the direction of eigenvector $e_2 = \frac{1}{\sqrt{2}}(-1, 1)$ (see Fig. 6.4). Every point $A \in M(O, \Sigma)$ satisfies $d_m(O, A; \Sigma) = 1$, i.e. M-circle $M(O, \Sigma)$ is the unit circle

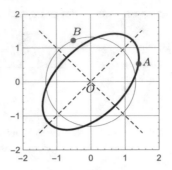

Fig. 6.4 M-circle $M(O, \Sigma)$ originated from the circle of radius $r = \sqrt[4]{\det \Sigma}$ centered at the origin

with respect to the distance-like function d_m, and every point $B \in \mathcal{C}(O, \sqrt[4]{3})$ has the same Euclidean distance to the origin, equal to $r = \sqrt[4]{3} = 1.31607$ (see Fig. 6.4).

6.3   Mahalanobis Distance Induced by a Set in the Plane

To keep things simple, we proceed considering again the problem in $\mathbb{R}^2$. For the given data set $\mathcal{A} = \{a_i = (x_i, y_i) : i = 1, \ldots, m\} \subset \mathbb{R}^2$ with weights $w_1, \ldots, w_m > 0$, we are looking for the main directions of elongation. For that purpose we consider the covariance matrix

$$\Sigma = \tfrac{1}{W} B^T D B = \begin{bmatrix} \frac{1}{W} \sum\limits_{i=1}^{m} w_i (x_i - \bar{x})^2 & \frac{1}{W} \sum\limits_{i=1}^{m} w_i (x_i - \bar{x})(y_i - \bar{y}) \\ \frac{1}{W} \sum\limits_{i=1}^{m} w_i (x_i - \bar{x})(y_i - \bar{y}) & \frac{1}{W} \sum\limits_{i=1}^{m} w_i (y_i - \bar{y})^2 \end{bmatrix}, \tag{6.16}$$

where the matrix B is given by (6.8), and $\bar{a} = (\bar{x}, \bar{y})$ is the centroid of the set $\mathcal{A}$

$$\bar{x} = \tfrac{1}{W} \sum_{i=1}^{m} w_i x_i, \quad \bar{y} = \tfrac{1}{W} \sum_{i=1}^{m} w_i y_i, \quad W = \sum_{i=1}^{m} w_i.$$

As we already noticed, the direction of the TLS-line, thus also the main direction of data (first principal component), is the direction of the eigenvector belonging to the larger eigenvalue of the matrix Σ. The ancillary direction (second principal component) is taken to be perpendicular to the main direction, i.e. it is the direction of the eigenvector belonging to the smaller eigenvalue of the same matrix. Therefore, looking for the most significant directions in the set $\mathcal{A}$, we will look in directions of the eigenvectors of the matrix Σ.

Remark 6.18 The covariance matrix Σ of a data set $\mathcal{A}$ with the centroid $\bar{a} = \text{mean}(\mathcal{A})$ can be written as the *Kronecker product*:

$$\Sigma = \text{cov}(\mathcal{A}) = \frac{1}{W} \sum_{i=1}^{m} w_i (\bar{a} - a_i)^T (\bar{a} - a_i). \tag{6.17}$$

The following lemma shows that the covariance matrix Σ defined by (6.16), respectively by (6.17), is positive definite if the data set $\mathcal{A} \subset \mathbb{R}^2$ does not lie on a straight line through the centroid (cf. Exercise 6.6).

Lemma 6.19 *Let $\mathcal{A} = \{a_i = (x_i, y_i) : i = 1, \ldots, m\} \subset \mathbb{R}^2$ be a set in the plane, and let $\bar{a} = (\bar{x}, \bar{y})$ be its centroid. The covariance matrix Σ defined by (6.16), respectively by (6.17), is a positive definite matrix if and only if the vectors $(x_1 - \bar{x}, \ldots, x_m - \bar{x})$ and $(y_1 - \bar{y}, \ldots, y_m - \bar{y})$ are linearly independent.*

Proof The matrix Σ is obviously symmetric, and positive definiteness follows from the Cauchy–Schwarz–Bunyakovsky inequality (see e.g. [19]).   □

Exercise 6.20 Let $u, v \in \mathbb{R}$ be arbitrary real numbers and $a = (x^1, \ldots, x^m)$, $b = (y^1, \ldots, y^m) \in \mathbb{R}^m$ vectors.

(a) If the vectors a and b are linearly independent (dependent), do the vectors $(x^1 - u, \ldots, x^m - u)$ and $(y^1 - v, \ldots, y^m - v)$ also need to be linearly independent (dependent)?
(b) If the vectors $(x^1 - u, \ldots, x^m - u)$ and $(y^1 - v, \ldots, y^m - v)$ are linearly independent (dependent), do the vectors a and b also need to be linearly independent (dependent)?

Example 6.21 Eigenvalues of the symmetric matrix $\Sigma = \begin{bmatrix} 2 & 1 \\ 1 & 2 \end{bmatrix}$ are $\lambda_1 = 3$ and $\lambda_2 = 1$, and the corresponding unit eigenvectors are $u_1 = \frac{\sqrt{2}}{2}(1, 1)$ and $u_2 = \frac{\sqrt{2}}{2}(-1, 1)$. Using this matrix, we are going to generate, in a neighborhood of the origin, the set $\mathcal{A}$ with $m = 500$ random points from bivariate normal distribution with expectation $0 \in \mathbb{R}^2$ and the covariance matrix Σ (see Fig. 6.5).

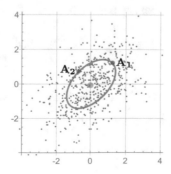

Fig. 6.5 Set of points generated by the covariance matrix Σ

```
In[1]:= SeedRandom[23];
        m=500; O={0,0}; Cov={{2,1},{1,2}};
        RandomReal[MultinormalDistribution[O, Cov],m];
```

According to (6.16), the centroid $\bar{a} = (\bar{x}, \bar{y})$ and the covariance matrix are

$$\bar{a} = (0.015, -0.051), \qquad \text{Cov} = \begin{bmatrix} 1.960 & 0.975 \\ 0.975 & 1.933 \end{bmatrix}.$$

Eigenvalues of the matrix Cov are $\lambda_1 = 2.921$ and $\lambda_2 = 0.972$, and the corresponding eigenvectors are $u_1 = (0.712, 0.702)$ (the main direction—the first principal component) and $u_2 = (-0.702, 0.712)$ (the ancillary direction—the second principal component).

Take two points on the M-circle $M(O, \text{Cov})$: $A_1 = (1.20855, 1.20855)$ on the main direction which is $d_2(A_1, O) \approx 1.71$ distant from the origin, and $A_2 = (-0.697, 0.697)$ on the ancillary direction which is $d_2(A_2, O) \approx 1$ distant from the origin. The M-distance to the origin of both points equals 1, i.e. $d_m(O, A_i, \text{Cov}) = 1$, $i = 1, 2$, (see Fig. 6.5).

Theorem 6.22 *Let $\mathcal{A} = \{a_i = (x_i, y_i) : i = 1, \ldots, m\} \subset \mathbb{R}^2$ be a data set in the plane, with weights $w_i > 0$, and let $\Sigma \in \mathbb{R}^{2 \times 2}$ be a symmetric positive definite matrix. The centroid of the set $\mathcal{A}$ coincides with the M-centroid of $\mathcal{A}$.*

Proof In accordance with Definition 2.4, the M-centroid of $\mathcal{A}$ is the point

$$\bar{c}_m = \arg\min_{x \in \mathbb{R}^2} \sum_{i=1}^m w_i \, d_m(x, a_i; \Sigma) = \arg\min_{x \in \mathbb{R}^2} \sum_{i=1}^m w_i \, (x - a_i) \, \Sigma^{-1} (x - a_i)^T,$$

where the function

$$F_m(x) = \sum_{i=1}^m w_i \, d_m(x, a_i; \Sigma) = \sum_{i=1}^m w_i \, (x - a_i) \, \Sigma^{-1} (x - a_i)^T,$$

attains its minimum. The stationary point c of the function F_m is determined by

$$\nabla F_m(c) = 2 \sum_{i=1}^m w_i \, \Sigma^{-1} (c - a_i)^T = 0.$$

Since the Hessian $\nabla^2 F_m = 2W\Sigma^{-1}$, $W = \sum_{i=1}^m w_i$, is positive definite, the function F_m attains its global maximum at the point

$$\bar{c}_m := \frac{1}{W} \sum_{i=1}^{m} w_i \, a_i.$$

Hence, the M-centroid of the set $\mathcal{A}$ coincides with the (ordinary) centroid of $\mathcal{A}$.   □

6.3.1   *Mahalanobis Distance Induced by a Set of Points in $\mathbb{R}^n$*

Considering the general case, let $\mathcal{A} = \{a_i = (a_i^1, \ldots, a_i^n) : i = 1, \ldots, m\} \subset \mathbb{R}^n$ be a data set with weights $w_1, \ldots, w_m > 0$. The centroid of the set $\mathcal{A}$ is the point

$$\bar{a} = (\bar{a}_1, \ldots, \bar{a}_n) = \underset{x \in \mathbb{R}^n}{\arg\min} \sum_{i=1}^{m} w_i \, d_m(x, a_i; \Sigma) = \frac{1}{W} \sum_{i=1}^{m} w_i \, a_i, \quad W = \sum_{i=1}^{m} w_i.$$

The Mahalanobis distance-like function $d_m : \mathbb{R}^n \times \mathbb{R}^n \to \mathbb{R}_+$ is defined by

$$d_m(x, y; \Sigma) = (x - y) \, \Sigma^{-1} (x - y)^T,$$

where the covariance matrix Σ is the $n \times n$ matrix

$$\Sigma = \frac{1}{W} \begin{bmatrix} \sum w_i(\bar{a}_1 - a_1^i)^2 & \cdots & \sum w_i(\bar{a}_1 - a_1^i)(\bar{a}_n - a_n^i) \\ \sum w_i(\bar{a}_2 - a_2^i)(\bar{a}_1 - a_1^i) & \cdots & \sum w_i(\bar{a}_2 - a_2^i)(\bar{a}_n - a_n^i) \\ \vdots & \ddots & \vdots \\ \sum w_i(\bar{a}_n - a_n^i)(\bar{a}_1 - a_1^i) & \cdots & \sum w_i(\bar{a}_n - a_n^i)^2 \end{bmatrix}, \quad W = \sum_{i=1}^{m} w_i,$$

which can be written as $\Sigma = \frac{1}{W} B^T D B$, where

$$B = \begin{bmatrix} \bar{a}_1 - a_1^1 & \cdots & \bar{a}_n - a_n^1 \\ \vdots & \ddots & \vdots \\ \bar{a}_1 - a_1^m & \cdots & \bar{a}_n - a_n^m \end{bmatrix}, \quad \text{and } D = \mathrm{diag}(w_1, \ldots, w_m),$$

or, using the Kronecker product (see Remark 6.18), as

$$\Sigma = \frac{1}{W} \sum_{i=1}^{m} w_i \, (\bar{a} - a_i)^T (\bar{a} - a_i).$$

6.4 Methods to Search for Optimal Partition with Ellipsoidal Clusters

Consider a set $\mathcal{A} \subset \mathbb{R}^n$ with $m = |\mathcal{A}|$ elements which we want to group into k clusters. In order to search for optimal partitions with ellipsoidal clusters, we are going to construct appropriate modifications of the k-means and of the incremental algorithms.

Let $\Pi = \{\pi_1, \ldots, \pi_k\}$ be a partition of the set $\mathcal{A}$, and for each cluster π_j let $c_j = \mathrm{mean}(\pi_j)$ be its centroid and $\Sigma_j = \frac{1}{|\pi_j|} \sum_{a \in \pi_j} (c_j - a)^T (c_j - a)$ the corresponding covariance matrix. If the eigenvalues of the covariance matrix differ significantly, it is without doubt the reason to look for partitions of $\mathcal{A}$ consisting of ellipsoidal clusters. Therefore, we will use M-distance like function.

To ensure that values of the objective function always decrease, when implementing the Mahalanobis k-means Algorithm 6.28, we will use the *normalized Mahalanobis distance-like function* $d_M : \mathbb{R}^n \times \mathbb{R}^n \to \mathbb{R}_+$

$$d_M(u, v; \Sigma) := \sqrt[n]{\det \Sigma} \, (u - v) \Sigma^{-1} (u - v)^T =: \|u - v\|_\Sigma^2. \tag{6.18}$$

Definition 6.23 Let $\Sigma \in \mathbb{R}^{n \times n}$ be a symmetric positive definite matrix, and $C \in \mathbb{R}^n$ a point.

- The set $M(C; \Sigma) = \{x \in \mathbb{R}^n : (x - C) \Sigma^{-1} (x - C)^T = 1\}$ is called the *Mahalanobis sphere (M-sphere)* centered at C.
- The set $M_N(C; \Sigma) = \{x \in \mathbb{R}^n : \sqrt[n]{\det \Sigma} \, (x - C) \Sigma^{-1} (x - C)^T = 1\}$ is called the *normalized Mahalanobis sphere (M$_N$-sphere)* centered at C.

Remark 6.24 In the plane, i.e. for $n = 2$, Mahalanobis sphere and M-sphere are called as before, Mahalanobis circle and M-circle, respectively.

Lemma 6.25 *Let $\Sigma \in \mathbb{R}^{n \times n}$ be a symmetric positive definite matrix with eigenvalues $\lambda_1 \geq \cdots \geq \lambda_n > 0$. Then:*

(i) *the M-sphere $M(C; \Sigma) \subset \mathbb{R}^n$ is an $(n-1)$-dimensional ellipsoid centered at C with main semi-axes $\sqrt{\lambda_1} \geq \cdots \geq \sqrt{\lambda_n} > 0$ in directions of eigenvectors of the matrix Σ;*

(ii) *the normalized M-sphere $M_N(C; \Sigma) \subset \mathbb{R}^n$ is an $(n-1)$-dimensional ellipsoid centered at C with main semi-axes $\dfrac{\sqrt{\lambda_1}}{\sqrt[2n]{\det \Sigma}} \geq \cdots \geq \dfrac{\sqrt{\lambda_n}}{\sqrt[2n]{\det \Sigma}} > 0$ in directions of eigenvectors of the matrix Σ. The normalized factor $\sqrt[n]{\det \Sigma}$ ensures the constant volume of the ellipsoid: $\displaystyle\prod_{i=1}^{n} \dfrac{\sqrt{\lambda_i}}{\sqrt[2n]{\det \Sigma}} \pi = \pi$.*

Proof The proof of (i) follows from the previous discussion (see also [111]).

To prove (ii), note that equation of the M_N-sphere can be written as

$$(x - C)\Big(\frac{1}{\sqrt[n]{\det \Sigma}} \Sigma\Big)^{-1}(x - C)^T = 1,$$

from which, using (i) for eigenvalues $\dfrac{\lambda_i}{\sqrt[n]{\det \Sigma}}$, we get (ii).   □

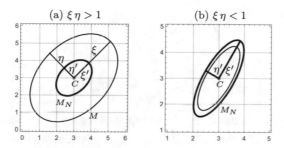

Fig. 6.6 M-circle and the normalized M_N-circle

Example 6.26 Let

$$\Sigma = \begin{bmatrix} \frac{13}{2} & \frac{5}{2} \\ \frac{5}{2} & \frac{13}{2} \end{bmatrix} = Q \begin{bmatrix} 3^2 & 0 \\ 0 & 2^2 \end{bmatrix} Q^T, \quad \text{where} \quad Q = \begin{bmatrix} \frac{\sqrt{2}}{2} & -\frac{\sqrt{2}}{2} \\ \frac{\sqrt{2}}{2} & \frac{\sqrt{2}}{2} \end{bmatrix}$$

be the eigenvalue decomposition of the positive definite symmetric matrix Σ.

The M-circle $M(C, \Sigma)$ is an ellipse centered at $C = 3$ with semi-axes $\xi = 3$ and $\eta = 2$ in directions of eigenvectors, i.e. columns of the orthogonal matrix Q, (the bigger ellipse in Fig. 6.6a). Since

$$\frac{1}{(\det \Sigma)^{1/2}} \Sigma = Q \begin{bmatrix} \big(\frac{3}{(\det \Sigma)^{1/4}}\big)^2 & 0 \\ 0 & \big(\frac{2}{(\det \Sigma)^{1/4}}\big)^2 \end{bmatrix} Q^T,$$

and $\sqrt[4]{\det \Sigma} = \sqrt{6}$, the normalized M-circle $M_N(C, \Sigma)$ in an ellipse centered at $C = (3, 3)$ with semi-axes $\xi' = \frac{3}{\sqrt{6}}$ and $\eta' = \frac{2}{\sqrt{6}}$ in directions of the eigenvectors, i.e. the columns of the orthogonal matrix Q, (the smaller ellipse in Fig. 6.6a). Notice that $\xi' \eta' = 1$ and that the area of M-circle $M(C, \Sigma)$ equals $\xi \eta \pi = 6\pi$, and that the area of the normalized M-circle $M_N(C, \Sigma)$ equals $\xi' \eta' \pi = \pi$.

Example 6.27 The matrix

$$\Sigma_2 = Q \begin{bmatrix} 1.4^2 & 0 \\ 0 & 0.5^2 \end{bmatrix} Q^T \approx \begin{bmatrix} 0.68 & 0.74 \\ 0.74 & 1.53 \end{bmatrix}, \quad \text{where} \quad Q = \begin{bmatrix} 1/2 & -\sqrt{3}/2 \\ \sqrt{3}/2 & 1/2 \end{bmatrix}$$

defines the M-circle $M(C, \Sigma_2)$ centered at $C = (3, 3)$, with semi-axes $\xi = 1.4$ and $\eta = 0.5$ in directions of eigenvectors of Σ_2 (the smaller ellipse in Fig. 6.6b).

The corresponding normalized M-circle $M_N(C, \Sigma_2)$ is the ellipse centered at C, with semi-axes $\xi' = \frac{1.4}{r}$ and $\eta' = \frac{0.5}{r}$, where $r = \sqrt[4]{\det \Sigma_2} \approx 0.84$ (the bigger ellipse in Fig. 6.6b). Again, $\xi'\eta' = 1$.

Using the normalized Mahalanobis distance-like function, the objective function can be defined in the following two ways:

$$\mathcal{F}_M(\Pi) = \sum_{j=1}^{k} \sum_{a \in \pi_j} d_M^{(j)}(c_j, a; \Sigma_j), \tag{6.19}$$

or

$$F_M(c_1, \ldots, c_k) = \sum_{i=1}^{m} \min_{1 \le j \le k} d_M^{(j)}(c_j, a_i; \Sigma_j). \tag{6.20}$$

If the covariance matrix $S_j \approx I$ (all eigenvalues are ≈ 1), then the cluster π_j has an almost spherical shape. If $\lambda_{\max}^{(j)} \gg \lambda_{\min}^{(j)}$, the cluster π_j has a very elongated elliptical shape in direction of the eigenvector with the largest eigenvalue. Numerical problems can occur if $\text{cond}\,(S_j) = \lambda_{\max}^{(j)}/\lambda_{\min}^{(j)}$ is very large (say 10^{20}). How to act in such cases is discussed in [7].

For searching for optimal k-partitions with ellipsoidal clusters we are going to present the Mahalanobis k-means algorithm and the Mahalanobis incremental algorithm (see [116]).

6.4.1   Mahalanobis k-Means Algorithm

The Mahalanobis k-means algorithm is initialized in Step 0 after which steps A and B are successively repeated.

Algorithm 6.28 (Mahalanobis k-means algorithm)

Step 0:   Given a finite set $\mathcal{A} \subset \mathbb{R}^n$ and choosing k distinct points $z_1, \ldots, z_k$ in $\mathbb{R}^n$, for every $j \in J := \{1, \ldots, k\}$ determine the following:
clusters: $\pi_j = \{a_i \in \mathcal{A} : \|z_j - a_i\| \le \|z_s - a_i\|, \forall s \in J\}$;   [by using the minimal distance principle];
centroids: $c_j := \text{mean}(\pi_j)$;
covariance matrices: $\Sigma_j := \frac{1}{|\pi_j|} \sum_{a \in \pi_j} (c_j - a)^T (c_j - a)$;

M-distance like functions: $d_M^{(j)}(x, y; \Sigma_j) := \sqrt[n]{\det \Sigma_j} (x-y) \Sigma_j^{-1} (x-y)^T$.

Step A: (Assignment step). For centroids $c_j \in \mathbb{R}^n$, covariance matrices Σ_j and M-distance like functions $d_M^{(j)}$, using the minimal distance principle, determine new clusters

$$\pi_j = \{a_i \in \mathcal{A} : d_M^{(j)}(c_j, a_i; \Sigma_j) \le d_M^{(s)}(c_s, a_i; \Sigma_s), \ \forall s \in J\}, \ j \in J;$$

Step B: (Update step). For the partition $\Pi = \{\pi_1, \ldots, \pi_k\}$ determine the following:

centroids: $c_j := \mathrm{mean}(\pi_j)$;

covariance matrices: $\Sigma_j := \frac{1}{|\pi_j|} \sum_{a \in \pi_j} (c_j - a)^T (c_j - a)$;

M-distance like functions: $d_M^{(j)}(x, y; \Sigma_j) := \sqrt[n]{\det \Sigma_j}\,(x - y)\Sigma_j^{-1}(x - y)^T$.

Notice that in Step A, clusters $\pi_1, \ldots, \pi_k$ are formed in such a way that an element $a_i \in \mathcal{A}$ is added to that cluster π_j for which the internal distance $d_M^{(j)}(c_j, a_i; \Sigma_j)$ is the smallest.

Since the sequence of function values obtained in each iteration is monotonically decreasing (see [176]), the stopping criterion for Algorithm 6.28 can be the same as in (4.11). Also, as in Theorem 4.8, one can show that Algorithm 6.28 finds a LOPart.

Example 6.29 The following synthetic data set is constructed similarly as in [184]. Take two points $C_1 = (3, 2)$ and $C_2 = (8, 6)$ in $\Delta = [0, 10]^2$. In neighborhood of each point C_j generate 100 random points from the bivariate normal distribution with expectation $C_j \in \mathbb{R}^2$ and the covariance matrices $\Sigma_1 = \frac{1}{2}\begin{bmatrix} 2 & -1 \\ -1 & 0.9 \end{bmatrix}$ and $\Sigma_2 = \frac{1}{3}\begin{bmatrix} 2 & 1 \\ 1 & 1 \end{bmatrix}$, respectively. Next, choose three straight line segments $\ell_1 = [(1, 9), (6, 9)]$, $\ell_2 = [(3, 5), (5, 9)]$, and $\ell_3 = [(3, 6), (7, 2)]$, and on each one generate 100 uniformly distributed points and then to each point add a random error from the bivariate normal distribution with expectation $0 \in \mathbb{R}^2$ and the covariance matrix $\sigma^2 I$, where $\sigma^2 = 0.1$ (see Fig. 6.7a). This defines 5 clusters $\pi_1, \ldots, \pi_5$ and the set $\mathcal{A} = \bigcup \pi_j \subset \Delta$ with $m = 500$ points, and the partition $\Pi = \{\pi_1, \ldots, \pi_5\}$ of the set $\mathcal{A}$.

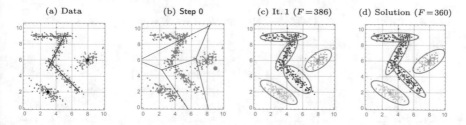

Fig. 6.7 Mahalanobis k-means algorithm

Using the Mahalanobis k-means Algorithm 6.28 we will search for an optimal 5-partition. First, according to Step 0, choose 5 initial centers $z_0 = \{(2, 2), (9, 5), (3, 9), (4, 7), (5, 4)\}$ (orange points in Fig. 6.7b) and by the minimal distance principle determine initial clusters $\pi_1^{(0)}, \ldots, \pi_5^{(0)}$ (Voronoi diagram in

Fig. 6.7b), corresponding covariance matrices, and the M-distance like functions. The result of the first run of Algorithm 6.28 is shown in Fig. 6.7c, and the M-optimal 5-partition $\Pi^\star$ obtained after four iterations, in Fig. 6.7d. Depicted ellipses encompass 95% of points in each cluster.

The Rand index 0.989, Jaccard index 0.801, and the confusion matrix

$$
S(\Pi^0, \Pi^\star) = \begin{bmatrix} 100 & 0 & 0 & 0 & 0 \\ 0 & 100 & 0 & 0 & 0 \\ 0 & 0 & 97 & 3 & 0 \\ 0 & 0 & 12 & 87 & 1 \\ 0 & 1 & 0 & 13 & 86 \end{bmatrix}
$$

defined in Sect. 5.2.1 show a high degree of agreement between the original and the obtained partition, owing to the fact that the confusion matrix is almost diagonal.

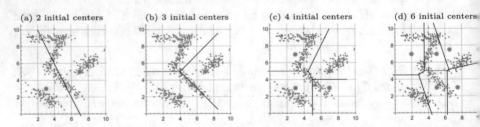

Fig. 6.8 Various choices of initial centers

Example 6.30 We could have started the Mahalanobis k-means Algorithm 6.28 with less or with more initial centers. Figure 6.8 shows Voronoi diagrams for the set $\mathcal{A}$ in Example 6.29 for two, three, four, and six initial centers, and Fig. 6.9 shows the corresponding Mahalanobis k-LOParts. Depicted ellipses encompass 95% of points in each cluster.

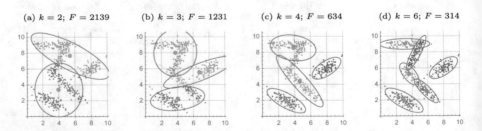

Fig. 6.9 Results of the Mahalanobis k-means algorithm

One of the things that can be enquired is whether any of this Mahalanobis k-LOParts is at the same time the Mahalanobis MAPart.

Remark 6.31 One can show that the Mahalanobis k-means Algorithm 6.28 coincides considerably with the *Generalized Mixture Decomposition Algorithmic Scheme* [184] as a special case of the *Expectation Maximization* algorithm (see [201, p. 31]), but the efficiency of Algorithm 6.28, comparing CPU-times, is considerably better (see Sect. 6.4.3).

Exercise 6.32 Construct the Algorithm 6.28 for data with weights $w_i > 0$.

Remark 6.33 The corresponding *Mathematica*-module MWKMeans[] is described in Sect. 9.2, and the link to appropriate *Mathematica*-code is supplied.

6.4.2 Mahalanobis Incremental Algorithm

The next algorithm to search for Mahalanobis k-LOPart which we are going to present is the *Mahalanobis incremental algorithm*. Let $\mathcal{A} \subset \mathbb{R}^n$ be a data set. The incremental algorithm starts by choosing the initial center $\hat{c}_1 \in \mathbb{R}^n$, for instance, the centroid of $\mathcal{A}$. The next center $\hat{c}_2$ is found by solving the GOP for the function $\Phi \colon \mathbb{R}^n \to \mathbb{R}$:

$$c_2 \in \underset{x \in \mathbb{R}^n}{\arg\min}\ \Phi(x), \quad \Phi(x) := \sum_{i=1}^{m} \min\{\|c_1 - a_i\|^2, \|x - a_i\|^2\}.$$

Applying the Mahalanobis k-means Algorithm 6.28 to centers $\hat{c}_1, \hat{c}_2$, we obtain centers $c_1^\star, c_2^\star$ of 2-LOPart $\Pi^{(2)}$.

In general, knowing the k centers $\hat{c}_1, \ldots, \hat{c}_k$, the next one, $\hat{c}_{k+1}$, will be the solution to the GOP for the function $\Phi \colon \mathbb{R}^n \to \mathbb{R}$:

$$\hat{c}_{k+1} \in \underset{x \in \mathbb{R}^n}{\arg\min}\ \Phi(x), \quad \Phi(x) := \sum_{i=1}^{m} \min\{\delta_k^i, \|x - a_i\|^2\}, \tag{6.21}$$

where $\delta_k^i = \min\limits_{1 \le s \le k} \|\hat{c}_s - a_i\|^2$. Here one can apply just a few, say 10 iterations of the algorithm DIRECT.

After that, the Mahalanobis k-means Algorithm 6.28 will find centers $c_1^\star, \ldots, c_{k+1}^\star$ of the $(k+1)$-LOPart $\Pi^{(k+1)}$.

Remark 6.34 Notice that we could have started the algorithm in a similar way with several initial centers.

Example 6.35 Let us start the Mahalanobis incremental algorithm for the data set $\mathcal{A}$ from Example 6.29 starting from the centroid of $\mathcal{A}$.

Figure 6.10 shows the resulting Mahalanobis k-LOPart for $k = 2, 3, 4, 5, 6$ with the corresponding objective function values. Depicted ellipses encompass 95% of points in each cluster.

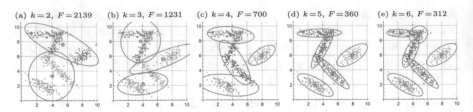

(a) $k=2$, $F=2139$ (b) $k=3$, $F=1231$ (c) $k=4$, $F=700$ (d) $k=5$, $F=360$ (e) $k=6$, $F=312$

Fig. 6.10 Incremental Mahalanobis algorithm

Notice that for each k, neither the obtained partitions nor the function values are the same as those obtained using the Mahalanobis k-means algorithm in Example 6.30. Again, one can enquire whether any of this Mahalanobis k-partitions is at the same time the Mahalanobis `MAPart`.

Remark 6.36 In Sect. 9.2 the corresponding *Mathematica*-module `MInc[]` is described, and the link to appropriate *Mathematica*-code is supplied.

6.4.3 Expectation Maximization Algorithm for Gaussian Mixtures

Now we will give a survey of an algorithm, in the literature known as the *expectation maximization algorithm for Gaussian mixtures* [10, 23]. It turns out that a special case of this algorithm, which we will call here *expectation maximization algorithm for normalized Gaussian mixtures*, is closely related to the Mahalanobis k-means algorithm.

Let A be an absolutely continuous random vector in $\mathbb{R}^n$ whose distribution depends on the parameter vector $\Theta \in \mathcal{P}$, where $\mathcal{P}$ is the parameter space, and the corresponding density function is $a \mapsto f_A(a; \Theta)$ (see e.g. [10, 183]).

Assume that the elements of the data set $\mathcal{A} = \{a_i : i = 1 \ldots, m\} \subset \mathbb{R}^n$ represent independent realizations $(a_1, \ldots, a_m)$ of a random sample $(A_1, \ldots, A_m)$ from a distribution of A. Using the *maximum likelihood principle* (see e.g. [183]), we want to estimate, based on the set $\mathcal{A}$, the parameter vector $\Theta \in \mathcal{P}$. To do so, we define the criterion function $\Theta \mapsto L(\Theta) := \prod_{i=1}^m f_A(a_i; \Theta)$, and call it the *likelihood function*. The parameter $\hat{\Theta} \in \mathcal{P}$ such that

$$\hat{\Theta} = \underset{\Theta \in \mathcal{P}}{\arg\max}\, L(\Theta) \tag{6.22}$$

is called the *maximum likelihood estimate* (MLE). Intuitively, MLE is the parameter vector for which the data set $\mathcal{A}$ is the "most probable" realization of the random vector A.

Since the function $t \mapsto \ln t$ is strictly increasing, to solve the problem (6.22), instead of the function $\Theta \mapsto L(\Theta)$, it is worthy to consider the function $\Theta \mapsto \ell(\Theta) := \ln L(\Theta) = \sum_{i=1}^{m} \ln f_A(a_i; \Theta)$, called the *log-likelihood function*, so the MED can be obtained as the solution to the following problem:

$$\hat{\Theta} = \underset{\Theta \in \mathcal{P}}{\arg\max} \, \ell(\Theta). \tag{6.23}$$

As illustrated by the following example, in certain cases the MED can be calculated explicitly.

Example 6.37 Let $A \sim \mathcal{N}(c, \Sigma)$ be a multivariate normal random (or multivariate Gaussian) vector in $\mathbb{R}^n$ with expectation $c \in \mathbb{R}^n$ and covariance matrix $\Sigma \in \mathbb{R}^{n \times n}$ for which we assume to be symmetric and positive definite. The corresponding density function is of the following form:

$$f_A(a; \Theta) = \frac{1}{(2\pi)^{\frac{n}{2}} \sqrt{\det \Sigma}} e^{-\frac{1}{2}(a-c)\Sigma^{-1}(a-c)^T}, \quad \Theta = (c, \Sigma^{-1}).$$

Based on given realizations $(a_1, \ldots, a_m)$ from a distribution of A, one has to estimate the MED $\hat{\Theta} = (\hat{c}, \hat{\Sigma}^{-1})$.

The corresponding log-likelihood function has the following form:

$$\ell(\Theta) = \sum_{i=1}^{m} \ln \left(\frac{1}{(2\pi)^{\frac{n}{2}} \sqrt{\det \Sigma}} e^{-\frac{1}{2}(a_i-c)\Sigma^{-1}(a_i-c)^T} \right)$$

$$= -\frac{mn}{2} \ln(2\pi) - \frac{m}{2} \ln(\det \Sigma) - \frac{1}{2} \sum_{i=1}^{m} (a_i - c)\Sigma^{-1}(a_i - c)^T.$$

In order to search for the MED $\hat{\Theta} = (\hat{c}, \hat{\Sigma}^{-1})$ we have to solve the global optimization problem (6.23). Since the objective function $\Theta \mapsto \ell(\Theta)$ is differentiable, we will use standard methods from calculus, i.e. we will determine the MED by solving the following system of equations:

$$\nabla \ell(\Theta) = \left(\frac{\partial \ell(\Theta)}{\partial c}, \frac{\partial \ell(\Theta)}{\partial \Sigma^{-1}} \right) = 0.$$

Since Σ^{-1} is a positive definite matrix, from

$$\frac{\partial \ell(\Theta)}{\partial c} = \sum_{i=1}^{m} \Sigma^{-1}(c - a_i)^T = 0$$

we obtain $\hat{c} = \frac{1}{m} \sum_{i=1}^{m} a_i$.

To estimate the covariance matrix $\hat{\Sigma}$ we will need a few auxiliary identities (see e.g. [39]):

(i) Let M_1, M_2, and M_3 be matrices such that the products $M_1 M_2 M_3$, $M_2 M_1 M_3$, and $M_2 M_3 M_1$ are defined. Then

$$\text{tr}(M_1 M_2 M_3) = \text{tr}(M_2 M_1 M_3) = \text{tr}(M_2 M_3 M_1).$$

(ii) If $x \in \mathbb{R}^n$ and $M \in \mathbb{R}^{n \times n}$, then $x M x^T = \text{tr}(x M x^T) = \text{tr}(x\, x^T M)$.

(iii) If M_1 and M_2 are matrices such that the product $M_1 M_2$ is defined, then $\frac{\partial}{\partial M_1} \text{tr}(M_1 M_2) = M_2^T$.

(iv) If M is a regular matrix, then $\frac{\partial}{\partial M} \ln \det(M) = (M^{-1})^T$.

(v) If M is a regular matrix, then $\frac{\partial}{\partial M} \det(M) = \det(M)\, (M^{-1})^T$.

Using the facts (i)–(iv), the function $\Theta \mapsto \ell(\Theta)$ can be written as

$$\ell(\Theta) = -\tfrac{mn}{2} \ln(2\pi) + \tfrac{m}{2} \ln(\det \Sigma^{-1}) - \tfrac{1}{2} \sum_{i=1}^{m} \text{tr}\left((a_i - c)(a_i - c)^T \Sigma^{-1}\right)$$

and hence

$$\frac{\partial \ell(\Theta)}{\partial \Sigma^{-1}} = \frac{m}{2} \Sigma - \frac{1}{2} \sum_{i=1}^{m} (a_i - c)^T (a_i - c).$$

Finally, from $\dfrac{\partial \ell(\Theta)}{\partial \Sigma^{-1}} = 0$, we obtain

$$\hat{\Sigma} = \frac{1}{m} \sum_{i=1}^{m} (a_i - \hat{c})^T (a_i - \hat{c}).$$

This shows that $\hat{\Theta} = (\hat{c}, \hat{\Sigma}^{-1})$ is the unique stationary point of the function $\Theta \mapsto \ell(\Theta)$. Besides, it is not difficult to show that $-\nabla^2 \ell(\hat{\Theta})$ is a positive definite matrix, meaning that $\hat{\Theta} = (\hat{c}, \hat{\Sigma}^{-1})$ is indeed the MED.

Next we will consider random vectors $B_j \sim \mathcal{N}(c_j, \Sigma_j)$, $j = 1, \ldots, k$. Define the new random vector A whose density function is a convex combination of density functions for the random vectors B_j, $j = 1, \ldots, k$

$$f_A(a; \Theta) = \sum_{j=1}^{k} \frac{p_j}{(2\pi)^{\frac{n}{2}} \sqrt{\det \Sigma_j}} e^{-\frac{1}{2}(a - c_j)\Sigma_j^{-1}(a - c_j)^T}, \quad \Theta = (p_1, c_1, \Sigma_1^{-1}, \ldots, p_k, c_k, \Sigma_k^{-1}),$$

where $\sum_{j=1}^{k} p_j = 1$. The random vector A is called a *Gaussian mixture*.

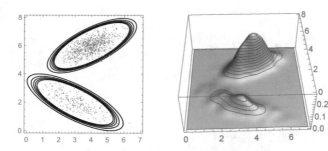

Fig. 6.11 Data set $\mathcal{A}$ and `ContourPlot` of probability density function for B_1 and B_2

Example 6.38 Figure 6.11 shows a data set $\mathcal{A}$, $|\mathcal{A}| = 1000$, generated from Gaussian mixture of two Gaussian vectors $B_1 \sim \mathcal{N}\left((3,2), \frac{1}{2}\begin{bmatrix} 2 & -1 \\ -1 & 0.9 \end{bmatrix}\right)$ and $B_1 \sim \mathcal{N}\left((4,6), \frac{1}{3}\begin{bmatrix} 2 & 1 \\ 1 & 1 \end{bmatrix}\right)$, with $p_1 = \frac{1}{4}$ and $p_2 = \frac{3}{4}$, and the `ContourPlot` of probability density function of B_1 and B_2.

If $(a_1, \ldots, a_m)$ are realizations from distribution of a Gaussian mixture Λ, then the MED $\hat{\Theta} = (\hat{p}_1, \hat{c}_1, \hat{\Sigma}_1^{-1}, \ldots, \hat{p}_k, \hat{c}_k, \hat{\Sigma}_k^{-1})$ is obtained by solving the following problem:

$$\hat{\Theta} = \underset{\Theta \in \mathcal{P}}{\arg\max}\, \ell(\Theta), \tag{6.24}$$

where

$$\ell(\Theta) = \sum_{i=1}^{m} \ln\left(\sum_{j=1}^{k} \frac{p_j}{(2\pi)^{\frac{n}{2}} \sqrt{\det \Sigma_j}} e^{-\frac{1}{2}(a_i - c_j)\Sigma_j^{-1}(a_i - c_j)^T} \right),$$

and $\mathcal{P} = \{(p_1, c_1, \Sigma_1^{-1}, \ldots, p_k, c_k, \Sigma_k^{-1}) : c_j \in \mathbb{R}^n, \Sigma_j \in \mathbb{R}^{n \times n}, p_j > 0, \sum_{i=1}^{k} p_j = 1\}$, where Σ_j are symmetric positive definite matrices.

Since this is an optimization problem with constrains, we define the Lagrange function

$$\Phi(\Theta, \lambda) = \ell(\Theta) + \lambda\left(\sum_{j=1}^{m} p_j - 1 \right), \ \lambda \in \mathbb{R}.$$

Contrary to the case in Example 6.37 where $k = 1$, this problem cannot be solved in a closed form. Instead, to estimate the parameter values, we will construct a numerical iterative method. From

$$\frac{\partial \Phi(\Theta, \lambda)}{\partial c_j} = \sum_{i=1}^{m} \gamma_{ij}(a_i - c_j) = 0, \ j = 1, \ldots, k,$$

where

$$\gamma_{ij} = \frac{\frac{p_j}{\sqrt{\det \Sigma_j}} e^{-\frac{1}{2}(a_i - c_j)\Sigma_j^{-1}(a_i - c_j)^T}}{\sum_{l=1}^{k} \frac{p_l}{\sqrt{\det \Sigma_l}} e^{-\frac{1}{2}(a_i - c_l)\Sigma_l^{-1}(a_i - c_l)^T}}, \quad i = 1, \ldots, m, \ j = 1, \ldots, k,$$

we obtain

$$c_j = \frac{\sum_{i=1}^{m} \gamma_{ij} a_i}{\sum_{i=1}^{m} \gamma_{ij}}, \quad j = 1, \ldots, k.$$

Also, from

$$\frac{\partial \Phi(\Theta, \lambda)}{\partial \Sigma_j^{-1}} = \frac{1}{2} \sum_{i=1}^{m} \gamma_{ij} \Sigma_j - \frac{1}{2} \sum_{i=1}^{m} \gamma_{ij} (a_i - c_j)^T (a_i - c_j) = 0, \quad j = 1, \ldots, k,$$

we obtain

$$\Sigma_j = \frac{\sum_{i=1}^{m} \gamma_{ij} (a_i - c_j)^T (a_i - c_j)}{\sum_{i=1}^{m} \gamma_{ij}}, \quad j = 1, \ldots, k.$$

Finally, from

$$\frac{\partial \Phi(\Theta, \lambda)}{\partial p_j} = \sum_{i=1}^{m} \frac{\gamma_{ij}}{p_j} + \lambda = 0, \quad j = 1, \ldots, k,$$

follows

$$p_j = -\frac{1}{\lambda} \sum_{i=1}^{m} \gamma_{ij}, \quad j = 1, \ldots, k,$$

which, together with $\sum_{j=1}^{k} p_j = 1$, gives $\lambda = -\sum_{j=1}^{k} \sum_{i=1}^{m} \gamma_{ij} = -\sum_{i=1}^{m} \sum_{j=1}^{k} \gamma_{ij} = -m$,

i.e. $p_j = \frac{1}{m} \sum_{i=1}^{m} \gamma_{ij}, \ j = 1, \ldots, k.$

Based on the preceding calculation, we can construct an iterative procedure, in literature known as the *Expectation Maximization (EM) algorithm for Gaussian mixtures*. After inputting initial data in Step 0, the following two steps alternate: Step A (*Expectation step*—E step) and Step B (*Maximization step*—M step). One can show (see e.g. [23, 184]) that the algorithm converges to a local maximum of the problem (6.24).

Algorithm 6.39 (EM algorithm)

Step 0: Input the data set $\mathcal{A} \subset \mathbb{R}^n$ and initial parameters (c_j, Σ_j, p_j), $j = 1, \ldots, k, k \geq 1$.

Step A: (E step). Determine

$$
\gamma_{ij} = \frac{\dfrac{p_j}{\sqrt{\det \Sigma_j}} e^{-\frac{1}{2}(a_i - c_j)\Sigma_j^{-1}(a_i - c_j)^T}}{\sum_{l=1}^{k} \dfrac{p_l}{\sqrt{\det \Sigma_l}} e^{-\frac{1}{2}(a_i - c_l)\Sigma_l^{-1}(a_i - c_l)^T}}, \quad i = 1, \ldots, m, \ j = 1, \ldots, k.
$$

Step B: (M step). Determine:

$$
c_j = \frac{\sum_{i=1}^{m} \gamma_{ij} a_i}{\sum_{i=1}^{m} \gamma_{ij}}, \quad j = 1, \ldots, k \quad \text{[centroids]};
$$

$$
\Sigma_j = \frac{\sum_{i=1}^{m} \gamma_{ij}(a_i - c_j)^T(a_i - c_j))}{\sum_{i=1}^{m} \gamma_{ij}}, \quad j = 1, \ldots, k \quad \text{[covariance matrices]};
$$

$$
p_j = \frac{1}{m} \sum_{i=1}^{m} \gamma_{ij}, \quad j = 1, \ldots, k \quad \text{[probabilities]}.
$$

Remark 6.40 In Sect. 9.2 the corresponding *Mathematica*-module EM[] is described, and the link to appropriate *Mathematica*-code is supplied.

6.4.4 Expectation Maximization Algorithm for Normalized Gaussian Mixtures and Mahalanobis k-Means Algorithm

Now we are going to consider multivariate Gaussians vectors $B_j \sim \mathcal{N}(c_j, S_j)$, $j = 1, \ldots, k$, with covariance matrices S_j such that $\det S_j = 1$, $j = 1, \ldots, k$ (thus the particular notation). Define the new random vector A with the density function

$$
f_A(a; \Theta) = \frac{1}{k} \sum_{j=1}^{k} \frac{1}{(2\pi)^{\frac{n}{2}}} e^{-\frac{1}{2}(a - c_j)S_j^{-1}(a - c_j)^T}, \quad \Theta = (c_1, S_1^{-1}, \ldots, c_k, S_k^{-1}).
$$

Such a random vector A will be called *Normalized Gaussian mixture*.

Similarly as in the previous section, based on realizations $(a_1, \ldots, a_m)$ from a distribution of normalized Gaussian mixture A, one has to estimate the MED $\hat{\Theta} = (\hat{c}_1, \hat{S}_1^{-1}, \ldots, \hat{c}_k, \hat{S}_k^{-1})$.

In order to do so, one has to maximize the log-likelihood function

$$
\ell(\Theta) = \sum_{i=1}^{m} \ln \left(\frac{1}{k} \sum_{j=1}^{k} \frac{1}{(2\pi)^{\frac{n}{2}}} e^{-\frac{1}{2}(a_i - c_j)S_j^{-1}(a_i - c_j)^T} \right)
$$

on the set $\mathcal{P} = \{(c_1, S_1, \ldots, c_k, S_k) : c_j \in \mathbb{R}^n, S_j \in \mathbb{R}^{n \times n}, \det S_j = 1\}$, where S_j are symmetric positive definite matrices. Since this is an optimization problem with constrains, we define the Lagrange function

$$\Psi(\Theta, \lambda_1, \ldots, \lambda_k) = \ell(\Theta) + \sum_{j=1}^{k} \lambda_j \left(\sqrt[n]{\det S_j} - 1 \right), \quad \lambda_j \in \mathbb{R}.$$

From

$$\frac{\partial \Psi(\Theta, \lambda_1, \ldots, \lambda_k)}{\partial c_j} = \sum_{i=1}^{m} \gamma_{ij} (a_i - c_j) = 0, \ j = 1, \ldots, k,$$

where

$$\gamma_{ij} = \frac{e^{-\frac{1}{2}(a_i - c_j) S_j^{-1} (a_i - c_j)^T}}{\sum_{l=1}^{k} e^{-\frac{1}{2}(a_i - c_l) S_l^{-1} (a_i - c_l)^T}}, \quad i = 1, \ldots, m, \quad j = 1, \ldots, k,$$

we obtain

$$c_j = \frac{\sum_{i=1}^{m} \gamma_{ij} a_i}{\sum_{i=1}^{m} \gamma_{ij}}, \quad j = 1, \ldots, k.$$

From

$$\frac{\partial \Phi(\Theta, \lambda_1, \ldots, \lambda_k)}{\partial S_j^{-1}} = \frac{\sum_{i=1}^{m} \gamma_{ij}}{2} S_j - \frac{1}{2} \sum_{i=1}^{m} \gamma_{ij} (a_i - c_j)^T (a_i - c_j) + \frac{1}{2n} \lambda_j \sqrt[n]{\det S_j^{-1}} \, S_j$$

follows

$$S_j \left(\sum_{i=1}^{m} \gamma_{ij} + \frac{\lambda_j}{n} \right) = \sum_{i=1}^{m} \gamma_{ij} (a_i - c_j)^T (a_i - c_j), \tag{6.25}$$

and therefore

$$\det\left(S_j \left(\sum_{i=1}^{m} \gamma_{ij} + \frac{\lambda_j}{n} \right) \right) = \det\left(\sum_{i=1}^{m} \gamma_{ij} (a_i - c_j)^T (a_i - c_j) \right),$$

i.e.

$$\left(\sum_{i=1}^{m} \gamma_{ij} + \frac{\lambda_j}{n} \right)^n = \det\left(\sum_{i=1}^{m} \gamma_{ij} (a_i - c_j)^T (a_i - c_j) \right),$$

$$\sum_{i=1}^{m} \gamma_{ij} + \frac{\lambda_j}{n} = \sqrt[n]{\det\left(\sum_{i=1}^{m} \gamma_{ij} (a_i - c_j)^T (a_i - c_j) \right)},$$

$$\lambda_j = n \sqrt[n]{\det\left(\sum_{i=1}^{m} \gamma_{ij} (a_i - c_j)^T (a_i - c_j) \right)} - n \sum_{i=1}^{m} \gamma_{ij}.$$

Together with (6.25), this gives

$$S_j = \frac{\sum_{i=1}^{m} \gamma_{ij}(a_i - c_j)^T (a_i - c_j)}{\sqrt[n]{\det\left(\sum_{i=1}^{m} \gamma_{ij}(a_i - c_j)^T (a_i - c_j)\right)}},$$

i.e.

$$S_j^{-1} = \sqrt[n]{\det\left(\sum_{i=1}^{m} \gamma_{ij}(a_i - c_j)^T (a_i - c_j)\right)} \left(\sum_{i=1}^{m} \gamma_{ij}(a_i - c_j)^T (a_i - c_j)\right)^{-1}$$

$$= \sqrt[n]{\det\left(\frac{\sum_{i=1}^{m} \gamma_{ij}(a_i - c_j)^T (a_i - c_j)}{\sum_{i=1}^{m} \gamma_{ij}}\right)} \left(\frac{\sum_{i=1}^{m} \gamma_{ij}(a_i - c_j)^T (a_i - c_j)}{\sum_{i=1}^{m} \gamma_{ij}}\right)^{-1}$$

$$= \sqrt[n]{\det \Sigma_j} \; \Sigma_j^{-1},$$

where $\Sigma_j = \frac{\sum_{i=1}^{m} \gamma_{ij}(a_i - c_j)^T (a_i - c_j)}{\sum_{i=1}^{m} \gamma_{ij}}$.

The EMN algorithm is initialized in Step 0, and then successively alternates steps A and B.

Algorithm 6.41 (EMN algorithm for normalized Gaussian Mixture)

Step 0: Input the set $\mathcal{A} \subset \mathbb{R}^n$ and initial parameters (c_j, Σ_j), $j = 1, \ldots, k, k \geq 1$.

Step A: (E step). Determine

$$\gamma_{ij} = \frac{e^{-\frac{1}{2}\sqrt[n]{\det(\Sigma_j)}(a_i - c_j)\Sigma_j^{-1}(a_i - c_j)^T}}{\sum_{l=1}^{k} e^{-\frac{1}{2}\sqrt[n]{\det(\Sigma_l)}(a_i - c_l)\Sigma_l^{-1}(a_i - c_l)^T}}, \quad i = 1, \ldots, m, \; j = 1, \ldots, k.$$

Step B: (M step). Determine:

$$c_j = \frac{\sum_{i=1}^{m} \gamma_{ij} a_i}{\sum_{i=1}^{m} \gamma_{ij}}, \quad j = 1, \ldots, k; \quad \text{[centroids]}$$

$$\Sigma_j = \frac{\sum_{i=1}^{m} \gamma_{ij}(a_i - c_j)^T (a_i - c_j)}{\sum_{i=1}^{m} \gamma_{ij}}, \quad j = 1, \ldots, k. \; \text{[covariance matrices (non-}$$
normalized)]

Notice that for a reasonable approximation

$$\gamma_{ij} = \frac{e^{-\frac{1}{2}d_M^{(j)}(c_j, a_i; \Sigma_j)}}{\sum_{l=1}^{k} e^{-\frac{1}{2}d_M^{(l)}(c_l, a_i; \Sigma_l)}} \approx \begin{cases} 1, \; j = \underset{l=1,\ldots,k}{\arg\min} \, d_M^{(l)}(c_l, a_i; \Sigma_l) \\ 0, \; \text{otherwise}, \end{cases}$$

the E step of Algorithm 6.41 coincides with the Assignment step and the M step coincides with the Update step in the Mahalanobis k-means algorithm.

Example 6.42 In Example 6.29 we defined the data set $\mathcal{A}$ shown in Fig. 6.6, to which we applied the Mahalanobis k-means algorithm. It required 0.125 s of CPU-

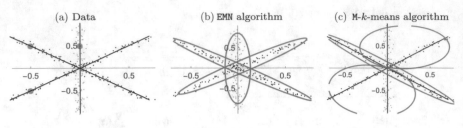

(a) Data   (b) EMN algorithm   (c) M-*k*-means algorithm

Fig. 6.12

time. The EMN algorithm applied to the same problem, finished after 10 iterations
and used 3.45 seconds of CPU-time.

Example 6.43 Take three straight line segments $\ell_1 = [(0, -1), (0, 1)]$, $\ell_2 = [-\frac{\sqrt{2}}{2}(1, 1), \frac{\sqrt{2}}{2}(1, 1)]$, and $\ell_3 = \frac{\sqrt{2}}{2}[(-1, 1), \frac{\sqrt{2}}{2}(1, -1)]$ in the plane, and on
each generate 100 uniformly distributed points. Then to each point add a random
error from bivariate normal distribution with expectation $0 \in \mathbb{R}^2$ and the covariance
matrix $\sigma^2 I$, where $\sigma^2 = 0.001$ (see Fig. 6.12a).

Starting with initial centers $c_1 = (0, 0.5)$, $c_2 = (-0.5, -0.5)$, and $c_3 = (-0.5, 0.5)$ (see also Fig. 6.12a), the EMN algorithm produced after 10 iterations
the solution shown in Fig. 6.12b. The confusion matrix $\begin{bmatrix} 88 & 2 & 10 \\ 3 & 90 & 7 \\ 2 & 0 & 98 \end{bmatrix}$ and the Rand index
0.97 show a high degree of agreement with the original partition. The CPU-time
was 0.97 s.

On the other hand, the Mahalanobis k-means algorithm obviously does not
provide an acceptable solution (see Fig. 6.12c), which is confirmed by the confusion
matrix $\begin{bmatrix} 44 & 38 & 18 \\ 45 & 37 & 18 \\ 0 & 0 & 100 \end{bmatrix}$ and also by the Rand index 0.92. The CPU-time was 0.14 s.

The two previous examples confirm the assertion that the EMN algorithm is less
efficient than the Mahalanobis k-means algorithm, but it does find a better MAPart.

6.5 Choosing a Partition with the Most Appropriate Number of Ellipsoidal Clusters

Similarly as for spherical clusters, see Chap. 5, one can pose the following question:

> Given a data set $\mathcal{A}$, into how many ellipsoidal clusters should one group the elements of $\mathcal{A}$,
> i.e. how to choose a Mahalanobis partition with the most appropriate number of clusters
> (Mahalanobis MAPart)?

For that purpose we are going to define indexes generalizing those in Chap. 5
(see [116]). Let $\Pi = \{\pi_1, \ldots, \pi_k\}$ be a partition of the set $\mathcal{A} \in \mathbb{R}^n$ with
$m = |\mathcal{A}|$ elements. For each cluster π_j let $c_j = \text{mean}(\pi_j)$ be its centroid,

$\Sigma_j = \frac{1}{|\pi_j|} \sum\limits_{a \in \pi_j} (c_j - a)^T (c_j - a)$ the corresponding covariance matrix, and $d_M^{(j)}$ the M-distance like function defined by (6.18).

- Mahalanobis Calinski–Harabasz index is defined as

$$\mathrm{MCH}(k) = \frac{\frac{1}{k-1} \sum\limits_{j=1}^{k} |\pi_j| d_M^{(j)}(c, c_j; \Sigma_j)}{\frac{1}{m-k} \sum\limits_{j=1}^{k} \sum\limits_{a \in \pi_j} d_M^{(j)}(c_j, a; \Sigma_j)}, \qquad c = \mathrm{mean}(\mathcal{A}); \qquad (6.26)$$

- Mahalanobis Davies–Bouldin index is defined as

$$\mathrm{MDB}(k) = \frac{1}{k} \sum\limits_{j=1}^{k} \max_{s \neq j} \frac{V(\pi_j) + V(\pi_s)}{d_M^{(j)}(c_j, c_s; \Sigma_j + \Sigma_s)}, \quad V(\pi_j) = \frac{1}{|\pi_j|} \sum\limits_{a \in \pi_j} d_M^{(j)}(c_j, a; \Sigma_j);$$

$$(6.27)$$

- To define Mahalanobis Simplified Silhouette Width Criterion (MSSWC), first for all $a_i \in \pi_r, i = 1, \ldots, m, r = 1, \ldots, k$, calculate the numbers

$$\alpha_{ir} = d_M^{(r)}(c_r, a_i; S_r), \quad \beta_{ir} = \min_{q \neq r} d_M^{(q)}(c_q, a_i; S_q), \quad s_i = \frac{\beta_{ir} - \alpha_{ir}}{\max\{\alpha_{ir}, \beta_{ir}\}},$$

and then the MSSWC index is defined as the average

$$\mathrm{MSSWC}(k) = \frac{1}{m} \sum\limits_{i=1}^{m} s_i. \qquad (6.28)$$

- Mahalanobis Area index is defined analogously to the Fuzzy Hypervolume index in [67]:

$$\mathrm{MArea}[k] = \sum\limits_{j=1}^{k} \sqrt{\det \Sigma_j}, \qquad (6.29)$$

where $\sqrt{\det \Sigma_j}$ is proportional to the area of the ellipse determined by the covariance matrix Σ_j.

The partition with the highest MCH index, or the lowest MDB index, or the highest MSSWC index, or the lowest MArea index will be called the *Mahalanobis partition with the most acceptable number of clusters* (Mahalanobis MAPart).

Example 6.44 Table 6.1 lists the values of MCH, MDB, MSSWC, and MArea indexes for partitions from Example 6.35, which are based on data from Example 6.29. All four indexes indicate that the 5-partition is the most acceptable one (the Mahalanobis MAPart).

Table 6.1 Mahalanobis indexes for partitions in Example 6.35

Index	$\Pi^{(2)}$	$\Pi^{(3)}$	$\Pi^{(4)}$	$\Pi^{(5)}$	$\Pi^{(6)}$
MCH	1166	1117	2266	**4742**	3029
MDB	0.300	0.222	0.189	**0.125**	0.277
MSSWC	0.851	0.865	0.861	**0.879**	0.862
MArea	0.037	0.026	0.016	**0.0065**	0.0068

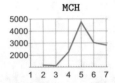 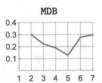 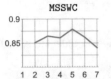 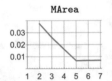

Fig. 6.13 All four Mahalanobis indexes indicate that the 5-partition in Example 6.35 is the Mahalanobis MAPart

Remark 6.45 MCH, MDB, MSSWC, and MArea indexes can be calculated using corresponding *Mathematica*-modules MCH[], MDB[], MSSWC[], and MArea[] which are described in Sect. 9.2, and links to appropriate *Mathematica*-codes are supplied.

Chapter 7
Fuzzy Clustering Problem

In previous chapters, we considered hard clustering methods. In hard clustering approach, each data point either belongs completely to some cluster or not.

In cases when one expects that some elements of the set A could belong to two or more clusters, one should apply *fuzzy clustering*. In the literature, fuzzy clustering is considered to be a sort of *soft clustering*. Fuzzy clustering is used in analyzing images and signals, in medical diagnostics, tomography, analysis of gene expression data, astronomy, market segmentation, credit scoring, speech recognition, in ranking projects, traffic volume data estimation, environmental science, etc. (see e.g. [21, 29, 34, 75, 77, 99, 168, 181, 184, 202]). As we are going to see in Sect. 8.4, such a problem occurs also when defining constituencies in a specified area (like in the capital of Croatia).

Suppose we want to group elements of a data set $A \subset \mathbb{R}^n$ into clusters $\pi_1, \ldots, \pi_k$ which are not defined precisely as in Definition 3.1, but with possibility that some elements $a_i \in A$ end up in several clusters "up to some degree." In such situations the elements u_{ij} of the membership matrix U (see Sect. 4.2.1) have to be numbers between 0 and 1, i.e. $u_{ij} \in [0, 1]$. According to [20, 21, 184], the degree to which we consider that the element a_i belongs to the cluster π_j, sometimes called the *degree of belongingness*, is determined by $u_{ij}^q(c)$, $q > 1$. The function $c \mapsto u_{ij}^q(c)$ is called the *membership function* and q is called the *fuzzifier* (note that q is the exponent, not an upper index). From now on, for the membership function value $u_{ij}^q(c)$ we will simply write u_{ij}^q.

Knowing the distance-like function $d \colon \mathbb{R}^n \times \mathbb{R}^n \to \mathbb{R}_+$, the objective function (4.8) becomes (see e.g. [21, 81, 168, 178, 184])

$$\Phi(c, U) = \sum_{i=1}^{m} \sum_{j=1}^{k} u_{ij}^q \, d(c_j, a_i), \qquad (7.1)$$

with conditions

© The Author(s), under exclusive license to Springer Nature Switzerland AG 2021
R. Scitovski et al., *Cluster Analysis and Applications*,
https://doi.org/10.1007/978-3-030-74552-3_7

$$\sum_{j=1}^{k} u_{ij}(c) = 1, \quad i = 1, \ldots, m, \tag{7.2}$$

$$\sum_{i=1}^{m} u_{ij}(c) > 0, \quad j = 1, \ldots, k. \tag{7.3}$$

Note that c in (7.1) is the concatenation of points $c_1, \ldots, c_k$, i.e. $c = (c_1^1, \ldots, c_1^n, c_2^1, \ldots, c_2^n, \ldots, c_k^1 \ldots, c_k^n) \in \mathbb{R}^{kn}$, and $U \in [0, 1]^{m \times k}$ is the matrix whose elements are $u_{ij}(c)$, i.e. values of the membership functions u_{ij}.

7.1   Determining Membership Functions and Centers

Knowing the distance-like function $d \colon \mathbb{R}^n \times \mathbb{R}^n \to \mathbb{R}_+$, and given a data set $\mathcal{A} \subset \mathbb{R}^n$, $|\mathcal{A}| = m$, the objective function Φ is defined by the membership functions $u_{ij} \colon \mathbb{R}^n \to \mathbb{R}_+$ and centers $c_1, \ldots, c_k$.

7.1.1   Membership Functions

We are going to determine the membership functions u_{ij} in such a way that the function Φ, meeting the conditions (7.2), attains its minimum. Thus, we define the Lagrange function [63, 184]

$$\mathcal{J}(c, U, \lambda) := \sum_{i=1}^{m} \sum_{j=1}^{k} u_{ij}^q d(c_j, a_i) - \sum_{i=1}^{m} \lambda_i \left(\sum_{j=1}^{k} u_{ij} - 1 \right). \tag{7.4}$$

Assuming $a_r \neq c_s$, the partial derivative

$$\frac{\partial \mathcal{J}(c, U, \lambda)}{\partial u_{rs}} = q \, u_{rs}^{q-1} \, d(c_s, a_r) - \lambda_r \tag{7.5}$$

equals 0 for

$$u_{rs}(c) = \left(\frac{\lambda_r}{q \, d(c_s, a_r)} \right)^{\frac{1}{q-1}}. \tag{7.6}$$

Substituting (7.6) into (7.2) we obtain

$$\sum_{j=1}^{k} \left(\frac{\lambda_r}{q \, d(c_j, a_r)} \right)^{\frac{1}{q-1}} = 1,$$

and therefore

$$\lambda_r = \frac{q}{\left(\sum_{j=1}^{k} \left(\frac{1}{d(c_j,a_r)} \right)^{1/(q-1)} \right)^{q-1}}. \tag{7.7}$$

Finally, substituting (7.7) into (7.6), we obtain

$$u_{rs}(c) = \frac{1}{\sum_{j=1}^{k} \left(\frac{d(c_s,a_r)}{d(c_j,a_r)} \right)^{1/(q-1)}}, \qquad a_r \neq c_s. \tag{7.8}$$

Remark 7.1 Notice that it can happen that some datum $a_i \in \mathcal{A}$ coincides with some center c_j. If this happens, the function (7.8) is not defined. In [80] is therefore suggested that the membership functions u_{ij} should be defined as

$$u_{ij}(c) = \begin{cases} \dfrac{1}{\sum_{s=1}^{k} \left(\frac{d(c_j,a_i)}{d(c_s,a_i)} \right)^{1/(q-1)}} & \text{if } I_i = \emptyset \\[2ex] \dfrac{1}{|I_i|}, & \text{if } I_i \neq \emptyset \text{ and } j \in I_i \\[1ex] 0, & \text{if } I_i \neq \emptyset \text{ and } j \notin I_i \end{cases}, \tag{7.9}$$

where $I_i = \{s : c_s = a_i\} \subseteq \{1, \ldots, k\}$. Note that this automatically ensures the condition (7.3).

7.1.2 Centers

Similarly as in the previous section, we are going to find centers $c_1, \ldots, c_k$ by requiring that the function Φ, assuming conditions (7.2), attains its minimum. To find the center c_j, equalize the partial derivative of the Lagrange function to 0

$$\frac{\partial \mathcal{J}(c, U, \lambda)}{\partial c_j} = \sum_{i=1}^{m} u_{ij}^{q} \frac{\partial d(c_j, a_i)}{\partial c_j} = 0, \tag{7.10}$$

and substitute (7.8). This will give an equation whose solution is c_j. In general, this way of finding the centers $c_1, \ldots, c_k$ boils down to solving a system of nonlinear equations.

In particular, for the LS distance-like function $d(c_j, a_i) := \|c_j - a_i\|^2$, Eq. (7.10) becomes

$$2 \sum_{i=1}^{m} u_{ij}^{q} (c_j - a_i) = 0,$$

from which, using (7.8), we obtain

$$c_j = \Big(\sum_{i=1}^{m} u_{ij}^{\,q} \Big)^{-1} \sum_{i=1}^{m} u_{ij}^{\,q} \, a_i, \quad j = 1, \dots, k. \tag{7.11}$$

Similarly, for the ℓ_1-metric function $d(x, a_i) := \|x - a_i\|_1$, centers $c_1, \dots, c_k$ are the weighted medians [72, 144, 149, 187] of data $a_1, \dots, a_m \in \mathbb{R}^n$ with weights $u_{ij}^{\,q}, i = 1, \dots, m$:

$$c_j = \operatorname*{med}_{i=1,\dots,m} (u_{ij}^{\,q}, a_i), \quad j = 1, \dots, k. \tag{7.12}$$

Exercise 7.2 Show, using Exercise 2.21, that the Itakura–Saito centers c_j for the data set $\mathcal{A} = \{a_i \in \mathbb{R} : i = 1, \dots, m\}$ are given by

$$c_j = \sum_{i=1}^{m} u_{ij}^{\,q} \Big(\sum_{i=1}^{m} \frac{u_{ij}^{\,q}}{a_i} \Big)^{-1}, \quad j = 1, \dots, k.$$

What would be the generalization in case $\mathcal{A} \subset \mathbb{R}^n$?

Exercise 7.3 Show, using Exercise 2.22, that the Kullback–Leibler centers c_j for the data set $\mathcal{A} = \{a_i \in \mathbb{R} : i = 1, \dots, m\}$ are given by

$$c_j = \Big(\prod_{i=1}^{m} a_i^{u_{ij}^{\,q}} \Big)^{1 / \sum_{i=1}^{m} u_{ij}^{\,q}}, \quad j = 1, \dots, k.$$

What would be the generalization in case $\mathcal{A} \subset \mathbb{R}^n$?

Remark 7.4 Let $D \subseteq \mathbb{R}^n$ be a convex set and let $\phi \colon D \to \mathbb{R}_+$ be a strictly convex twice continuously differentiable function on $\operatorname{int} D \neq \emptyset$, and let $d_\phi \colon D \times \operatorname{int} D \to \mathbb{R}_+$ be the Bregman divergence. Show that, using the distance-like function $D_\phi \colon \operatorname{int} D \times D \to \mathbb{R}_+$ given by $D_\phi(x, y) = d_\phi(y, x)$, the centers of the set $\mathcal{A} = \{a_i \in D : i = 1, \dots, m\}$ are

$$c_j = \frac{1}{\sum_{i=1}^{m} u_{ij}^{\,q}} \sum_{i=1}^{m} u_{ij}^{\,q} \, a_i, \quad j = 1, \dots, k.$$

Remark 7.5 Searching for the optimal value of the fuzzifier q for which the objective function (7.1) attains its global minimum is a very complex process, and there are numerous papers about this problem (see e.g., [152, 196]). In applied research (see e.g. [21, 196]), the most commonly used is $q \in [1.5, 2.5]$. In our examples and applications we will use $q = 2$.

Hereafter we will use only the LS distance-like function, and the membership function will be denoted by u_{ij}.

7.2   Searching for an Optimal Fuzzy Partition with Spherical Clusters

Searching for an optimal fuzzy partition is quite a complex GOP because the objective function (7.1) can have a rather large number of variables, it does not need to be either convex or differentiable, and it usually has several local minima. On the other hand, the objective function (7.1) is a symmetric Lipschitz-continuous function [69, 149] and therefore we could try to apply directly some of the known global optimization methods. However, due to the large number of independent variables, this problem cannot be directly solved, not even by using some specialized methods for symmetric Lipschitz-continuous functions [55, 69, 89, 154].

Therefore, various simplifications are often proposed in the literature which would find only a locally optimal partition for which we usually do not know how close it is to the globally optimal one. An overview of the most popular methods to search for a fuzzy (and hard) optimal partition can be found in [184].

In cases when coordinates of data do not belong to similar intervals, it is advisable to first *normalize* these intervals and transform the data set $\mathcal{A} \subset \mathbb{R}^n$ into $[0, 1]^n$, as it was shown at the beginning of Chap. 4.

7.2.1   Fuzzy c-Means Algorithm

The most popular algorithm used to search for a fuzzy locally optimal partition is the well known *fuzzy c-means algorithm* (see e.g. [21, 63, 168, 184]) using the LS distance-like function and producing clusters of spherical shape. Similarly as with the k-means Algorithm 4.7, this algorithm can be described in two steps which successively alternate.

Algorithm 7.6 (Fuzzy c-means algorithm)

Step A: Given a finite subset $\mathcal{A} \subset \mathbb{R}^n$ and k distinct points $z_1, \ldots, z_k \in \mathbb{R}^n$, determine the membership matrix $U \in [0, 1]^{m \times k}$ according to (7.9).

Step B: Given a membership matrix $U \in [0, 1]^{m \times k}$, determine the corresponding cluster centers $c_1, \ldots, c_k \in \mathbb{R}^n$ according to (7.11), and calculate the objective function value $\Phi(c, U)$ according to (7.1).
Set $z_j := c_j, j = 1, \ldots, k$.

Similarly as with the standard k-means algorithm (see Sect. 4.2.3), or with the Mahalanobis k-means algorithm (see Sect. 6.4.1), this algorithm also produces monotonically decreasing sequence of objective function values. Therefore the literature suggests (see e.g. [10, 21, 184]) that, given an $\epsilon_{fcm} > 0$, the iterative process stops when

$$\frac{\Phi_{j-1} - \Phi_j}{\Phi_j} < \epsilon_{fcm}, \quad \Phi_j = \Phi(c^{(j)}, U^{(j)}), \tag{7.13}$$

or when

$$\|U^{(j)} - U^{(j-1)}\| < \epsilon_{\text{fcm}}. \tag{7.14}$$

This method heavily depends on the initial approximation and thus numerous heuristic approaches can be found in the literature (see e.g. [100]).

Remark 7.7 The corresponding *Mathematica*-module cmeans[] is described in Sect. 9.2, and the link to appropriate *Mathematica*-code is supplied.

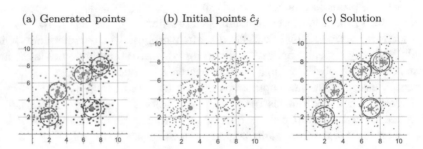

(a) Generated points   (b) Initial points $\hat{c}_j$ (c) Solution

Fig. 7.1 c-means algorithm

Example 7.8 Similarly to [134], choose $k = 5$ points $C_1 = (2, 2)$, $C_2 = (3, 5)$, $C_3 = (6, 7)$, $C_4 = (7, 3)$, and $C_5 = (8, 8)$ in the square $[0, 10]^2 \subset \mathbb{R}^2$. In the neighborhood of each point C_j generate a set π_j with 100 random points from bivariate normal distribution $\mathcal{N}(C_j, \sigma^2 I)$ with expectation $C_j \in \mathbb{R}^2$ and the covariance matrix $\sigma^2 I$, where $\sigma^2 = 1$. Notice that the obtained partition $\Pi = \{\pi_1, \ldots, \pi_k\}$ is very close to the globally optimal partition of the set $\mathcal{A} = \bigcup_{j=1}^{k} \pi_j$, $m = |\mathcal{A}| = 500$. The region of the highest point concentration in cluster π_j is determined by its *main circle* $\mathcal{C}(C_j, \sigma)$ (see Sect. 5.1.2 and the red circles in Fig. 7.1a). The whole data set $\mathcal{A}$ is depicted in Fig. 7.1b. The same figure shows initial centers $\hat{c}_j$, $j = 1, \ldots, k$, at which the c-means algorithm was started. The fuzzy c-means algorithm produced new centers (Fig. 7.1c)

$$c_1^{\star} = (1.76, 2.16), \quad c_2^{\star} = (3.06, 4.8), \quad c_3^{\star} = (6.08, 7.06),$$

$$c_4^{\star} = (7.05, 2.85), \quad c_5^{\star} = (8.31, 7.99).$$

The fuzzy c-means Algorithm 7.6 used 6.39 s of CPU-time.

The cluster with center $c_j^{\star}$ consists of all data from $\mathcal{A}$ with weights $u_{ij}^{\star q}$, $i = 1, \ldots, m$. Let

$$\sigma_j^{\star 2} := \frac{1}{\sum_{i=1}^{m} u_{ij}^{\star q}} \sum_{i=1}^{m} u_{ij}^{\star q} \, \|c_j^{\star} - a_i\|^2, \quad j = 1, \ldots, k. \tag{7.15}$$

Table 7.1 Confusion matrix and its percentage structure

Cluster	$\pi_1^\star$	$\pi_2^\star$	$\pi_3^\star$	$\pi_4^\star$	$\pi_5^\star$	Σ	$\pi_1^\star$	$\pi_2^\star$	$\pi_3^\star$	$\pi_4^\star$	$\pi_5^\star$	Σ
π_1	65.0	16.9	4.6	7.5	2.8	96.8	67.1	17.5	4.8	7.7	2.9	100
π_2	16.5	64.3	11.2	8.6	5.1	105.6	15.6	60.8	10.6	8.1	4.9	100
π_3	4.8	11.4	56.6	9.3	18.6	100.7	4.8	11.3	56.2	9.2	18.4	100
π_4	7.1	9.5	9.7	62.3	7.1	95.6	7.4	9.9	10.2	65.1	7.4	100
π_5	2.9	5.4	23.2	6.8	62.9	101.2	2.9	5.3	22.9	6.7	62.1	100
Σ	96.2	107.4	105.3	94.5	96.5	500	97.7	104.9	104.7	97.0	95.8	500

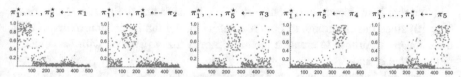

Fig. 7.2 Dispersion of elements of original clusters

The numbers $\sigma_j^{\star 2}$, $j = 1, \ldots, k$, are approximately equal to σ^2, and the circles $\mathcal{C}^\star(c_j^\star, \sigma_j^\star)$, $j = 1, \ldots, k$, (gray circles in Fig. 7.1c) are the main circles which almost coincide with $\mathcal{C}(C_j, \sigma)$, $j = 1, \ldots, k$.

Hausdorff distance between the sets of original and new centers is 0.097.

Table 7.1 shows the corresponding confusion matrix (5.18) and its percentage structure. According to Exercise 5.14, the confusion matrix can be determined easily. The j-th row of Table 7.1 shows how are the elements of the j-th cluster, formed around center C_j, arranged: most of them belong to the new j-th cluster. This is also illustrated by Fig. 7.2.

Exercise 7.9 Using the module `cmeans[]` from Sect. 9.2, apply the c-means algorithm to the data sets from Test-examples 9.2 and 9.3, and analyze the results.

7.2.2   Fuzzy Incremental Clustering Algorithm (FInc)

Assume now that the most appropriate number of clusters is not known in advance. Based on the method developed in [8, 9, 11, 13–15, 116, 166], we will construct an incremental algorithm to search for an optimal partition with $2, 3, \ldots$ clusters. In order to be able to decide which of these partitions is the most appropriate one, we will use the common procedure applying different validity indexes.

First notice that, similarly as in Sect. 3.4, we can define the objective function

$$\Psi(c, U) = \sum_{i=1}^{m} \min_{1 \le j \le k} u_{ij}^{\,q} \, \|c_j - a_i\|^2, \qquad (7.16)$$

which is related to the function Φ in a similar way as are related the functions $\mathcal{F}$ and F given by (3.5) and (3.40), respectively.

Our algorithm will start by choosing the initial center $\hat{c}_1 \in \mathbb{R}^n$. For example, it can be the mean of the data set $\mathcal{A}$. An approximation of the next center $\hat{c}_2$ can be obtained by solving the following GOP:

$$\underset{x \in \mathbb{R}^n}{\arg\min} \, \Psi_2(x), \quad \Psi_2(x) := \sum_{i=1}^{m} \min\{\|\hat{c}_1 - a_i\|^2, \|x - a_i\|^2\}. \tag{7.17}$$

Note that it suffices to carry out just a few iterations of the DIRECT algorithm (say 10) since this is enough to obtain a sufficiently good initial approximation for the c-means algorithm to create a locally optimal fuzzy partition with two clusters with centers $c_1^\star$ and $c_2^\star$.

In general, knowing a fuzzy partition with $r-1$ centers $\hat{c}_1, \ldots, \hat{c}_{r-1}$, the next, the r-th center $\hat{c}_r$, is obtained by solving the following GOP:

$$\hat{c}_r \in \underset{x \in \mathbb{R}^n}{\arg\min} \, \Psi_r(x), \quad \Psi_r(x) := \sum_{i=1}^{m} \min\{\delta_{r-1}^i, \|x - a_i\|^2\}, \tag{7.18}$$

where $\delta_{r-1}^i = \min_{1 \le j \le r-1} u_{is}^q \|\hat{c}_s - a_i\|^2$. Then, with obtained centers $\hat{c}_1, \ldots, \hat{c}_r$, the c-means algorithm gives centers $c_1^\star, \ldots, c_r^\star$ of the locally optimal r-partition. The number K of partitions obtained in this way can be set in advance. Alternatively, using the fact that increasing the number of clusters, the objective function values decrease (Theorem 3.7), generating new partitions can be stopped at the r-th stage if, for a given $\epsilon_{\text{Bagi}} > 0$, the following is achieved [10, 11]:

$$\frac{\Phi_{r-1} - \Phi_r}{\Phi_r} < \epsilon_{\text{Bagi}}, \quad \Phi_j = \Phi(c^{(j)}, U^{(j)}). \tag{7.19}$$

Based on the above considerations, the *fuzzy incremental algorithm* FInc is set up similarly as the hard clustering incremental Algorithm 2.

Remark 7.10 In Sect. 9.2 the corresponding *Mathematica*-module FInc[] is described, and the link to appropriate *Mathematica*-code is supplied.

Example 7.11 For the data set $\mathcal{A}$ in Example 7.8 we perform $K = 5$ iterations of FInc algorithm starting with the center $\hat{c}_1 = \text{mean}(\mathcal{A}) \approx (5.21, 5.00)$. In order to solve the problem (7.17), we use 10 iterations of the DIRECT algorithm to obtain the next center $\hat{c}_2$. Using the c-means algorithm with initial centers $\hat{c}_1$ and $\hat{c}_2$ we obtain optimal centers $c_1^\star$ and $c_2^\star$ (Fig. 7.3a).

In a similar way, solving the problem (7.18) for $r = 3$ using 10 iterations of the DIRECT algorithm, we obtain the third center $\hat{c}_3$ (the red dot in Fig. 7.3a). Then, using the c-means algorithm, we obtain optimal fuzzy 3-partition (Fig. 7.3b). Repeating the procedure we obtain the other fuzzy partitions shown in Fig. 7.3. The used CPU-time was 27.4 s, which was mostly used for the c-means algorithm.

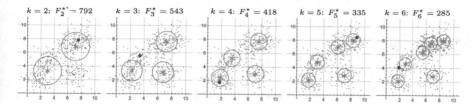

$k = 2: F_2^{\star} = 792 \quad k = 3: F_3^{\star} = 543 \quad k = 4: F_4^{\star} = 418 \quad k = 5: F_5^{\star} = 335 \quad k = 6: F_6^{\star} = 285$

Fig. 7.3 Fuzzy incremental algorithm

Of course, the problem of choosing the most suitable partition, remains.

Remark 7.12 Note that, if needed, the Algorithm 2 (p. 77) can also be run starting from several initial centers.

7.2.3 Choosing the Most Appropriate Number of Clusters

To be able to decide which of the obtained partitions is the most appropriate one, we are going to use the customary procedure to apply various indexes. Some of these indexes arouse by adjusting already known indexes, while some were specially designed to find the most acceptable fuzzy optimal partitions (see [10, 21, 168, 188, 197]).

Let $\Pi^{\star} = \{\pi_1^{\star}, \ldots, \pi_k^{\star}\}$ be a k-optimal fuzzy partition with membership matrix $U^{\star}$ and corresponding cluster centers $c_1^{\star}, \ldots, c_k^{\star}$. We will briefly mention the following indexes for fuzzy spherical clusters:

- **Xie–Beni fuzzy index** is defined as (see e.g. [168, 197])

$$\text{FXB}(k) = \frac{1}{m \min_{i \neq j} \|c_i^{\star} - c_j^{\star}\|^2} \sum_{i=1}^{m} \sum_{j=1}^{k} u_{ij}^{\star q} \|c_j^{\star} - a_i\|^2, \tag{7.20}$$

and the minimal FXB value indicates the most acceptable partition.

- **Calinski–Harabasz fuzzy index** can be defined as (see e.g. [21, 168, 197])

$$\text{FCH}(k) = \frac{\frac{1}{k-1} \sum_{j=1}^{k} \kappa_j^{\star} \|c_A - c_j^{\star}\|^2}{\frac{1}{m-k} \Phi(c^{\star}, U^{\star})}, \quad c_A = \text{mean } \mathcal{A}, \quad \kappa_j^{\star} = \sum_{i=1}^{m} u_{ij}^{\star q}, \tag{7.21}$$

and the largest FCH value indicates the most appropriate partition.

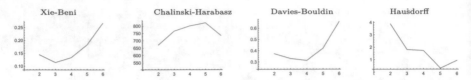

Fig. 7.4 Fuzzy indexes and Hausdorff distances between the sets of original and calculated centers

- **Davies–Bouldin fuzzy index** is defined as (see e.g. [21, 168])

$$\text{FDB}(k) = \frac{1}{k}\sum_{j=1}^{k}\max_{s\neq j}\frac{V(\pi_j^\star) + V(\pi_s^\star)}{\|c_j^\star - c_s^\star\|^2}, \quad V(\pi_j^\star) = \frac{1}{\sum\limits_{i=1}^{m} u_{ij}^{\star q}}\sum_{i=1}^{m} u_{ij}^{\star q}\|c_j^\star - a_i\|^2,$$

$$(7.22)$$

and the smallest FDB value indicates the most appropriate partition.

Example 7.13 In Example 7.11 we have run $K = 5$ iterations of the FInc algorithm on the data set from Example 7.8, and we obtained fuzzy optimal c-partitions for $c = 2, 3, 4, 5, 6$, and for each partition we calculated the values of the aforementioned fuzzy indexes (Fig. 7.4). The FXB index suggests that the most appropriate partition should be the 3-partition, the FCH index shows that the most appropriate partition is the 5-partition, whereas the FDB index suggests the 4-partition. Looking at Fig. 7.3, one can see why this happened. The data are obviously very dispersed around their original centers, and using the above indexes made the real reconstruction difficult.

Notice that the Hausdorff distances between sets of original and calculated centers confirm that the fuzzy 5-partition is the most appropriate one.

Exercise 7.14 Using the FInc[] module from Sect. 9.2, apply the FInc algorithm to the data sets from Test-examples 9.2 and 9.3, and, similarly as in Example 7.13, analyze the results.

Remark 7.15 FXB, FCH, and FDB indexes can be calculated by using corresponding *Mathematica*-modules FXB[], FCH[], and FDB[] which are described in Sect. 9.2, and links to appropriate *Mathematica*-codes are supplied.

7.3 Methods to Search for an Optimal Fuzzy Partition with Ellipsoidal Clusters

We are going to present two most prevailing methods found in the literature for searching for an optimal fuzzy partition with ellipsoidal clusters. Analogously to the fuzzy c-means Algorithm 7.6, we will consider an adjustment of the Mahalanobis k-means algorithm (see Sect. 6.4.1) for the case of fuzzy clustering—the well known

Gustafson–Kessel c-means (GKc-means) algorithm. Furthermore, analogously to the FInc algorithm (see Sect. 7.2.2), we will consider an adjustment of the Mahalanobis incremental algorithm (see Sect. 6.4.2) to the case of fuzzy clustering.

7.3.1   Gustafson–Kessel c-Means Algorithm

The well known GKc-means algorithm was developed to search for a fuzzy optimal partition with clusters of ellipsoidal shape when the number of clusters is set beforehand (see e.g. [21, 61, 63, 73, 95, 102, 173, 184]).

Let $\mathcal{A} = \{a_i \in \mathbb{R}^n : i = 1, \ldots, m\}$ be a data set. If the distinct centers $c_1, \ldots, c_k \in \mathbb{R}^n$ are known, the membership matrix $U \in [0, 1]^{m \times k}$ with properties (7.2) and (7.3) is defined by (7.9), and the corresponding covariance matrices are

$$S_j = \frac{1}{\sum\limits_{i=1}^{m} u_{ij}{}^q} \sum_{i=1}^{m} u_{ij}{}^q (c_j - a_i)(c_j - a_i)^T, \quad j = 1, \ldots, k. \tag{7.23}$$

Conversely, if the membership matrix $U \in [0, 1]^{m \times k}$ with properties (7.2) and (7.3) is known, then, using (7.11), one can define centers $c_1, \ldots, c_k \in \mathbb{R}^n$.

The cluster π_j is determined by the j-th column of the membership matrix U such that $\{(a_i, u_{ij}) : a_i \in \mathcal{A}, i = 1, \ldots, m\}$ is a weighted set of data of length $|\pi_j| = \sum\limits_{i=1}^{m} u_{ij}$. Note that because of (7.2),

$$\sum_{j=1}^{k} |\pi_j| = \sum_{j=1}^{k} \sum_{i=1}^{m} u_{ij} = \sum_{i=1}^{m} \left(\sum_{j=1}^{k} u_{ij} \right) = \sum_{i=1}^{m} 1 = m.$$

Knowing the locally covariance matrices, for each cluster π_j we can define the locally Mahalanobis distance-like function (see also [7, 116, 176]):

$$d_M^{(j)}(x, y; S_j) := \sqrt[n]{\det S_j}\,(x - y)S_j^{-1}(x - y)^T. \tag{7.24}$$

The factor $\sqrt[n]{\det S_j}$ is the geometric mean of eigenvalues of the matrix S_j, and it is introduced here in order to ensure a steady decrease of the objective function values obtained by the c-means algorithm (see e.g. [176]).

If an eigenvalue of the covariance matrix S_j vanishes or if $\lambda_{\max}^{(j)} \gg \lambda_{\min}^{(j)}$, then S_j is a nearly singular matrix, and the condition number $\text{cond}(S_j) := \lambda_{\max}^{(j)}/\lambda_{\min}^{(j)}$ is very large, e.g. 10^{20}. In [7] one can see what to do in such cases.

The problem of searching for a globally optimal fuzzy k-partition is related to the following GOP:

$$\underset{c\in\mathbb{R}^{kn},\, U\in[0,1]^{m\times k}}{\arg\min}\mathcal{F}_M(c,U), \quad \mathcal{F}_M(c,U) = \sum_{i=1}^{m}\sum_{j=1}^{k} u_{ij}^{\,q}\, d_M(c_j,a_i;S_j). \tag{7.25}$$

Similarly to the fuzzy spherical clustering case, no global optimization method can be directly applied in this case, either.

It can be easily checked that, similarly as in Sect. 7.1.1, we obtain the membership function

$$u_{rs} = \left(\sum_{j=1}^{k} \left(\frac{d_M(c_s,a_r;S_s)}{d_M(c_j,a_r;S_j)} \right)^{1/(q-1)} \right)^{-1}, \quad a_r \neq c_s, \tag{7.26}$$

and, similarly as in Sect. 7.1.2, the centers

$$c_j = \left(\sum_{i=1}^{m} u_{ij}^{\,q} \right)^{-1} \sum_{i=1}^{m} u_{ij}^{\,q}\, a_i, \quad j = 1,\ldots,k. \tag{7.27}$$

The GKc-means algorithm[1] to search for an approximate solution to the global optimization problem (7.25) produces an approximation of the solution depending on the quality of the initial approximation. Therefore, numerous heuristic approaches can be found in the literature (see e.g. [7, 100, 173]).

The algorithm is formally written in two steps, performing either one or the other, depending on which data were initially supplied:

Algorithm 7.16 (Gustafson–Kessel c-means algorithm)

Step A: Given a finite subset $\mathcal{A} \subset \mathbb{R}^n$ and k distinct points $z_1,\ldots,z_k \in \mathbb{R}^n$, apply the minimal distance principle to determine the membership matrix $U \in [0,1]^{m\times k}$ and the covariance matrix $S_j \in \mathbb{R}^{n\times n}$, $j = 1,\ldots,k$,

$$u_{ij} = \left(\sum_{s=1}^{k} \left(\frac{d_M(z_j,a_i;S_j)}{d_M(z_s,a_i;S_s)} \right)^{1/(q-1)} \right)^{-1}, \tag{7.28}$$

$$S_j = \left(\sum_{i=1}^{m} u_{ij}^{\,q} \right)^{-1} \sum_{i=1}^{m} u_{ij}^{\,q}\, (z_j - a_i)^T (z_j - a_i). \tag{7.29}$$

Step B: Given the membership matrix $U \in [0,1]^{m\times k}$, determine the corresponding cluster centers $c_1,\ldots,c_k \in \mathbb{R}^n$ and calculate the objective function value $\mathcal{F}_M(c,U)$,

---

[1]The Gustafson–Kessel c-means algorithm and the corresponding MATLAB-code are available in [7].

$$c_j = \left(\sum_{i=1}^{m} u_{ij}^{\,q} \right)^{-1} \sum_{i=1}^{m} u_{ij}^{\,q} \, a_i, \quad j = 1, \ldots, k, \tag{7.30}$$

$$\mathcal{F}_M(c, U) = \sum_{i=1}^{m} \sum_{j=1}^{k} u_{ij}^{\,q} \, d_M(c_j, a_i; S_j); \tag{7.31}$$

Set $z_j := c_j$, $j = 1, \ldots, k$.

Note that in (7.28) it would be better to use Remark 7.1.

Step A of the GKc-means algorithm finds a solution to the problem (7.25) assuming that c is constant and Step B assuming that U is constant. If we are able to find a good enough initial cluster center, then the GKc-means algorithm runs the Step A, and if we are able to find a good enough initial membership matrix, then the GKc-means algorithm runs the Step B.

The algorithm produces a monotonically decreasing sequence of objective function values. Therefore the literature suggests (see e.g. [7, 10, 21, 184]), that with an $\epsilon_{\text{GKcm}} > 0$ set in advance, the iterative process should be stopped when

$$\frac{\mathcal{F}_M^{(j-1)} - \mathcal{F}_M^{(j)}}{\mathcal{F}_M^{(j-1)}} < \epsilon_{\text{GKcm}}, \quad \mathcal{F}_M^{(j)} = \mathcal{F}_M(c^{(j)}, U^{(j)}), \tag{7.32}$$

or when

$$\|U^{(j)} - U^{(j-1)}\| < \epsilon_{\text{GKcm}}. \tag{7.33}$$

Remark 7.17 The corresponding *Mathematica*-module GKcmeans [] is described in Sect. 9.2, and the link to appropriate *Mathematica*-code is supplied.

Example 7.18 Choose $k = 5$ points $C_1 = (2, 2)$, $C_2 = (6.5, 5.5)$, $C_3 = (4, 4)$, $C_4 = (2.5, 8)$, and $C_5 = (7.5, 2)$ in the square $[0, 10]^2 \subset \mathbb{R}^2$. To each point C_j we add 200 random points from the bivariate normal distribution with expectation $C_j \in \mathbb{R}^2$ and the covariance matrix

$$\text{Cov}_j = Q(\vartheta_j) \, \text{diag}(\xi_j^2, \eta_j^2) \, Q^T(\vartheta_j), \quad j = 1, \ldots, 5,$$

where $Q(\vartheta_j)$ is the rotation matrix given by (6.11), $\vartheta_j = 0, -\frac{15}{180}\pi, -\frac{72}{180}\pi, -\frac{15}{180}\pi, \frac{30}{180}\pi$, and $(\xi_j, \eta_j) = (0.5, 0.5), (1.0, 1.5), (1.7, 0.3), (1.0, 0.5), (1.0, 0.5)$.

The largest concentrations of data are inside M_N-circles (red ellipses) given by equations

$$d_M(x, C_j, \text{Cov}_j) = 1, \quad j = 1, \ldots, 5,$$

(Fig. 7.5a). The whole data set $\mathcal{A}$, $m = |\mathcal{A}| = 1000$, is shown in Fig. 7.1b. The same figure renders also the initial centers $\hat{c}_j$, $j = 1, \ldots, 5$, at which the GKc-means

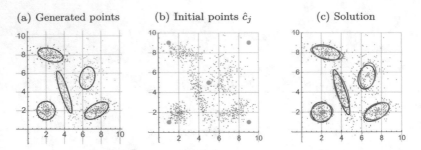

Fig. 7.5 GKc-means algorithm

Table 7.2 Confusion matrix and the percentage structure

Cluster	$\pi_1^\star$	$\pi_2^\star$	$\pi_3^\star$	$\pi_4^\star$	$\pi_5^\star$	Σ	$\pi_1^\star$	$\pi_2^\star$	$\pi_3^\star$	$\pi_4^\star$	$\pi_5^\star$	Σ
π_1	176.3	2.4	3.8	0.8	3.6	187	94.3	1.3	2.0	0.5	1.9	100
π_2	10.6	141.7	18.1	25.6	22.0	218	4.9	65.0	8.3	11.8	10.1	100
π_3	17.2	10.6	159.5	9.7	13.0	210	8.2	5.0	75.9	4.6	6.2	100
π_4	3.0	4.8	16.3	156.6	1.3	182	1.7	2.6	9.0	86.0	0.7	100
π_5	6.1	11.5	9.1	1.9	174.5	203	3.0	5.7	4.5	0.9	86.0	100
Σ	213.2	170.9	206.8	194.7	214.5	1000	112	79.6	99.7	103.8	104.9	500

algorithm was started. The new centers obtained are

$$c_1^\star = (2.01, 1.95), \ c_2^\star = (6.64, 5.68), \ c_3^\star = (4.00, 4.14),$$
$$c_4^\star = (2.58, 7.96), \ c_5^\star = (7.55, 2.09),$$

and the corresponding normalized Mahalanobis circles (gray ellipses) contain the largest concentrations of the data (Fig. 7.1c). The CPU-time used by the GKc-means algorithm was 42 s.

Hausdorff distance between the sets of original and new centers is 0.225.

Table 7.2 shows the corresponding confusion matrix (5.18), which can easily be determined (see Exercise 5.14), and its percentage structure. The j-th row of the table shows how the elements of the j-th cluster, obtained around the center C_j, are arranged: most of them are in the new j-th cluster.

Exercise 7.19 Using the module GKcmeans[] from Sect. 9.2, apply the GKc-means algorithm to data sets from Test-examples 9.6 and 9.10, and analyze the results

7.3.2   Mahalanobis Fuzzy Incremental Algorithm (MFInc)

Assume now that the most appropriate number of clusters is not known in advance. Based on the method developed in [8, 10, 11, 13–15, 116, 166], we will devise an incremental algorithm to search for an optimal partition with $2, 3, \dots$ clusters. In order to be able to conclude which of these partitions is the most appropriate one, we will use the common procedure applying different validity indexes. Some of the known fuzzy indexes shall in fact be adjusted to our situation, that is to search for a fuzzy optimal partition with clusters of ellipsoidal shape.

Our algorithm will start by choosing an initial center $\hat{c}_1 \in \mathbb{R}^n$. For example, it can be the mean or the median of the data set $\mathcal{A}$. An approximation of the next center $\hat{c}_2$ can be obtained by solving the following GOP:

$$\underset{x \in \mathbb{R}^n}{\arg\min} \, \Phi_2(x), \quad \Phi_2(x) := \sum_{i=1}^{m} \min\{\|\hat{c}_1 - a_i\|^2, \|x - a_i\|^2\}. \tag{7.34}$$

Note that it suffices to determine just a few, say 10 iterations of the DIRECT algorithm, since this results in a sufficiently good initial approximation for the GKc-means algorithm, producing an optimal fuzzy partition with two clusters with centers $c_1^\star$ and $c_2^\star$.

In general, knowing $r - 1$ centers $\hat{c}_1, \dots, \hat{c}_{r-1}$, the next one, i.e. the r-th center $\hat{c}_r$, will be found by solving the following GOP:

$$\underset{x \in \mathbb{R}^n}{\arg\min} \, \Phi(x), \quad \Phi(x) := \sum_{i=1}^{m} \min\{\delta_{r-1}^i, \|x - a_i\|^2\}, \tag{7.35}$$

where $\delta_{r-1}^i = \underset{1 \leq s \leq r-1}{\min} \, d_M(\hat{c}_s, a_i; S_j)$. Again, to solve this GOP, it suffices to perform just a few, say 10 iterations of the DIRECT algorithm, which is enough to obtain a sufficiently good initial approximation for the GKc-means algorithm.

After that, the GKc-means algorithm finds centers $c_1^\star, \dots, c_r^\star$ of the optimal r-partition.

The number K of partitions produced in this way can be set in advance. Alternatively, since increasing the number of partitions decreases the corresponding objective function values (Theorem 3.7), setting an $\epsilon_{\text{Bagi}} > 0$ in advance, generating new partitions can be stopped at the r-th step if [10, 11]

$$\frac{\Phi_{r-1} - \Phi_r}{\Phi_{r-1}} < \epsilon_{\text{Bagi}}, \quad \Phi_j = \Phi(c^{(j)}, U^{(j)}). \tag{7.36}$$

Remark 7.20 The corresponding *Mathematica*-module MFInc[] is described in Sect. 9.2, and the link to appropriate *Mathematica*-code is supplied.

Example 7.21 We implemented the MFInc algorithm for the data set $\mathcal{A}$ from Example 7.18, starting at the initial center $\hat{c}_1 = \text{mean}(\mathcal{A}) \approx (4.73, 2.31)$. After

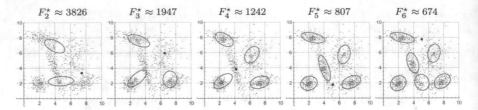

$F_2^\star \approx 3826$ $F_3^\star \approx 1947$ $F_4^\star \approx 1242$ $F_5^\star \approx 807$ $F_6^\star \approx 674$

Fig. 7.6 Fuzzy incremental Mahalanobis algorithm

applying 10 iterations of the DIRECT algorithm to the GOP (7.34), producing the center $\hat{c}_2$, we applied the GKc-means algorithm to obtain two centers $c_1^\star, c_2^\star$ and the corresponding M_N-circles, shown in Fig. 7.6a. The same figure shows the new center $\hat{c}_3$, the red dot, which was obtained by applying 10 iterations of the DIRECT algorithm to the GOP (7.35) for $r = 3$. Next, the GKc-means algorithm produced the fuzzy optimal 3-partition shown in Fig. 7.6b, and so on. Increasing the number of clusters, the value of the objective function decreases (Theorem 3.7), as illustrated in Fig 7.6. The algorithm was stopped after $K = 5$ iterations. The same would have been achieved using the criterion (7.36) with $\epsilon_{\text{Bagi}} = 0.05$. In the same way we found optimal partitions with $2, \ldots, 6$ clusters. The total CPU-time was 98.4 s, out of which only 0.1 s were used by the DIRECT algorithm, while the rest was used by the GKc-means algorithm.

However, the problem of choosing the most suitable partition remains.

Remark 7.22 Note that, when needed, this algorithm can also be run from more than one initial center.

7.3.3 Choosing the Most Appropriate Number of Clusters

The most appropriate number of clusters in a fuzzy partition is usually determined based on several validity indexes. Some of the most frequently mentioned and well known such indexes used for fuzzy clustering (see e.g., [21, 32, 188, 197, 203]), will be customized to our situation.

Let $\Pi^\star = \{\pi_1^\star, \ldots, \pi_k^\star\}$ be an optimal fuzzy partition with the membership matrix $U^\star$, the corresponding cluster centers $c_1^\star, \ldots, c_k^\star$, and covariance matrices $S_1^\star, \ldots, S_k^\star$. The following validity indexes will be used:

The **Mahalanobis Xie–Beni fuzzy index** is defined as

$$\text{MFXB}(k) = \frac{1}{m \min\limits_{i \neq j} d_M(c_i^\star, c_j^\star; S_i^\star + S_j^\star)} \sum_{i=1}^{m} \sum_{j=1}^{k} u_{ij}^{\star\, q}\, d_M(c_j^\star, a_i, S_j^\star). \qquad (7.37)$$

The more compact and better separated clusters in an optimal partition, the smaller the MFXB(k) number.

The *Mahalanobis Calinski–Harabasz fuzzy index* can be defined as (cf. (5.3), (6.26), and (7.21))

$$\text{MFCH}(k) = \frac{\frac{1}{k-1}\sum_{j=1}^{k}\kappa_j^{\star}\,d_M(c_A, c_j^{\star}, S_j^{\star}+S)}{\frac{1}{m-k}\mathcal{F}_M(c^{\star}, U^{\star})}, \quad S = \frac{1}{m}\sum_{i=1}^{m}(c_A-a_i)^T(c_A-a_i),\; c_A = \frac{1}{m}\sum_{i=1}^{m}a_i,$$

(7.38)

where $\kappa_j^{\star} = |\pi_j^{\star}| = \sum_{i=1}^{m}u_{ij}^{\star}$. The more compact and better separated clusters in an optimal partition, the larger the MFCH(k) number.

The *Mahalanobis Davies–Bouldin fuzzy index* is defined as (cf. (5.9), (6.27), and (7.22))

$$\text{MFDB}(k) = \frac{1}{k}\sum_{j=1}^{k}\max_{s\neq j}\frac{V(\pi_j^{\star})+V(\pi_s^{\star})}{d_M(c_j^{\star}, c_s^{\star}; S_j^{\star}+S_s^{\star})}, \quad V(\pi_j^{\star}) = \frac{1}{\sum_{i=1}^{m}u_{ij}^{\star q}}\sum_{i=1}^{m}u_{ij}^{\star q}\,d_M(c_j^{\star}, a_i; S_j^{\star}).$$

(7.39)

The more compact and better separated clusters in an optimal partition, the smaller the MFDB(k) number.

The *Mahalanobis fuzzy hyper-volume index*

$$\text{MFHV}(k) = \sum_{j=1}^{k}\sqrt{\det S_j^{\star}}$$

(7.40)

represents the sum of fuzzy hyper-volumes of all clusters. The more compact and better separated clusters in an optimal partition, the smaller the MFHV(k) number.

Example 7.23 We calculated the corresponding values of aforementioned indexes for the data set $\mathcal{A}$ from Example 7.18 and partitions with $2, \ldots, 6$ clusters given in Example 7.21. Table 7.3 and Fig. 7.7 show that, according to all indexes, 5 is the most appropriate number of clusters.

Table 7.3 Index values for optimal partitions from Example 7.21 (indexes of the optimal partition are in bold)

	$k = 2$	$k = 3$	$k = 4$	$k = 5$	$k = 6$
MFXB	0.10	0.09	0.07	**0.06**	0.12
MFCH	1258	1591	1864	**2313**	2119
MFDB	0.24	0.22	0.24	**0.18**	0.31
MFHV	4.81	4.01	3.75	**2.96**	3.38

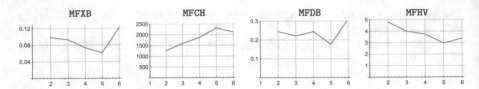

Fig. 7.7 Mahalanobis fuzzy indexes for optimal partitions from Example 7.23

Exercise 7.24 Using the module `MFInc[]` from Sect. 9.2, apply the `MFinc` algorithm to data sets from Test-examples 9.6–9.10, and analyze the results.

Remark 7.25 MFXB, MFCH, MFDB, MFHV indexes can be calculated by using corresponding *Mathematica*-modules `MFXB[]`, `MFCH[]`, `MFDB[]`, `MFHV[]` which are described in Sect. 9.2, and links to appropriate *Mathematica*-codes are supplied.

7.4 Fuzzy Variant of the Rand Index

In Sect. 5.2 we compared two partitions whose clusters were well separated. Now we want to investigate how to compare two fuzzy partitions $\Pi^{(1)} = \{\pi_1^{(1)}, \ldots, \pi_k^{(1)}\}$ and $\Pi^{(2)} = \{\pi_1^{(2)}, \ldots, \pi_\ell^{(2)}\}$ of a set $\mathcal{A} \subset \mathbb{R}^n$.

In case when the clusters are well separated, looking at the membership matrix U [145], one can see that a pair (a_r, a_s) is paired (Definition 5.11), if the scalar product $\langle U(r), U(s) \rangle$ of the corresponding matrix rows equals 1, and the pair is not paired if this scalar product equals 0. By analogy to [62], in case of fuzzy clustering, for the matrix U whose elements u_{rs} are defined by (7.9), one defines the **fuzzy Rand index**, cf. (5.16),

$$\mathrm{FR}(\Pi^{(1)}, \Pi^{(2)}) := \frac{a+d}{a+b+c+d}, \tag{7.41}$$

where

$$a = \sum_{(a_r, a_s) \in \mathcal{C}} \psi^{(1)}(a_r, a_s)\, \psi^{(2)}(a_r, a_s),$$

$$b = \sum_{(a_r, a_s) \in \mathcal{C}} \psi^{(1)}(a_r, a_s)\big(1 - \psi^{(2)}(a_r, a_s)\big),$$

$$c = \sum_{(a_r, a_s) \in \mathcal{C}} \big(1 - \psi^{(1)}(a_r, a_s)\big)\, \psi^{(2)}(a_r, a_s),$$

$$d = \sum_{(a_r, a_s) \in \mathcal{C}} \big(1 - \psi^{(1)}(a_r, a_s)\big)\big(1 - \psi^{(2)}(a_r, a_s)\big),$$

with

$$\psi^{(\kappa)}(a_r, a_s) = \sum_{(a_r, a_s) \in \mathcal{C}} U^{(\kappa)}(a_r) U^{(\kappa)}(a_s), \quad \kappa = 1, 2,$$

where $U^{(\kappa)}(a_r)$ denotes the row of the membership matrix $U^{(\kappa)}$ for partition $\Pi^{(\kappa)}$, corresponding to the element $a_r \in \mathcal{A}$. More precisely, if $U^{(\kappa)}(r)$ is the r-th row of the membership matrix $U^{(\kappa)}$ showing the dispersion of element a_r over clusters of partition $\Pi^{(\kappa)}$, then we can write

$$\psi^{(\kappa)}(a_r, a_s) = \sum_{r=1}^{m-1} \sum_{s=r+1}^{m} \langle U^{(\kappa)}(r), U^{(\kappa)}(s) \rangle, \quad \kappa = 1, 2. \tag{7.42}$$

Remark 7.26 In a similar way one can also define the *fuzzy Jaccard index* (FJ).

The fuzzy Rand and fuzzy Jaccard indexes can be calculated fast using the *Mathematica*-module `RandFrigue[]` which is described in Sect. 9.2, and the link to appropriate *Mathematica*-code is supplied.

Short overviews of existing techniques to compare fuzzy partitions which have been proposed in the literature can be found in [82, 83].

7.4.1 Applications

Now we will apply the aforementioned indexes to recognize the partition which is closest to the globally optimal one.

Example 7.27 We compared the partition Π of the data set $\mathcal{A}$ from Example 7.8, with partitions $\Pi^{(2)}, \ldots, \Pi^{(6)}$ obtained using the FInc algorithm in Example 7.11. Table 7.4 shows the FR and FJ indexes and the Hausdorff distances between the sets of centers. The best agreement was for the $\Pi^{(5)}$ partition, meaning that the $\Pi^{(5)}$ partition is the one closest to the globally optimal partition.

Table 7.4 FR and FJ indexes, and Hausdorff distances between sets of centers

	$\Pi^{(2)}$	$\Pi^{(3)}$	$\Pi^{(4)}$	$\Pi^{(5)}$	$\Pi^{(6)}$
FR	0.573	0.691	0.742	0.773	**0.777**
FJ	0.241	0.269	0.275	**0.276**	0.244
H	3.878	1.810	1.726	**0.312**	0.944

Example 7.28 The partition Π of the data set $\mathcal{A}$ from Example 7.18 was compared with partitions $\Pi^{(2)}, \dots, \Pi^{(6)}$ obtained using MFInc algorithm in Example 7.21. Table 7.5 shows the values of the FR and FJ indexes, and the Hausdorff distances between the sets of centers. The best agreement was for the $\Pi^{(5)}$ partition, meaning that $\Pi^{(5)}$ is the partition closest to the globally optimal one.

Table 7.5 FR indexes, FJ indexes, and Hausdorff distances between the sets of centers

	$\Pi^{(2)}$	$\Pi^{(3)}$	$\Pi^{(4)}$	$\Pi^{(5)}$	$\Pi^{(6)}$
FR	0.59	0.73	0.8	**0.87**	0.85
FJ	0.26	0.33	0.39	**0.50**	0.43
H	2.9	3.07	2.34	**0.27**	2.20

Chapter 8
Applications

As we already mentioned in the Introduction, there are numerous applications of cluster analysis in various areas of science and applications. In this chapter we are going to analyze in more details some applications previously investigated in our published papers, and which originated from different real-world problems. For each application which we will consider, we developed corresponding *Mathematica* programs used to execute the necessary calculations.

8.1 Multiple Geometric Objects Detection Problem and Applications

In this section we are going to look at the problem of recognizing several geometric objects (curves) of the same type in the plane: circles, ellipses, generalized circles, straight lines, straight line segments, etc., but whose number may not be known in advance.

Let $\Delta = [a, b] \times [c, d] \subset \mathbb{R}^2$, $a < b$, $c < d$, be a rectangle in the plane, and let $\mathcal{A} = \{a_i = (x_i, y_i) \in \Delta : i = 1, \ldots, m\}$ be a set of data points which are assumed to be scattered along multiple unknown geometric objects of the same type and which should be reconstructed or detected. Although the geometric objects are of the same type, their parameters and sometimes even their numbers are not known.

Furthermore, we will assume that the data points coming from geometric objects satisfy the *homogeneity property*, i.e. we assume that these points are scattered around geometric objects as if they were obtained by adding noise to points uniformly distributed along the geometric object contained in the rectangle Δ. We also say that such data are *homogeneously scattered* around the geometric object.

We start with the following definition.

Definition 8.1 Let $\pi = \pi(\gamma)$ be a cluster set with $|\pi(\gamma)|$ data points originating from a geometric object (curve) $\gamma \subset \Delta$ of length $|\gamma|$. The number $\rho(\pi) = \frac{|\pi(\gamma)|}{|\gamma|}$ is called the *local density* of the cluster π.

8.1.1   The Number of Geometric Objects Is Known in Advance

Let us first consider the problem of recognizing a known number k of geometric objects in the plane. All objects are of the same type, each one specified by r parameters which have to be determined based on the data set $\mathcal{A}$. Such a problem will be called an MGOD-problem (*Multiple Geometric Object Detection problem*), for short.

Denote by $\gamma_j(t_j)$, $j = 1, \ldots, k$, these geometric objects, where $t_j \in \mathbb{R}^r$, are the parameter-vectors. The MGOD-problem will be considered as a center-based clustering problem, where the cluster centers are the geometric objects $\gamma_j(t_j)$, $j = 1, \ldots, k$ (γ-cluster-centers).

Once the distance $\mathfrak{D}(a, \gamma)$ from a point $a \in \mathcal{A}$ to the geometric object γ is well defined, the problem of finding the optimal k-partition $\Pi = \{\pi_1, \ldots, \pi_k\}$ whose cluster centers are the geometric objects $\gamma_1(t_1), \ldots, \gamma_k(t_k)$ can be formulated in the following two ways:

$$\underset{\Pi}{\arg\min}\, \mathcal{F}(\Pi), \quad \mathcal{F}(\Pi) = \sum_{j=1}^{k} \sum_{a_i \in \pi_j} \mathfrak{D}(a_i, \gamma_j(t_j)), \tag{8.1}$$

$$\underset{t \in \mathbb{R}^{kr}}{\arg\min}\, F(t), \quad F(t) = \sum_{i=1}^{m} \min_{1 \leq j \leq k} \mathfrak{D}(a_i, \gamma_j(t_j)), \quad t = (t_1, \ldots, t_k). \tag{8.2}$$

In spite of the fact that the objective function (8.2) for all aforementioned MGOD-problems is Lipschitz-continuous (see e.g. [162]), solving the GOP (8.2) directly using one of the global optimization algorithms (for example, the DIRECT algorithm) is not acceptable since one deals with a symmetric function with rk variables arranged into k parameter-vectors $t_1, \ldots, t_k \in \mathbb{R}^r$, and thus the function attains at least $k!$ different global minima. By its construction, the DIRECT algorithm would search for all global minima, thus being extremely inefficient.

To solve the GOP (8.2) one can proceed as follows (see [161, 162]):

(i) either use just a few, say 10 iterations of the DIRECT algorithm, and find a sufficiently good initial approximation of γ-cluster-centers, or use the DBSCAN algorithm and find a sufficiently good initial partition;

(ii) find an optimal solution using the k-means algorithm in such a way that the cluster centers are the given geometric objects.

The modified k-means algorithm in the case of γ-cluster-centers can be set up in two steps which alternate consecutively.

Algorithm 8.2 (Modification of the k-means algorithm for γ-cluster-centers (KCG))

Step A: (Assignment step) Given a finite set $\mathcal{A} \subset \mathbb{R}^2$ and k distinct geometric objects $\gamma_1(t_1), \ldots, \gamma_k(t_k)$, apply the minimal distance principle to determine the clusters

$$\pi_j := \{a \in \mathcal{A} : \mathfrak{D}(a, \gamma_j(t_j)) \le \mathfrak{D}(a, \gamma_s(t_s)), \ \forall s \ne j\}, \ j = 1, \ldots k, \qquad (8.3)$$

to get the partition $\Pi = \{\pi_1, \ldots, \pi_k\}$.

Step B: (Update step) Given the partition $\Pi = \{\pi_1, \ldots, \pi_k\}$ of the set $\mathcal{A}$, determine the corresponding γ-cluster-centers $\hat{\gamma}_j(\hat{t}_j)$, $j = 1, \ldots, k$, by solving the following GOPs:

$$\hat{\gamma}_j(\hat{t}_j) \in \arg\min_{t \in \mathbb{R}^r} \sum_{a \in \pi_j} \mathfrak{D}(a, \gamma_j(t)), \quad j = 1, \ldots, k, \qquad (8.4)$$

and calculate the objective function value $F(\hat{t}_1, \ldots, \hat{t}_k)$.
Set $\gamma_j(t_j) := \hat{\gamma}_j(\hat{t}_j)$, $j = 1, \ldots, k$.

Similarly as in the case of ordinary spherical or elliptical clusters (see Theorem 4.8), one can show that the KCG algorithm finds in a finite number of steps an optimal partition, and that the sequence (F_n) of function values defined by (8.2) is monotonously decreasing. Therefore the stopping criterium for the algorithm may be

$$\frac{F_{n-1} - F_n}{F_{n-1}} < \epsilon_{\text{KCG}}, \qquad (8.5)$$

for some small $\epsilon_{\text{KCG}} > 0$ (e.g. 0.005).

8.1.2   The Number of Geometric Objects Is Not Known in Advance

When the number of clusters is unknown, i.e. it is not known in advance, one can search for optimal partitions with $k = 1, 2, \ldots, K$ clusters, and then, in order to recognize the most optimal partition, try to use one of the known indexes adapted for the case of γ-cluster-centers.

We are going to modify the general incremental algorithm introduced in Sect. 4.3 like this. The algorithm can start by determining the best representative of the set $\mathcal{A}$

$$\hat{\gamma}_1(\hat{t}_1) \in \arg\min_{t \in \mathbb{R}^r} \sum_{a \in \mathcal{A}} \mathfrak{D}(a, \gamma(t)) \qquad (8.6)$$

but can also start with some other initial object.

Furthermore, if the γ-cluster-centers $\hat{\gamma}_1, \ldots, \hat{\gamma}_{k-1}$ are known, the next center $\hat{\gamma}_k$ is found by solving the GOP

$$\underset{t \in \mathbb{R}^r}{\arg\min}\, \Phi(t), \quad \Phi(t) = \sum_{i=1}^{m} \min\{\delta_{k-1}^{(i)}, \mathfrak{D}(a_i, \gamma(t))\}, \tag{8.7}$$

where $\delta_{k-1}^{(i)} = \min\{\mathfrak{D}(a_i, \hat{\gamma}_1), \ldots, \mathfrak{D}(a_i, \hat{\gamma}_{k-1})\}$.

Next, applying the KCG algorithm one obtains optimized γ-cluster-centers $\gamma_1^\star, \ldots, \gamma_{k-1}^\star, \gamma_k^\star$ of the optimal k-LOPart $\Pi^{(k)}$. In such a way the incremental algorithm generates a sequence of k-LOParts.

Similarly as in Sect. 4.3, the incremental algorithm can be stopped (see also [12]) when the relative value of the objective function (8.2) for some k becomes smaller than a predefined number $\epsilon_B > 0$ (e.g. 0.005)

$$\frac{\Phi_{k-1} - \Phi_k}{\Phi_{k-1}} < \epsilon_B. \tag{8.8}$$

Note that the described algorithm could just as well start with more than one initial geometric object.

It was demonstrated (see [143, 146, 156, 163]) that solving the MGOD-problem, the stopping criterion for the incremental algorithm can be better defined using the parameters MinPts and $\epsilon(\mathcal{A})$ which were used in the DBSCAN algorithm to estimate the ϵ-density of the set $\mathcal{A}$ (see Sect. 4.5.1).

Let $\Pi = \{\pi_1, \ldots, \pi_k\}$ be the k-LOPart obtained in the k-th step of the incremental algorithm. For every cluster $\pi_j \in \Pi$ with γ-cluster-center γ_j define the set

$$\mu_j := \{\mathfrak{D}_1(a, \gamma_j) : a \in \pi_j\}, \tag{8.9}$$

where $\mathfrak{D}_1(a, \gamma_j)$ denotes the orthogonal distance from the point a to the γ-cluster-center γ_j. For instance, for a circle this distance is given by (8.19), for an ellipse by (8.39), and for a straight line by (8.59).

Denote by $\mathrm{QD}(\pi_j)$ the 90 % quantile of the set μ_j, and define

$$\mathrm{QD}(\Pi) = \max_{\pi \in \Pi} \mathrm{QD}(\pi). \tag{8.10}$$

We expect that the k-partition Π is close to k-LOPart, if

$$\mathrm{QD}(\Pi) < \epsilon(\mathcal{A}). \tag{8.11}$$

Therefore the stopping criterion (8.8) for the incremental algorithm can be replaced by the better criterion (8.11).

Remark 8.3 The corresponding Mathematica-module DBC[] calculating the Density-Based Clustering index according to (8.10) is described in Sect. 9.2, and the link to appropriate *Mathematica*-code is supplied.

Notice that (8.11) will be satisfied easier if $QD(\pi_j)$ is defined as $p\%$, $(p < 90)$, quantile of the set μ_j. This way the algorithm will stop earlier and the number K of geometric objects obtained will be smaller. In general, varying the number p one can affect the number of geometric objects produced by the incremental algorithm.

8.1.3 Searching for MAPart and Recognizing Geometric Objects

The incremental algorithm, but also some other approaches, like using the RANSAC method (see Sect. 8.1.8), produce, as a rule, several k-LOParts, $k \geq 1$. When solving MGOD-problems, the most important question is the following one: does there exist among obtained k-LOParts the one whose γ-cluster-centers coincides with the original geometric objects based on which the data originated, and if so, how to detect it? Such k-LOPart is called the ***Most Appropriate Partition*** (MAPart). We will devote special attention to define criteria for recognizing MAPart.

Since we tackle the MGOD-problem as a center-based clustering problem, one could try to solve the detection problem of MAPart by modifying some of the known indexes from Chap. 5.

Let us state only modifications of CH and DB indexes. Let $m = |\mathcal{A}|$, let $\Pi^\star = \{\pi_1^\star, \ldots, \pi_\kappa^\star\}$ be the κ-LOPart given by γ-cluster-centers $\gamma_1^\star, \ldots, \gamma_\kappa^\star$, and let $\gamma^\star$ be the representative of the whole set $\mathcal{A}$. Then (see [68, 158])

$$\text{CHG}(\kappa) := \frac{\frac{1}{\kappa-1} \sum_{j=1}^{\kappa} |\pi_j^\star| d_H(\gamma_j^\star, \gamma^\star)}{\frac{1}{m-\kappa} \sum_{j=1}^{\kappa} \sum_{a \in \pi_j^\star} \mathfrak{D}(a, \gamma_j^\star)}, \tag{8.12}$$

$$\text{DBG}(\kappa) := \frac{1}{\kappa} \sum_{j=1}^{\kappa} \max_{s \neq j} \frac{\sigma_j + \sigma_s}{d_H(\gamma_j^\star, \gamma_s^\star)}, \quad \sigma_j^2 = \frac{1}{|\pi_j^\star|} \sum_{a \in \pi_j^\star} \mathfrak{D}(a, \gamma_j^\star), \tag{8.13}$$

where $\mathfrak{D}$ is the distance-like function used to define the distance from the point a to the curve $\gamma_j^\star$, and d_H is the Hausdorff distance between the respective curves. However, since curves are not finite sets, one cannot use the formula (4.20) directly but has to replace min by inf and max by sup—therefore the notation d_H instead of D_H. In practice this might not be carried out easily. In such situations, on each curve choose a finite, but not too small set of points, preferably uniformly distributed along the curves and take the Hausdorff distance D_H between these two sets.

Since the incremental algorithm is of local character, it will, as a rule, produce more curves γ than expected regarding the configuration of the set $\mathcal{A}$ (as can be seen e.g. in multiple straight line segments detection problems in Examples 8.33 and 8.34), but those from which the data stemmed may not be among them. This means that often the MAPart will not be among produced partitions, hence one cannot expect that using modified classical indexes will always detect MAPart.

Next, we will present an approach which uses specificities of the problem under consideration and which proved to be more effective (see [143, 146, 156, 163]).

Let $\Pi^\star = \{\pi_1^\star, \ldots, \pi_k^\star\}$ be a k-partition of the set $\mathcal{A}$. Recall that we assumed that the data points coming from some geometric object satisfy the *homogeneity property*. Since, for each $a \in \mathcal{A}$, the corresponding disc of radius $\epsilon(\mathcal{A})$ contains at least MinPts elements from the set $\mathcal{A}$, the lower bound of the local density for each cluster $\pi^\star \in \Pi^\star$ can be estimated by (see also Remark 8.5)

$$\rho(\pi) \geq \frac{\text{MinPts}}{2\epsilon(\mathcal{A})}. \tag{8.14}$$

Therefore the γ-cluster-centers for all clusters for which (8.14) does not hold should be dropped out, and from their data points formed the residual $\mathcal{R}$.

If $|\mathcal{R}| \leq \kappa$, where κ is some predefined threshold (e.g. $0.025|A|$), elements of the residual $\mathcal{R}$ should be distributed to remaining clusters by using the minimal distance principle. The LOPart obtained in this way will be considered to be MAPart (see Examples 8.12, 8.13, 8.22, and 8.23).

If $|\mathcal{R}| > \kappa$, then the residual $\mathcal{R}$ can be regarded as a new initial data set (see Example 8.34).

Now we are in position to define necessary conditions for the partition $\Pi^\star$ to be MAPart:

$\mathbf{A}_0 : \text{QD}(\Pi^\star) < \epsilon(\mathcal{A})$;

$\mathbf{A}_1 : \rho(\pi_j^\star) \geq \frac{\text{MinPts}}{2\epsilon(\mathcal{A})}$ for all clusters $\pi_j^\star \in \Pi^\star$; $\qquad\qquad$ (8.15)

$\mathbf{A}_2 :$ all clusters $\pi^\star \in \Pi^\star$ have similar local density $\rho(\pi^\star)$.

Remark 8.4 While solving MGOD-problems for circles, ellipses, straight line segments, etc., numerous numerical experiments showed that one would not err much considering these conditions also sufficient in order that r-LOPart be at the same time MAPart.

Remark 8.5 Another index designed specifically for MGOD-problems was defined in [159]. Let us briefly describe it. Let $\Pi^\star = \{\pi_1^\star(\gamma_1^\star), \ldots, \pi_\kappa^\star(\gamma_\kappa^\star)\}$ be a κ-LOPart with γ-cluster-centers $\gamma_1^\star, \ldots, \gamma_\kappa^\star$. For each cluster $\pi_j^\star$ find the local density $\rho_j(\pi_j^\star) = \frac{|\pi_j^\star|}{|\gamma_j^\star|}$ as in Definition 8.1. Notice that the cluster whose γ-cluster-center is close to a curve from which the data originated has a higher local density.

Next, find the variance

$$\mathrm{Var}(\rho) = \frac{1}{\kappa-1} \sum_{j=1}^{\kappa} (\rho_j - \bar{\rho})^2, \quad \bar{\rho} = \frac{1}{\kappa} \sum_{j=1}^{\kappa} \rho_j, \quad \kappa \geq 2, \tag{8.16}$$

for the set $\rho = \{\rho_1, \ldots, \rho_\kappa\}$, $\kappa \geq 2$, where $\rho_j := \rho_j(\pi_j^*)$ is the local density of the cluster π_j^*, and define the *Geometrical Objects* index (GO) by

$$\mathrm{GO}(\kappa) := \begin{cases} \mathrm{Var}(\rho_1, \ldots, \rho_\kappa), & \text{for } \kappa \geq 2, \\ \rho_1, & \text{for } \kappa = 1. \end{cases} \tag{8.17}$$

Note that for $\kappa = 1$ there is only one γ-cluster-center γ_1^* and the value of GO index equals $\rho_1 = \frac{|A|}{|\gamma_1^*|}$.

The smaller value of GO index is associated with the more appropriate partition.

8.1.4 Multiple Circles Detection Problem

We will start the discussion of MGOD-problems with the simplest case—the *Multiple circles detection problem* (MCD). Let $A = \{a_i - (x_i, y_i) : i = 1, \ldots, m\} \subset \mathbb{R}^2$ be a data set which originated from several unknown circles in the plane, and which are to be recognized or reconstructed. We will use either of the following two notations for a circle centered at the point $C = (p, q)$ with radius r:

(i) $\qquad K(C, r) = \{(x, y) \in \mathbb{R}^2 : (x - p)^2 + (y - q)^2 = r^2\},$

(ii) $\qquad K(C, r) = \{C + r(\cos \tau, \sin \tau) : \tau \in [0, 2\pi]\}.$

This problem comes up in numerous applications, e.g. pattern recognition and computer vision, image analysis in industrial applications such as automatic inspection of manufactured products and components, pupil and iris detection, investigation of animal behavior, etc., [2, 33, 35].

There are many different approaches in the literature for solving this problem, which are most often based on applying the LS-fitting [31, 59, 66], Hough transformation [120], RANSAC method [57], and different heuristic methods [2, 33, 35, 97].

We will consider this problem as a center-based clustering problem, where the cluster centers are circles (see [158]).

Circle as the Representative of a Data Set

First consider the problem of determining a circle representative of a data set $A \subset \Delta = [a, b] \times [c, d] \subset \mathbb{R}^2$, assuming that the data originated from a single circle.

Artificial Data Set Originating from a Single Circle

Artificial data set arising from a circle $K(C, r)$ centered at C with radius r can be constructed as follows: generate $m > 2$ uniformly distributed points $\tau_i \in [0, 2\pi]$, $i = 1, \ldots, m$, and to each point $T_i = C + r(\cos \tau_i, \sin \tau_i) \in K(C, r)$ add a random error from bivariate normal distribution $\mathcal{N}(0, \sigma^2 I)$ with expectation $0 \in \mathbb{R}^2$ and covariance matrix $\sigma^2 I$, I being the unit 2×2 matrix. This will ensure that the data in $\mathcal{A}$ coming from a circle satisfy the *homogeneity property*, i.e. that they are homogeneously spread around the circle. According to Definition 8.1, the local density of the set $\mathcal{A}$ is $\rho(\mathcal{A}) = \frac{m}{2r\pi}$.

Example 8.6 Let $K = K\big((2.5, 2.5), 2\big)$ be the circle centered at $(2.5, 2.5)$ with radius 2. In the neighborhood of K we generate $m = 100$ random points as was just described (see Fig. 8.1 and the *Mathematica*-program below).

```
In[1]:= A ={}; cen = {2.5,2.5}; r = 2; m = 100; sigma = .01;
        Do[
        t = RandomReal[{0, 2 Pi}];
        T = cen + r {Cos[t], Sin[t]};
        A=Append[A,RandomReal[
            MultinormalDistribution[T, sigma IdentityMatrix[2]], {1}][[1]]
            ];
        ,{ii,m}]
```

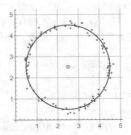

Fig. 8.1 Data coming from the circle $K\big((2.5, 2.5), 2\big)$

The Best Representative

The best representative of the set $\mathcal{A}$ is the circle $K\big((p, q), r\big)$ whose parameters p, q, r are obtained as the solution to the GOP

$$\underset{(p,q)\in\Delta,\, r\in[0,R]}{\arg\min} \sum_{i=1}^{m} \mathfrak{D}(a_i, K((p, q), r)), \quad R = \tfrac{1}{2}\max\{b - a, d - c\}, \quad (8.18)$$

where $\mathfrak{D}$ is a distance-like function, in the literature usually defined in one of the following three ways [31, 125, 158]:

$$\mathcal{D}_1(a_i, K((p,q), r)) := \left| \, \|a_i - (p,q)\| - r \, \right|, \tag{8.19}$$

$$\mathcal{D}_2(a_i, K((p,q), r)) := (\|a_i - (p,q)\| - r)^2, \tag{8.20}$$

$$\mathcal{D}_A(a_i, K((p,q), r)) := (\|a_i - (p,q)\|^2 - r^2)^2. \tag{8.21}$$

When the distance $\mathcal{D}_1$ is used ($\mathcal{D}_1$-approach) one says that the circle (8.18) was obtained in accordance with the LAD-principle, using $\mathcal{D}_2$ distance-like function ($\mathcal{D}_2$-approach)—in accordance with the TLS-principle, and using $\mathcal{D}_A$ distance-like function ($\mathcal{D}_A$-approach)—in accordance with the *Algebraic Criterion*.

It is the algebraic criterion, i.e. the $\mathcal{D}_A$-approach, which is the most used one in the literature and in applications. The reason for this is the fact that the minimizing function in (8.18) is differentiable. In this case, given a data set $\mathcal{A}$, in [156] the necessary and sufficient conditions for the existence of a global minimum of the GOP (8.18) are presented, and explicit formulas for the best parameters of the searched circle are given.

To solve the problem (8.18) for an arbitrary distance-like function $\mathcal{D}$ one can use not only the global optimization algorithm DIRECT but also some local optimization method (Newton method, quasi-Newton method, Nelder–Mead method, etc.) because in this case one can determine a good initial approximation of circle-center $\hat{C} = (\hat{p}, \hat{q})$ and radius $\hat{r}$.

For instance, using the $\mathcal{D}_A$ distance-like function, for $\hat{C}$ one can take the centroid of the set $\mathcal{A}$, and using the property (2.18) of the arithmetic mean

$$\sum_{i=1}^{m} \left(\|\hat{C} - a_i\|^2 - r^2 \right)^2 \geq \sum_{i=1}^{m} \left(\|\hat{C} - a_i\|^2 - \frac{1}{m} \sum_{s=1}^{m} \|\hat{C} - a_s\|^2 \right)^2 \tag{8.22}$$

one obtains an initial approximation for the radius

$$\hat{r}^2 = \frac{1}{m} \sum_{i=1}^{m} \|\hat{C} - a_i\|^2. \tag{8.23}$$

Using the $\mathcal{D}_2$ distance-like function, for $\hat{C}$ one can also take the centroid of the set $\mathcal{A}$, and similarly as when using the algebraic criterion, one gets an initial approximation for the radius

$$\hat{r} = \frac{1}{m} \sum_{i=1}^{m} \|\hat{C} - a_i\|. \tag{8.24}$$

And also when using the $\mathcal{D}_1$-distance, for $\hat{C}$ one can choose the centroid of the set $\mathcal{A}$, and using the property of the median (2.20)

$$\sum_{i=1}^{m} \left| \, \|\hat{C} - a_i\| - r \right| \geq \sum_{i=1}^{m} \left| \, \|\hat{C} - a_i\| - \operatorname*{med}_{a_s \in \mathcal{A}} \|\hat{C} - a_s\| \right|,$$

one obtains an initial approximation for the radius

$$\hat{r} = \operatorname*{med}_{a_s \in \mathcal{A}} \|\hat{C} - a_s\|. \tag{8.25}$$

Finally, let us mention that when there are outliers among data (e.g. real-world images usually contain errors or damages), in all three approaches it is advisable for $\hat{C}$ to choose the median of the set $\mathcal{A}$, and apply the $\mathfrak{D}_1$-approach.

Example 8.7 To the set $\mathcal{A}$ from Example 8.6 we will add 10 points (10% of $|\mathcal{A}|$) relatively far from the set $\mathcal{A}$ (see Fig. 8.2), consider these points as outliers, and denote the new data set by $\mathcal{A}_1$. This simulates the occurrence of errors in an image, something to be always reckoned with in real situations. We will try to reconstruct the original (gray) circle by solving the GOP (8.18).

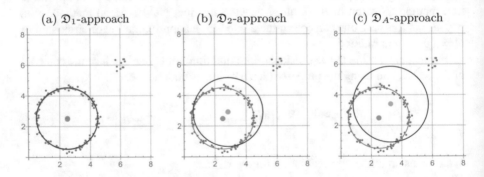

Fig. 8.2 Recognition of gray circle $K\big((2.5,\ 2.5), r\big)$ using different approaches

The result (red circle) obtained using $\mathfrak{D}_1$-approach with initial approximation $\hat{C} = \operatorname{med}(\mathcal{A}_1)$ and $\hat{r}$ given by (8.25) is shown in Fig. 8.2a.

The result (red circle) obtained using $\mathfrak{D}_2$-approach with initial approximation $\hat{C} = \operatorname{mean}(\mathcal{A}_1)$ and $\hat{r}$ given by (8.24) is shown in Fig. 8.2b.

The result (red circle) obtained using $\mathfrak{D}_A$-approach with initial approximation $\hat{C} = \operatorname{mean}(\mathcal{A}_1)$ and $\hat{r}$ given by (8.23) is shown in Fig. 8.2c.

As Fig. 8.2 clearly shows, the $\mathfrak{D}_A$-approach is the most sensitive one regarding outliers, and the $\mathfrak{D}_1$-approach is almost not affected by them. This is something to be taken into account in real-world applications.

Exercise 8.8 Check that a good approximation of the radius of the gray circle in previous example is $\hat{r} \approx \sqrt{\operatorname{tr}(\operatorname{cov}(\mathcal{A}))}$.

Exercise 8.9 Let $K(c, r) \subset \mathbb{R}$ be the "circle" centered at $c \in \mathbb{R}$ with radius $r > 0$. Generate a set $\mathcal{A} \subset \mathbb{R}$ with $m = 40$ data points originating from this circle, and solve the inverse problem using $\mathfrak{D}_1$, $\mathfrak{D}_2$, and $\mathfrak{D}_A$ approaches. Create corresponding graphical demonstrations. What does the `ContourPlot` for the appropriate minimizing function look like?

Multiple Circles Detection Problem in the Plane

Let $\mathcal{A} = \{a_i = (x_i, y_i) \in \mathbb{R}^2 : i = 1, \ldots, m\} \subset \Delta = [a, b] \times [c, d]$ be a data set originating from several unknown circles, i.e. which are not known in advance, and which have to be recognized or reconstructed. The corresponding artificial data set can be constructed in a similar way as was described for data originating for a single circle. Assume also that data coming from each circle satisfy the homogeneity property, i.e. that the data are mostly homogeneously scattered around that circle (see Definition 8.1).

The search for k-MAPart $\Pi^\star = \{\pi_1^\star, \ldots, \pi_k^\star\}$ where the cluster centers are circles $K_j^\star(C_j^\star, r_j^\star)$, $j = 1, \ldots, k$, boils down to finding optimal parameters $(p_j^\star, q_j^\star, r_j^\star)$, $j = 1, \ldots, k$, which are, in accordance with (8.2), solution to the following GOP (cf. (3.40)):

$$\underset{(p_j, q_j) \in \Delta, r_j \in [0, R]}{\arg\min} \sum_{i=1}^{m} \min_{1 \leq j \leq k} \mathfrak{D}\left(a_i, K_j((p_j, q_j), r_j)\right), \tag{8.26}$$

where the distance-like function $\mathfrak{D}$ has to be the same as the one used to find the best representative, and for R one can take $R = \frac{1}{2}\max\{b - a, d - c\}$.

In general, the function in (8.26) is non-convex and non-differentiable, but one can prove it to be Lipschitz-continuous (see [162]). Hence, one could try to solve the problem (8.26) using some globally optimizing method, e.g. the `DIRECT` algorithm, but numerically it would be an extremely demanding task.

The Number of Circles Is Known

The problem of searching for MAPart of a set $\mathcal{A}$ with circles as cluster centers, when the number k of clusters is known, can be solved using KCG algorithm (Algorithm 8.2), where γ-cluster-centers are circles, hence this algorithm will be called k-closest circles algorithm, KCC (see [108, 158]).

KCC Algorithm

In Step A of the KCC algorithm, the set $\mathcal{A}$ is split into k clusters using the Minimal Distance Principle and the same distance-like function $\mathfrak{D}$ to be used in the search for the best representative.

In Step B of the KCC algorithm, the best circle representative is determined for each cluster. How this can be done has previously been described.

The algorithm starts by choosing initial circle-centers or an initial partition. For initial circle-centers one can take k circles of the same radius, say 1, whose centers can be determined by solving the center-based clustering problem for the set $\mathcal{A}$ (see also [161]):

$$\underset{(p_j, q_j) \in \Delta}{\arg\min} \sum_{i=1}^{m} \min_{1 \le j \le k} \|(p_j, q_j) - a_i\|^2. \tag{8.27}$$

To do this it suffices to perform just a few, say 10 iterations of the DIRECT algorithm.

Example 8.10 The top row in Fig. 8.3 shows data originating from 2, 3, 4, and 5 circles and also circles of radius 1 with centers determined by solving the respective problems (8.27) as initial circle-centers for the KCC algorithm.

The resulting circles obtained after at most twelve (for $k = 3$ only six) iterations of the KCC algorithm are shown in the bottom row.

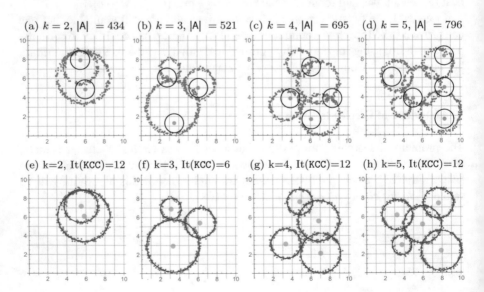

(a) $k = 2$, $|A| = 434$ (b) $k = 3$, $|A| = 521$ (c) $k = 4$, $|A| = 695$ (d) $k = 5$, $|A| = 796$

(e) k=2, It(KCC)=12   (f) k=3, It(KCC)=6   (g) k=4, It(KCC)=12   (h) k=5, It(KCC)=12

Fig. 8.3 Optimal partitions obtained by the KCC algorithm

The Number of Circles Is Not Known

If the number of circles from which the data originated is not known, we are going to search for optimal partitions with $k = 2, 3, \ldots, K$ clusters by adapting the incremental algorithm for circle-centers, and among them try to determine MAPart.

We will start the algorithm by choosing an initial circle-center $\hat{K}_1(\hat{C}_1, \hat{r}_1)$, for instance, the best representative of the set $\mathcal{A}$. Alternatively, instead of one, we can start with several initial circles. The approximation of the next circle-center $\hat{K}_2(\hat{C}_2, \hat{r}_2)$ is obtained by solving the GOP

$$\underset{(p,q)\in\Delta,\, r\in[0,R]}{\arg\min} \sum_{i=1}^{m} \min\{\mathfrak{D}(a_i, \hat{K}_1(C_1, r_1)), \mathfrak{D}(a_i, K((p, q), r))\}, \qquad (8.28)$$

where the distance-like function $\mathfrak{D}$ has to be the same as the one used to find the best representative.

Next, one has to apply the KCC algorithm to the circle-centers $\hat{K}_1$ and $\hat{K}_2$ to obtain a locally optimal 2-partition $\Pi^{(2)}$. Note that in solving the problem (8.28) it suffices to perform just a few, say 10 steps of the DIRECT algorithm to obtain a sufficiently good initial approximation for the KCC algorithm which will then provide a locally optimal 2-partition $\Pi^{(2)}$ with circle-centers $K_1^\star$ and $K_2^\star$.

In general, knowing $k-1$ circle-centers $\hat{K}_1, \ldots, \hat{K}_{k-1}$, the next circle-center is obtained by solving the following GOP

$$\underset{(p,q)\in\Delta,\, r\in[0,R]}{\arg\min} \sum_{i=1}^{m} \min\{\delta_{k-1}^i, \mathfrak{D}(a_i, K((p, q), r))\}, \qquad (8.29)$$

where $\delta_{k-1}^i = \min_{1\le s\le k-1} \mathfrak{D}(a_i, K(C_s, r_s))$. After that, using the KCC algorithm, we obtain the k-LOPart $\Pi^\star = \{\pi_1^\star, \ldots, \pi_k^\star\}$ with circle-centers $K_1^\star, \ldots, K_k^\star$.

The incremental algorithm is stopped by using the criterion (8.11).

The MAPart has to be chosen among K partitions obtained by the just described incremental algorithm. In order to do this one can calculate the CHG and DBG indexes using (8.12) and (8.13), respectively. To calculate the value of the DBG index we will use the explicit formula for the Hausdorff distance $d_H(K_1, K_2)$ between circles $K_1(C_1, r_1)$ and $K_2(C_2, r_2)$ (see [109, 158]):

$$d_H(K_1, K_2) = \|C_1 - C_2\| + |r_1 - r_2|. \qquad (8.30)$$

Nevertheless, it appears that it is better to use the specificity of this MGOD-problem and, in order that LOPart be MAPart as well, use the GO index or the necessary conditions (8.15).

Exercise 8.11 Prove that the Hausdorff distance $d_H(K_1, K_2)$ between circles $K_1(C_1, r_1)$ and $K_2(C_2, r_2)$ is indeed given by (8.30).

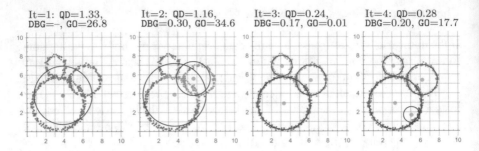

It=1: QD=1.33, It=2: QD=1.16, It=3: QD=0.24, It=4: QD=0.28
DBG=-, GO=26.8 DBG=0.30, GO=34.6 DBG=0.17, GO=0.01 DBG=0.20, GO=17.7

Fig. 8.4 Applying the incremental algorithm to the set from Fig. 8.3b

Example 8.12 Let $\mathcal{A}$ be the data set depicted in Fig. 8.3b. We will search for MAPart using the incremental algorithm.

We start the incremental algorithm with the circle $\hat{K}_1(\hat{C}_1, 1)$ where $\hat{C}_1 =$ mean($\mathcal{A}$), using the distance-like function $\mathfrak{D}_A$. The course of the algorithm is shown in Fig. 8.4. Since MinPts $= \lfloor \log |\mathcal{A}| \rfloor = 6$ and the ϵ-density of the set $\mathcal{A}$ based on 99 % quantile is $\epsilon(\mathcal{A}) = 0.369$, the incremental algorithm has to be stopped when QD < 0.369. As it can readily be seen, this happened in the third iteration. DBG and GO indexes also indicate that the 3-partition is the MAPart, and this is also confirmed by checking the necessary conditions (8.15).

Example 8.13 Let $\mathcal{A}$ be the data set depicted in Fig. 8.3c. We will search for MAPart using the incremental algorithm.

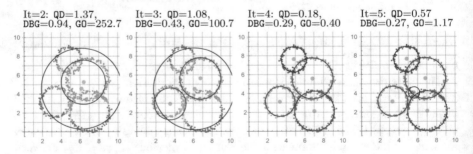

It=2: QD=1.37, It=3: QD=1.08, It=4: QD=0.18, It=5: QD=0.57
DBG=0.94, GO=252.7 DBG=0.43, GO=100.7 DBG=0.29, GO=0.40 DBG=0.27, GO=1.17

Fig. 8.5 Applying the incremental algorithm to the set from Fig. 8.3c

We start the incremental algorithm with the circle $\hat{K}_1(\hat{C}_1, 1)$ where $\hat{C}_1 =$ mean($\mathcal{A}$), using the distance-like function $\mathfrak{D}_A$. The course of the algorithm is shown in Fig. 8.5. Since MinPts $= \lfloor \log |\mathcal{A}| \rfloor = 6$ and the ϵ-density of the set $\mathcal{A}$ based on 99 % quantile is $\epsilon(\mathcal{A}) = 0.425$, the incremental algorithm has to be stopped when QD < 0.425. As it can readily be seen, this happened in the fourth iteration. The DBG index falsely suggests the 5-partition to be MAPart, whereas GO index correctly indicates that the 4-partition is the MAPart, which is also confirmed by checking the necessary conditions (8.15).

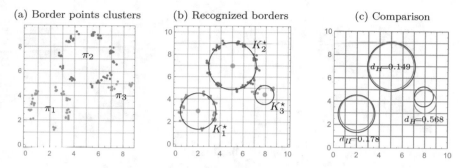

Fig. 8.6 Recognizing indistinct spherical data sets

Example 8.14 In Example 4.37 we constructed the data set $\mathcal{A}$ originating from three discs $K_1((2, 3), 1.5)$, $K_2((5, 7), 2)$, and $K_3((8, 4), 1)$ to which we added 2‰ outliers (see Fig 4.17a). Using the DBSCAN algorithm three clusters and the set $\pi_1 \cup \pi_2 \cup \pi_3$ of their border points were detected (see Fig 4.21c and also Fig. 8.6a). We will try to reconstruct boundaries of these discs, i.e. circles ∂K_1, ∂K_2, ∂K_3, based on these border points. Note that this is a problem of recognizing indistinct spherical objects, which has numerous applications in medicine (see e.g. [142]).

Using the $\mathfrak{D}_1$-approach, for each π_j, $j = 1, 2, 3$, we will find a circle $K_j^\star$ as the best representative in accordance with (8.18) (see Fig. 8.6b). Figure 8.6c shows comparison between the obtained red circles and original gray circles. Formula (8.30) for the Hausdorff distance between two circles gives: $d_H(\partial K_1, K_1^\star) = 0.178$, $d_H(\partial K_2, K_2^\star) = 0.149$, $d_H(\partial K_3, K_3^\star) = 0.568$.

Notice that although the sets π_1, π_2, and π_3 do not satisfy the homogeneity property, the proposed method still gives a very acceptable result.

Real-World Images

Let us consider the problem of detecting circles in real-world images. The problem of circle detection is most often observed in images from medicine, biology, agriculture, robotics, etc.

Figure 8.7 shows the case of a breast cancer. By detecting circles in the image one can precisely determine the size and the position of the objects.

For this purpose, first one has to carry out the Edge Detection procedure. In our case, this was done using the *Canny filter* which was implemented in module EdgeDetect[] in the *Mathematica* program [195] (see Fig. 8.7b). In this way, we obtained the data set $\mathcal{A}$ with $|\mathcal{A}| = 415$ points. Figure 8.7c shows both—the result of solving the multiple circle detection problem for the set $\mathcal{A}$ and the original image.

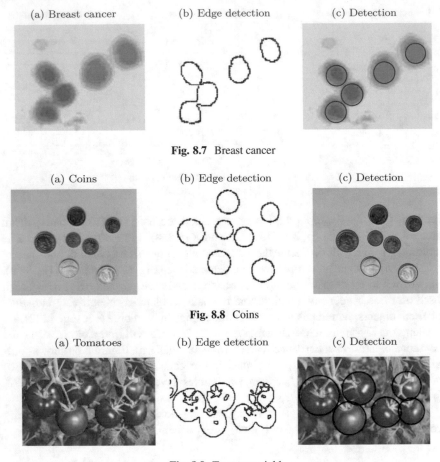

(a) Breast cancer   (b) Edge detection   (c) Detection

Fig. 8.7 Breast cancer

(a) Coins   (b) Edge detection   (c) Detection

Fig. 8.8 Coins

(a) Tomatoes   (b) Edge detection   (c) Detection

Fig. 8.9 Tomatoes yield

The images of 7 coins (Fig. 8.8) and of tomato yield (Fig. 8.9) have been analyzed in a similar way.

8.1.5 Multiple Ellipses Detection Problem

Let $\mathcal{A} = \{a_i = (x_i, y_i) : i = 1, \ldots, m\} \subset \mathbb{R}^2$ be a data set originating from several unknown ellipses in the plane, which have to be recognized or reconstructed.

This problem can be found in the literature as the *Multiple Ellipse Detection* (MED) problem. It appears in pattern recognition and computer vision [103], agriculture [93], anomaly detection in wireless sensor networks [119], astronomical and geological shape segmentation [198], applications in medicine [142], robotics, object detection, pupil tracking, and other image processing in industrial applications [2, 135].

The methods for solving MED-problem are mostly based on LS-fitting [59, 66], Hough transformations [2, 47, 120], etc. We are going to consider this problem as a center-based clustering problem with cluster centers being ellipses (see [68, 158]).

A Single Ellipse as the Representative of a Data Set

Let us first turn to the problem of determining a single ellipse as the best representative of the data set $\mathcal{A}$. Suppose $\mathcal{A}$ originated from a single ellipse $E(C, a, b, \vartheta) \subset \Delta = [a, b] \times [c, d] \subset \mathbb{R}^2$, centered at C, with semi-axes $a, b > 0$, and slanted, i.e. rotated about the center C, by the angle $\vartheta \in [0, \pi]$.

Artificial Data Set Originating from a Single Ellipse

An artificial data set originating from a single ellipse can be constructed like this:

First, on the ellipse $E(O, a, b, 0)$ centered at the origin O with semi-axes $a, b > 0$, and not slanted, i.e. $\vartheta = 0$, choose m uniformly allocated points according to [68] (see Fig. 8.10a), and then to each point add a randomly generated noise from bivariate normal distribution with expectation $0 \in \mathbb{R}^2$ and the covariance matrix $\sigma^2 I$, where I is the identity 2×2 matrix. This ensures that the data coming from the ellipse E satisfy the homogeneity property, i.e. that the data are mostly homogeneously scattered around this ellipse (see Fig. 8.10b). Denote by $\mathcal{A}$ the obtained set translated by C and rotated around C by the angle ϑ (see Fig. 8.10c). In accordance with Definition 8.1, the set $\mathcal{A}$ has local density $\rho(\mathcal{A}) = \frac{m}{|E|}$, where $|E|$ denotes the length of the ellipse, which can be estimated using the well known *Ramanujan approximation* (see [146])

$$|E| \approx \pi(a + b)\left(1 + \frac{3h}{10 + \sqrt{4 - 3h}}\right), \quad h = \frac{(a-b)^2}{(a+b)^2}. \tag{8.31}$$

Example 8.15 In the way which was just described, the following program will generate artificial data set $\mathcal{A}$, $|\mathcal{A}| = 100$, originating from the ellipse $E(C, a, b, \vartheta)$ with center $C = (4, 4)$, semi-axes $a = 4$, $b = 1$ and slanted by the angle $\vartheta = \frac{\pi}{4}$ (see Fig. 8.10).

```
In[1]:= a = 4; b = 1; theta = Pi/4; Cen = {4, 4}; m = 100; sigma = 0.08;
                    (* Points on ellipse *)
        h = (a-b)^2/(a+b)^2; dEl = Pi*(a+b)*(1+3 h/(10 + Sqrt[4-3 h]));
        tt = RandomVariate[ProbabilityDistribution[
                    a Sqrt[1-(1-b^2/a^2) Cos[x]^2]/dEl,{x,0,2 Pi}],m];
        toc1 = Transpose[{a Cos[tt], b Sin[tt]}];
                    (* Points around ellipse *)
        toc2 = toc1 + RandomVariate[NormalDistribution[0, sigma], {m, 2}];
                    (* Rotation *)
        Q = {{Cos[theta], -Sin[theta]}, {Sin[theta], Cos[theta]}};
        A = Table[Cen + Q.toc2[[i]], {i, Length[toc1]}];
```

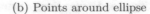

(a) Points on ellipse   (b) Points around ellipse   (c) Data set $\mathcal{A}$

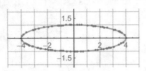
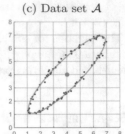

Fig. 8.10 Generating data originating from the ellipse $E(C, a, b, \vartheta)$, with center $C = (4, 4)$, semi-axes $a = 4$, $b = 1$, and slanted by the angle $\vartheta = \frac{\pi}{4}$

The Best Representative

The best representative of our set $\mathcal{A}$ is an ellipse $E(C, a, b, \vartheta)$ which we will regard as an M-circle $M(C, \Sigma)$ (see (6.14)) given by the equation

$$d_m(x, C; \Sigma) = 1, \tag{8.32}$$

where Σ is a symmetric positive definite matrix. Multiplying this equation by $r^2 :=$ $\sqrt{\det \Sigma}$ (see (6.15)), the equation for $M(C, \Sigma)$ becomes

$$d_M(x, C; \Sigma) = r^2, \tag{8.33}$$

with d_M being the normalized Mahalanobis distance-like function (see (6.18)), i.e. $M(C, \Sigma)$ can be regarded as the M-circle centered at C with radius r, covariance matrix $\Sigma = \begin{bmatrix} \alpha_1 & \alpha_2 \\ \alpha_2 & \alpha_3 \end{bmatrix}$ (see Definition 6.16 and Example 6.17) and will be denoted by $E(C, r, \Sigma)$.

The relation between these two notations of an ellipse is given by the following lemma.

Lemma 8.16 *The ellipse $E(C, \xi, \eta, \vartheta)$ centered at $C = (p, q)$, with semi-axes ξ, η, and slanted by the angle ϑ, which is defined parametrically by*

$$E(C, \xi, \eta, \vartheta) = \big\{ (\xi(\tau), y(\tau)) = (p, q) + (\xi \cos \tau, \eta \sin \tau) Q(-\vartheta) : \tau \in [0, \pi] \big\}, \tag{8.34}$$

where $Q(\vartheta) = \begin{bmatrix} \cos \vartheta & -\sin \vartheta \\ \sin \vartheta & \cos \vartheta \end{bmatrix}$ is the corresponding rotation matrix, can be represented as the M-circle [111]

$$E(C, r, \Sigma) = \{u \in \mathbb{R}^2 : d_M(C, u; \Sigma) = r^2\}, \quad \Sigma = \begin{bmatrix} \alpha_1 & \alpha_2 \\ \alpha_2 & \alpha_3 \end{bmatrix}, \tag{8.35}$$

where $r^2 = \sqrt{\det \Sigma} = \xi \eta$.

Conversely, using the eigenvalue decomposition, the M-circle $E(C, r, \Sigma)$ translates into the ellipse $E(C, \xi, \eta, \vartheta)$, where

$$\mathrm{diag}(\xi^2, \eta^2) = Q(\vartheta)\left(\frac{r^2}{\sqrt{\det \Sigma}} \Sigma\right) Q(\vartheta)^T, \quad Q(\vartheta) = \begin{bmatrix} \cos \vartheta & -\sin \vartheta \\ \sin \vartheta & \cos \vartheta \end{bmatrix}, \quad (8.36)$$

and the angle ϑ is given by

$$\vartheta = \frac{1}{2}\arctan\frac{2\alpha_2}{\alpha_1 - \alpha_3} \in \begin{cases} [0, \pi/4), & \alpha_2 \geq 0 \text{ and } \alpha_1 > \alpha_3 \\ [\pi/4, \pi/2), & \alpha_2 > 0 \text{ and } \alpha_1 \leq \alpha_3 \\ [\pi/2, 3\pi/4), & \alpha_2 \leq 0 \text{ and } \alpha_1 < \alpha_3 \\ [3\pi/4, \pi), & \alpha_2 < 0 \text{ and } \alpha_1 \geq \alpha_3 \end{cases}. \quad (8.37)$$

Proof The first assertion is proved by direct verification.

To prove the converse, write down the equation for the ellipse $E(C, r, \Sigma)$ like this:

$$\frac{1}{r^2} d_M(C, u; \Sigma) = 1 \implies \frac{\sqrt{\det \Sigma}}{r^2}(C - u) \Sigma^{-1}(C - u)^T = 1$$

$$\implies (C - u)\left(\frac{r^2}{\sqrt{\det \Sigma}} \Sigma\right)^{-1}(C - u)^T = 1,$$

and then the assertion is proved by using the eigenvalue decomposition of the matrix $\frac{r^2}{\sqrt{\det \Sigma}} \Sigma$. $\qquad\square$

Exercise 8.17 We are considering vectors as n-tuples, but if we write vectors as one-column matrices, show that the ellipse $E(C, \xi, \eta, \vartheta)$ centered at $C = (p, q)$, with semi-axes ξ, η, and slanted by the angle ϑ, can be defined parametrically by

$$E(C, \xi, \eta, \vartheta) = \left\{ \begin{bmatrix} \xi(\tau) \\ y(\tau) \end{bmatrix} = \begin{bmatrix} p \\ q \end{bmatrix} + Q(\vartheta) \begin{bmatrix} \xi \cos \tau \\ \eta \sin \tau \end{bmatrix} : \tau \in [0, \pi] \right\},$$

where $Q(\vartheta) = \begin{bmatrix} \cos \vartheta & -\sin \vartheta \\ \sin \vartheta & \cos \vartheta \end{bmatrix}$ is the corresponding rotation matrix.

The best representative of the set $\mathcal{A}$ is the M-circle $E(C, r, \Sigma)$, with parameters $p, q, r, \alpha_1, \alpha_2, \alpha_3$ being the solution to the GOP

$$\underset{(p,q)\in\Delta,\, r\in[0,R],\, \Sigma\in M_2}{\arg\min} \sum_{i=1}^{m} \mathfrak{D}(a_i, E((p, q), r; \Sigma)), \quad (8.38)$$

where $R = \frac{1}{2}\max\{b - a, d - c\}$, M_2 is the set of positive definite symmetric 2×2 matrices, and $\mathfrak{D}$ is the distance-like function which is in the literature usually defined in one of the following three ways (see [108, 125, 158]):

$$\mathfrak{D}_1(a_i, E((p, q), r; \Sigma)) = \big|\|a_i - (p, q)\|_\Sigma - r\big| \quad (8.39)$$

$$\mathfrak{D}_2(a_i, E((p,q),r;\Sigma)) = (\|a_i - (p,q)\|_\Sigma - r)^2 \tag{8.40}$$

$$\mathfrak{D}_A(a_i, E((p,q),r;\Sigma)) = (\|a_i - (p,q)\|_\Sigma^2 - r^2)^2 \tag{8.41}$$

and where $\|\cdot\|_\Sigma$ is defined in (6.18).

When the distance $\mathfrak{D}_1$ is used ($\mathfrak{D}_1$-approach) one says that the circle from (8.38) was obtained in accordance with the LAD-principle, using $\mathfrak{D}_2$ distance-like function ($\mathfrak{D}_2$-approach)—in accordance with the TLS-principle, and using $\mathfrak{D}_A$ distance-like function ($\mathfrak{D}_A$-approach)—in accordance with the *Algebraic Criterion*.

In order to solve the problem (8.38) one can always apply the global optimization algorithm DIRECT, but it is also possible to apply some local optimization method, like the Newton method, quasi-Newton method, Nelder-Mead method, etc., because in this situation one can determine a good initial approximation of the circle-center $\hat{C} = (\hat{p}, \hat{q})$ and of the radius $\hat{r}$.

Using the distance-like function $\mathfrak{D}_A$, $\hat{C}$ can be chosen to be the centroid of the set $\mathcal{A}$, and using the property (2.18) of the arithmetic mean

$$\sum_{i=1}^m \left(\|\hat{C} - a_i\|_\Sigma^2 - r^2\right)^2 \geq \sum_{i=1}^m \left(\|\hat{C} - a_i\|_\Sigma^2 - \frac{1}{m}\sum_{s=1}^m \|\hat{C} - a_s\|_\Sigma^2\right)^2, \tag{8.42}$$

one obtains the approximation for the radius

$$\hat{r}^2 = \frac{1}{m}\sum_{i=1}^m \|\hat{C} - a_i\|_\Sigma^2. \tag{8.43}$$

Also, using the distance-like function $\mathfrak{D}_2$, $\hat{C}$ can be chosen to be the centroid of the set $\mathcal{A}$. As in the case of algebraic criterion, one obtains the following initial approximation for the radius

$$\hat{r} = \frac{1}{m}\sum_{i=1}^m \|\hat{C} - a_i\|_\Sigma. \tag{8.44}$$

And again, using the $\mathfrak{D}_1$ distance, $\hat{C}$ can be chosen to be the centroid of the set $\mathcal{A}$, and using the property (2.20) of the median

$$\sum_{i=1}^m \left|\|\hat{C} - a_i\|_\Sigma - r\right| \geq \sum_{i=1}^m \left|\|\hat{C} - a_i\|_\Sigma - \underset{a_s \in \mathcal{A}}{\mathrm{med}} \|\hat{C} - a_s\|_\Sigma\right|,$$

one obtains the initial approximation for the radius:

$$\hat{r} = \underset{a_s \in \mathcal{A}}{\mathrm{med}} \|\hat{C} - a_s\|_\Sigma. \tag{8.45}$$

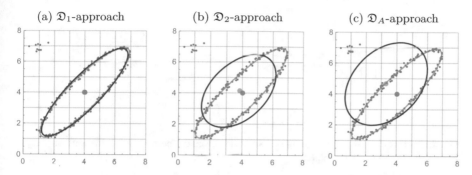

Fig. 8.11 Reconstruction of gray ellipse $E(C, a, b, \vartheta)$, with center $C = (4, 4)$, semi-axes $a = 4$, $b = 1$ and slanted by $\vartheta = \frac{\pi}{4}$, using $\mathfrak{D}_1$, $\mathfrak{D}_2$, and $\mathfrak{D}_A$ approaches

Finally note that when data contain outliers, and in the real-world images this is rarely not the case (errors, defects, damages), it is always good to choose $\hat{C}$ to be the median of $\mathcal{A}$ and apply the $\mathfrak{D}_1$-approach.

Example 8.18 To the set $\mathcal{A}$ from Example 8.15 we will add 10 points (10% of $|\mathcal{A}|$) relatively far from the set $\mathcal{A}$ (see Fig. 8.11), consider these points to be outliers, and denote the new data set by $\mathcal{A}_1$. This simulates the occurrence of errors in an image, something to be always reckoned with in real situations. We will try to reconstruct the original gray ellipse by solving the GOP (8.38).

The result (red ellipse) obtained using $\mathfrak{D}_1$-approach with initial approximation $\hat{C} = \mathrm{med}(\mathcal{A}_1)$ and $\hat{r}$ given by (8.45) is shown in Fig. 8.11a.

The result (red ellipse) obtained using $\mathfrak{D}_2$-approach with initial approximation $\hat{C} = \mathrm{mean}(\mathcal{A}_1)$ and $\hat{r}$ given by (8.44) is shown in Fig. 8.11b.

The result (red ellipse) obtained using $\mathfrak{D}_A$-approach with initial approximation $\hat{C} = \mathrm{mean}(\mathcal{A}_1)$ and $\hat{r}$ given by (8.43) is shown in Fig. 8.11c.

As Fig. 8.11 clearly shows, $\mathfrak{D}_2$-approach and in particular $\mathfrak{D}_A$-approach are sensitive regarding outliers, while the $\mathfrak{D}_1$-approach is almost not affected by them. This is something to be taken into account in real-world applications.

Remark 8.19 Determining the Euclidean distance from a point to an ellipse is not nearly as simple as determining the Euclidean distance from a point to a circle (see [186]).

In general, determining the Euclidean distance $d(T, E(C, a, b, \vartheta))$ between a point T and an ellipse $E(C, a, b, \vartheta)$ (see Fig. 8.12a), after a rotation and a translation, translates into a simpler problem of finding the distance between the point $T' = (T - C) Q(-\vartheta)$ and the ellipse $E'(O, a, b, 0)$ given by the equation $\frac{x^2}{a^2} + \frac{y^2}{b^2} = 1$, where $T' = (T - C) Q(-\vartheta)$ and Q is given by (8.36) (see Fig. 8.12b).

Let $P' = (\xi, \eta)$ be the orthogonal projection of $T' = (x_1, y_1)$ to the ellipse E'. The straight line through T' and P' is perpendicular to the tangent line to ellipse E' at P', and therefore $\frac{y_1 - \eta}{x_1 - \xi} = \frac{a^2 \eta}{b^2 \xi}$. Solving the system

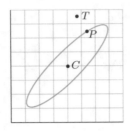
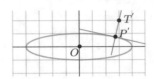

Fig. 8.12 Reducing the problem of finding the distance between a point and an ellipse $E(C, a, b, \vartheta)$ to a simpler problem

$$b^2 \xi (y_1 - \eta) = a^2 \eta (x_1 - \xi), \quad b^2 \xi^2 + a^2 \eta^2 = a^2 b^2, \qquad (8.46)$$

gives several real solutions for P', so one has to choose the one closest to T'. Then $d(T, E(C, a, b, \vartheta)) = d(T', P')$.

Exercise 8.20 Find the orthogonal projection of the point T to the ellipse $E(C, a, b, \vartheta)$.

Multiple Ellipses Detection Problem

Let $\mathcal{A} = \{a_i = (x_i, y_i) \in \mathbb{R}^2 : i = 1, \ldots, m\} \subset \Delta = [a, b] \times [c, d]$ be a data set originating from several unknown ellipses, which have to be recognized or reconstructed. The corresponding artificial data set can be constructed in a similar way as for data coming from a single ellipse. In addition, assume that data coming from each ellipse satisfy the homogeneity property, i.e. that data are mostly homogeneously scattered around that ellipse (see Definition 8.1).

Searching for k-MAPart $\Pi^\star = \{\pi_1^\star, \ldots, \pi_k^\star\}$, where the cluster centers $\pi_j^\star$ are ellipses $E_j^\star(C_j^\star, r_j^\star; \Sigma_j^\star)$, boils down to finding optimal parameters $(p_j^\star, q_j^\star, r_j^\star, \Sigma_j^\star)$, $j = 1, \ldots, k$, which are, in accordance with (8.2), the solution to the following GOP (cf. (3.40)):

$$\operatorname*{arg\,min}_{(p_j, q_j) \in \Delta, \, r_j \in [0, R], \, \Sigma_j \in M_2} \sum_{i=1}^{m} \mathfrak{D}(a_i, E((p_j, q_j), r_j; \Sigma_j)), \qquad (8.47)$$

where the distance-like function $\mathfrak{D}$ has to be same as the one used to find the best representative.

The Number of Ellipses Is Known in Advance

Searching for MAPart of a set $\mathcal{A}$ with ellipses as cluster centers when the number of clusters is known can be solved using the KCG algorithm (Algorithm 8.2), where, in this case, the γ-cluster-centers are ellipses, hence the algorithm will be called the *k-closest ellipses algorithm*, KCE (see [108, 158]).

KCE Algorithm

Choosing a distance-like function $\mathcal{D}$ and using the Minimal Distance Principle, Step A of the KCE algorithm divides the set $\mathcal{A}$ into k clusters.

In Step B of the KCE algorithm, for each of these k clusters one has to determine an ellipse which is the best representative of that cluster. Earlier, starting on page 182, we discussed how to do that.

The algorithm can be started by choosing M-circle-centers or by choosing an initial partition. For initial M-circle-centers one can take k circles of a given radius, say 1, with centers obtained by solving the center-based clustering problem (8.27) for the set $\mathcal{A}$. To do this, it will suffice to perform just a few, say 10 iterations of the DIRECT algorithm.

(a) $k = 2$, $|A| = 399$ (b) $k = 3$, $|A| = 552$ (c) $k = 4$, $|A| = 831$ (d) $k = 5$, $|A| = 1011$

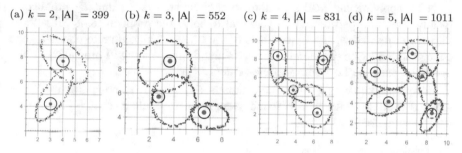

Fig. 8.13 Initial approximations obtained by 10 iteration of the DIRECT algorithm

(a) k=2, It(KCE)=5   (b) k=3, It(KCE)=2   (c) k=4, It(KCE)=7   (d) k=5, It(KCE)=10

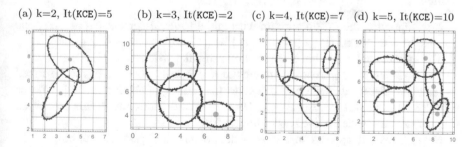

Fig. 8.14 Optimal partitions obtained by the KCE algorithm

Example 8.21 Figure 8.13 shows data originating from two, three, four, and five ellipses. For each of these data sets, first we obtained initial circle-centers for the KCC algorithm. The centers were obtained using the DIRECT algorithm to solve the corresponding center-based clustering problem (see also Chap. 3), and radii were taken to be 0.5. The results of the KCE algorithm are shown in Fig. 8.14.

The Number of Ellipses Is Not Known in Advance

If the number of ellipses from which the data originated is not known, we are going to search for optimal partitions with $k = 2, 3, \ldots, K$ clusters by adapting the incremental algorithm for M-circle-centers.

The algorithm starts by choosing an initial M-circle-center $\hat{E}_1(\hat{C}_1, \hat{r}_1, \hat{\Sigma}_1)$. This could be the best representative of the set $\mathcal{A}$, for instance. The approximation of the next M-circle-center will be found as the ordinary circle $\hat{E}_2(\hat{C}_2, \hat{r}_2, I_2)$, where I_2 is the 2×2 identity matrix, solving the following GOP:

$$\underset{(p,q)\in\Delta,\, r\in[0,R]}{\arg\min} \sum_{i=1}^m \min\{\mathfrak{D}(a_i, \hat{E}_1(\hat{C}_1, \hat{r}_1, \hat{\Sigma}_1)), \mathfrak{D}(a_i, E((p,q), r, I_2))\}. \tag{8.48}$$

Next, applying the KCE algorithm to M-circle-centers $\hat{E}_1$ and $\hat{E}_2$ produces a locally optimal 2-partition $\Pi^{(2)}$. Note that to solve the problem (8.48), it suffices to perform just a few, say 10 iterations of the DIRECT algorithm to obtain a good enough initial approximation for the KCE algorithm, which will produce a locally optimal 2-partition $\Pi^{(2)}$ with M-circle-centers $E_1^\star, E_2^\star$.

In general, knowing $k - 1$ M-circle-centers $\hat{E}_1, \ldots, \hat{E}_{k-1}$, the approximation of the next M-circle-center will be found as an ordinary circle $E(C, r, I_2)$ by solving the following GOP:

$$\underset{(p,q)\in\Delta,\, r\in[0,R]}{\arg\min} \sum_{i=1}^m \min\{\delta_{k-1}^{(i)}, \mathfrak{D}(a_i, E((p,q), r, I_2))\}, \tag{8.49}$$

where $\delta_{k-1}^{(i)} = \underset{1\le s\le k-1}{\min} \{\mathfrak{D}(a_i, \hat{E}_s(\hat{C}_s, \hat{r}_s, \hat{\Sigma}_s))\}$.

Next, applying the KCE algorithm with initial M-circle-centers $\hat{E}_1, \ldots, \hat{E}_k$, one obtains the k-GOPart $\Pi^\star = \{\pi_1^\star, \ldots, \pi_k^\star\}$ with M-circle-centers $E_1^\star, \ldots, E_k^\star$.

The incremental algorithm is stopped by using the criterion (8.11).

The MAPart has to be chosen among the K partitions obtained by the described incremental algorithm. One can calculate the values of CHG and DBG indexes using (8.12) and (8.13), respectively. Using the DBG index the distance between two ellipses can be determined according to [109].

Nevertheless, it appears that it is better to use the specificity of this MGOD problem and, in order that LOPart be MAPart as well, use the GO index or the necessary conditions (8.15).

Example 8.22 Let $\mathcal{A}$ be the data set as in Fig. 8.13b. We will search for MAPart using the incremental algorithm.

The incremental algorithm will be started with the circle $\hat{K}_1(\hat{C}_1, 0.5)$, where $\hat{C}_1 = \text{mean}(\mathcal{A})$ (see Fig. 8.15a). Applying the algebraic criterion we get the best representative of the set $\mathcal{A}$ (red ellipse in Fig. 8.15a). The course of the incremental algorithm can be followed in Fig. 8.15b–d. Since for MinPts $= \lfloor \log |\mathcal{A}| \rfloor = 6$, the ϵ-density of the set $\mathcal{A}$ based on the 99% quantile is $\epsilon(\mathcal{A}) = 0.349$, so the incremental

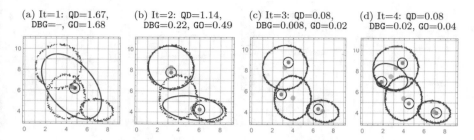

Fig. 8.15 Optimal partitions obtained by the KCE algorithm

algorithm has to be stopped as soon as QD < 0.349. As can readily be seen, this happened at the third iteration.

The DBG and GO indexes also indicate that the 3-partition is the MAPart, and this is also confirmed by checking the necessary conditions (8.15).

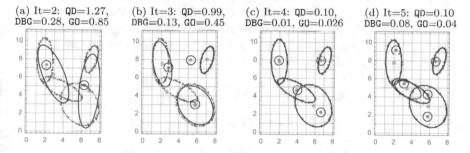

Fig. 8.16 Optimal partitions obtained by the KCE algorithm

Example 8.23 Let $\mathcal{A}$ be the data set as in Fig. 8.13c. We will search for MAPart using the incremental algorithm.

We will start the incremental algorithm with circle $\hat{K}_1(\hat{C}_1, 0.5)$, where $\hat{C}_1 = \text{mean}(\mathcal{A})$, and apply the algebraic criterion. The course of the incremental algorithm can be followed in Fig. 8.16b–d. Since for MinPts $= \lfloor \log |\mathcal{A}| \rfloor = 6$, the ϵ-density of the set $\mathcal{A}$ based on the 99 % quantile is $\epsilon(\mathcal{A}) = 0.299$, so the incremental algorithm has to be stopped as soon as QD < 0.299. As one can see, this happened at the fourth iteration.

The DBG index falsely suggests the 5-partition to be MAPart, whereas GO index correctly indicates that the 4-partition is the MAPart, which is also confirmed by checking the necessary conditions (8.15).

Example 8.24 In Example 4.41 we constructed a data set $\mathcal{A}$ derived from the region between two non-parallel ellipses, and from a disc, and to which we added 2‰ outliers (see Fig. 4.17c). Using the DBSCAN algorithm, two clusters were detected, as well as their border points (see Fig. 4.22b). Denote by π_1 the set of border points

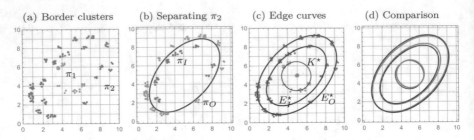

Fig. 8.17 Recognizing indistinct borders of ellipsoidal data sets

of the disc, and by π_2 the set of border points of the region between the two ellipses (see Fig. 8.17a). Based on π_1 and π_2 we will try to reconstruct the two ellipses and the circle from which the data originated.

Similarly as in Example 8.14, we will determine the circle $K^\star$, the best representative of the set π_1, by solving the GOP (8.18) using the $\mathfrak{D}_1$-approach.

The set π_2 carries data about two ellipses—the outer and the inner one. Therefore we will first find an ellipse $E_0(C_0, r_0, \Sigma_0)$ between these two as the best representative of the cluster π_2, by solving the GOP (8.38) using the $\mathfrak{D}_2$-approach. The ellipse $E_0(C_0, r_0, \Sigma_0)$ separates sets π_I and π_O of points originating from the inner, respectively, and outer ellipse (see Fig. 8.17b).

$$\pi_I = \{a_i \in \pi_2 : \sqrt{d_M(a_i, C_0, \Sigma_0)} < r_0\},$$

$$\pi_O = \{a_i \in \pi_2 : \sqrt{d_M(a_i, C_0, \Sigma_0)} \geq r_0\}.$$

The inner ellipse $E_I^\star$ and the outer ellipse $E_O^\star$ are now determined as the best representatives of sets π_I and π_O, respectively, by solving the corresponding GOP (8.38) using the $\mathfrak{D}_2$-approach.

All border curves detected are shown in Fig. 8.17c, while Fig. 8.17d shows the comparison between obtained (red) and the original (gray) curves.

Notice that although the sets π_1 and π_2 do not satisfy the homogeneity property, the proposed method still gives a very acceptable result.

Real-World Images

Let us consider the MED-problem in real-world images. By detecting ellipses in the image one can precisely determine the size and the position of objects.

We will show several examples of images in which we want to detect ellipses. The following figures consist of three subfigures: the first one is the image of original objects, the second one is obtained using *Mathematica* module EdgeDetect [], and the third one shows both the original image and the detected ellipses (Figs. 8.18–8.21).

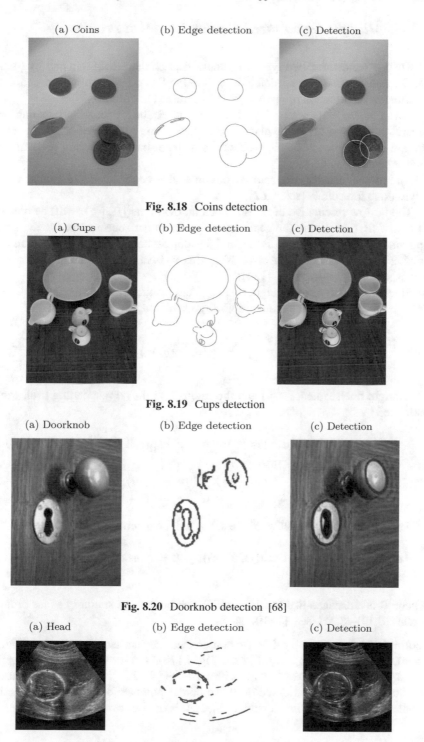

(a) Coins   (b) Edge detection   (c) Detection

Fig. 8.18 Coins detection

(a) Cups   (b) Edge detection   (c) Detection

Fig. 8.19 Cups detection

(a) Doorknob   (b) Edge detection   (c) Detection

Fig. 8.20 Doorknob detection [68]

(a) Head   (b) Edge detection   (c) Detection

Fig. 8.21 Ultrasound fetal head detection [104]

8.1.6   *Multiple Generalized Circles Detection Problem*

In Chap. 4 we defined two types of generalized circles—*oval*, defined by (4.28)–(4.29), and *circle-arc oval*, defined by (4.30)–(4.31). Let us briefly describe the multiple ovals detection problem as it is formulated in [163].

Let $\mathcal{A} = \{a_i = (x_i, y_i) : i = 1, \ldots, m\} \subset \mathbb{R}^2$ be a data set originating from several unknown ovals in the plane, which have to be recognized or reconstructed. We are going to consider this problem as a center-based clustering problem with cluster centers being ovals.

An artificial data set originating from an oval is constructed in a similar way as it was done for a circle (see [163]).

The best representative of the set $\mathcal{A}$ as an oval $\mathsf{Oval}([\mu, \nu], r)$ will be obtained like this: first determine the centroid $\bar{c}$ and the covariance matrix for the set $\mathcal{A}$, and find the unit eigenvectors v_1 and v_2 corresponding to the largest and the smallest eigenvalues of that matrix. Note that v_1 is also the direction vector of the corresponding TLS-line (see [124]).

Since the length of the straight line segment (as the center of the $\mathsf{Oval}([\mu, \nu], r)$ representing $\mathcal{A}$) can be according to [185], estimated by

$$\delta = \tfrac{2}{|\mathcal{A}|} \sum_{i=1}^{m} \left| \langle (a_i - \bar{c}), v_1 \rangle \right|, \tag{8.50}$$

the straight line segment $[\mu, \nu]$ and the radius r of the corresponding oval, can be estimated by

$$[\mu, \nu] \approx [\bar{c} - \tfrac{1}{2}\delta v_1, \bar{c} + \tfrac{1}{2}\delta v_1], \tag{8.51}$$

$$r \approx \tfrac{1}{|\mathcal{A}|} \sum_{i=1}^{m} \left| \langle (a_i - \bar{c}), v_2 \rangle \right|. \tag{8.52}$$

The best oval representative of the set $\mathcal{A}$ will be determined by solving the GOP

$$\underset{\mu, \nu \in \Delta, \, r \in [0, R]}{\arg\min} \sum_{a \in \pi_j} \mathfrak{D}(a, \mathsf{Oval}([\mu, \nu], r)), \quad R = \tfrac{1}{2} \max\{b - a, d - c\}, \tag{8.53}$$

where $\mathfrak{D}$ is a distance-like function which can be defined similarly as for ordinary circles (8.19)–(8.21) (see [163]).

Example 8.25 In Example 4.50 we constructed the data set $\mathcal{A}$ originating from the oval $\mathsf{Oval}([(4.5, 4.5), (8, 8)], 0.8)$ (see Fig. 4.17b), and the set of its border points was detected using the DBSCAN algorithm (see Fig 4.22a). We will try to reconstruct the boundary of the oval based on these border points. Notice that this is again a problem of recognizing an indistinct object, this time an oval.

Using the $\mathfrak{D}_2$-approach with (8.50)–(8.52), we obtain the initial cluster center approximation $\mathsf{Oval}([(5.2, 5.2), (7.5, 7.5)], 0.77)$ (see Fig. 8.22a). The best representative, $\mathsf{Oval}([(4.43491, 4.39389), (8.12537, 8.07639)], 0.83045)$, is obtained as the solution to the GOP (8.53) (see Fig. 8.22b). Figure 8.22c shows comparison between the obtained (red) and the original (gray) generalized circles.

Notice that although the set $\mathcal{A}$ does not satisfy the homogeneity property, the proposed method still gives a very acceptable result.

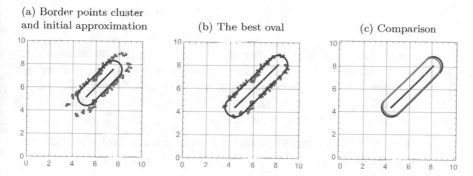

(a) Border points cluster and initial approximation   (b) The best oval   (c) Comparison

Fig. 8.22 Recognizing indistinct generalized circle border from Example 4.50

Remark 8.26 The corresponding *Mathematica*-module DGcir[] needed to solve the GOP (8.53) is described in Sect. 9.3, and the link to appropriate *Mathematica*-code is supplied.

The search for k-MAPart $\Pi^\star = \{\pi_1^\star, \ldots, \pi_k^\star\}$, where the cluster centers ovals are $\mathsf{Oval}_j^\star([\mu_j^\star, \nu_j^\star], r_j^\star)$, $j = 1, \ldots, k$, boils down to finding optimal parameters $(\mu_j^\star, \nu_j^\star, r_j^\star)$, $j = 1, \ldots, k$, which are, in accordance with (8.2), the solution to the following GOP (cf. (3.40)):

$$\underset{\mu_j, \nu_j \in \Delta, r_j \in [0, R]}{\arg\min} \sum_{i=1}^{m} \min_{1 \le j \le k} \mathfrak{D}(a_i, \mathsf{Oval}_j([\mu_j, \nu_j], r_j)), \tag{8.54}$$

with the distance-like function $\mathfrak{D}$ being the same as the one used to find the best representative.

Adapting the corresponding KCG algorithm is done similarly as in the case of MCD and MED problems. In this case, the initial approximation can be chosen as initial centers determined by k ordinary circles of a given radius, say 1, whose centers can be obtained by solving the center-based clustering problem (8.27) for the set $\mathcal{A}$, or as the initial partition by using the DBSCAN algorithm.

Remark 8.27 The corresponding *Mathematica*-module WKGC[] is described in Sect. 9.3, and the link to appropriate *Mathematica*-code is supplied.

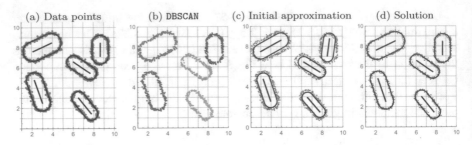

Fig. 8.23 Multiple ovals detection problem

Example 8.28 Let $\mathcal{A}$ be a data set originating from five ovals shown in Fig. 8.23a. The DBSCAN algorithm creates five clusters shown in Fig. 8.23b. Applying the described procedure to determine the best representative for each cluster, we obtain the initial approximation for the modified KCG algorithm (Fig. 8.23c), which produces the final solution shown in Fig. 8.23d.

The corresponding incremental algorithm is constructed similarly as in the case of MCD or MED problem.

Real-World Images

Let us consider a problem of detecting ovals in real-world images. We will take an example from biology—*Escherichia coli* detection. By detecting ovals in the image, one can determine precisely the size and the position of bacteria. The first figure is the original image, the second figure is obtained using *Mathematica* module EdgeDetect[] and the DBSCAN-algorithm for cluster detection, and the third figure shows both the original image and the detected ovals (Figs. 8.24 and 8.25).

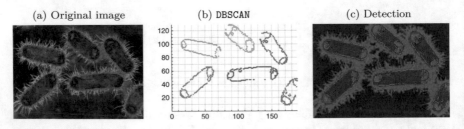

Fig. 8.24 *Escherichia coli*: $|A| = 887$, MinPts=6, $\epsilon = 7.07$, Time= 11.3 s

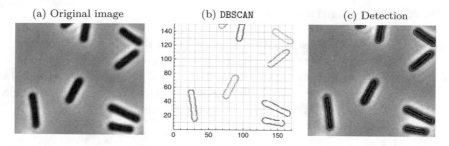

Fig. 8.25 *Escherichia coli*: $|A| = 498$, MinPts=6, $\epsilon = 5.385$, Time $= 5.9\,\mathrm{s}$

8.1.7 Multiple Lines Detection Problem

Since our data sets are finite, therefore bounded, looking at distances from data points to a straight line involves only a bounded part of that line. Hence, without abusing terminology and/or notation, by *line* we will always mean a bounded part, actually a segment, of a straight line.

Let $\mathcal{A} = \{a_i = (x_i, y_i) : i = 1, \ldots, m\} \subset \mathbb{R}^2$ be a data set originating from a possibly unknown number of lines in the plane, which have to be recognized or reconstructed.

Such problems appear in applications in various fields like computer vision and image processing [53, 105], robotics, laser range measurements [54], civil engineering and geodesy [105], crop row detection in agriculture [70, 190, 191], etc.

Methods for solving such problems often apply the Hough Transform [53, 91, 105, 120, 200] or LS-fitting [31].

We are going to consider this problem as a center-based clustering problem where the cluster centers are lines (see [143, 159]).

A Line as Representative of a Data Set

First let us have a look at the problem of determining a line as a representative of a data set $\mathcal{A}$. Suppose that the data set $\mathcal{A} \subset \Delta = [a, b] \times [c, d] \subset \mathbb{R}^2$ originates from a line. In Sect. 6.1 we have seen how to define and find the best representative of the set $\mathcal{A}$ as a TLS-line in normalized implicit form, and Example 6.12 shows how to construct an artificial data set originating from such a line.

The distance between the point $a_i = (x_i, y_i) \in \mathcal{A}$ and the line ℓ given in the normalized implicit form $\alpha x + \beta y + \gamma = 0$, $\alpha^2 + \beta^2 = 1$, equals $d(a_i, \ell) = |\alpha x_i + \beta y_i + \gamma|$ (see Exercise 6.2). This is not suitable for applying some global optimizing algorithm (e.g. the DIRECT algorithm), since the domain would be, because of the parameter γ, unbounded. To avoid this difficulty while solving the multiple lines detection problem, we will look for lines in the Hesse normal form.

The Best TLS-Line in Hesse Normal Form

Let $(O; (\vec{i}, \vec{j}))$ be the Cartesian coordinate system in the plane. A straight line ℓ
in the plane can be defined by a point $P_0 \in \ell$ and the unit normal vector $\vec{n}_0 =$
$\cos \vartheta \, \vec{i} + \sin \vartheta \, \vec{j}$ which is perpendicular to the line and points from P_0 away from
the origin O. The angle $\vartheta \in [0, 2\pi]$ is the angle between $\vec{i}$ and $\vec{n}_0$, i.e. the angle
between the positive direction of the x-axes and the vector $\vec{n}_0$.

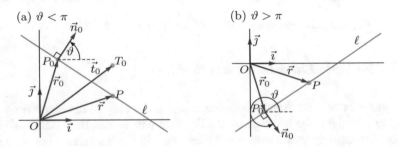

Fig. 8.26 Hesse normal form of the straight line equation in the plane

Denote by $\vec{r} = x\,\vec{i} + y\,\vec{j}$ the radius vector of an arbitrary point $P = (x, y) \in \ell$, and
by $\vec{r}_0$ the radius vector of the point P_0. The vector $\overrightarrow{P_0 P} = \vec{r} - \vec{r}_0$ is perpendicular to
the normal vector $\vec{n}_0$, and therefore

$$\langle \vec{r} - \vec{r}_0, \vec{n}_0 \rangle = 0, \tag{8.55}$$

resulting in equation of the straight line $\ell = \ell(\vartheta, \delta)$ in the *Hesse normal form*

$$\ell(\vartheta, \delta) = \{(x, y) \in \mathbb{R}^2 : x \cos \vartheta + y \sin \vartheta - \delta = 0\}, \tag{8.56}$$

where

$$\delta := \langle \vec{r}_0, \vec{n}_0 \rangle > 0$$

is the distance from the line ℓ to the origin O. The distance from an arbitrary point
$T_0 = (x_0, y_0)$ to the line ℓ equals (see Fig. 8.26a)

$$d(T_0, \ell) = |\langle \vec{t}_0, \vec{n}_0 \rangle - \langle \vec{r}_0, \vec{n}_0 \rangle| = |x_0 \cos \vartheta + y_0 \sin \vartheta - \delta|, \tag{8.57}$$

with $\vec{t}_0$ being the radius vector of the point T_0.

The Best Representative

To find a line as the best representative of the set $\mathcal{A}$, we will use the Hesse normal form when solving the GOP

$$\underset{\vartheta \in [0, 2\pi], \delta \in [0, M]}{\arg\min} \sum_{i=1}^{m} \mathfrak{D}(a_i, \ell(\vartheta, \delta)), \tag{8.58}$$

where $M = \max\limits_{i=1,\dots,m} \|a_i\|$, and $\mathfrak{D}$ is the distance-like function which is usually defined in one of the following two ways [31, 125, 158]:

$$\mathfrak{D}_1((x_i, y_i), \ell(\vartheta, \delta)) := |x_i \cos \vartheta + y_i \sin \vartheta - \delta|, \tag{8.59}$$

$$\mathfrak{D}_2((x_i, y_i), \ell(\vartheta, \delta)) := (x_i \cos \vartheta + y_i \sin \vartheta - \delta)^2. \tag{8.60}$$

Using the $\mathfrak{D}_2$ distance-like function ($\mathfrak{D}_2$-approach), because of the property (2.18) of the arithmetic mean (see also proof of Lemma 6.1), we have

$$\sum_{i=1}^{m} (x_i \cos \vartheta + y_i \sin \vartheta - \delta)^2 \geq \sum_{i=1}^{m} ((x_i - \bar{x}) \cos \vartheta + (y_i - \bar{y}) \sin \vartheta)^2,$$

hence the problem (8.58) boils down to the one-dimensional LS-optimization problem

$$\underset{\vartheta \in [0, 2\pi]}{\arg\min} \sum_{i=1}^{m} ((x_i - \bar{x}) \cos \vartheta + (y_i - \bar{y}) \sin \vartheta)^2, \tag{8.61}$$

and the resulting line is of the form

$$\ell(\vartheta) = \{(x, y) \in \mathbb{R}^2 : (x - \bar{x}) \cos \vartheta + (y - \bar{y}) \sin \vartheta = 0\}. \tag{8.62}$$

To solve the GOP (8.61) we can again use not only the DIRECT algorithm but also some other algorithm for one-dimensional global optimization, e.g. the *Pijavskij's broken lines algorithm* [132] or the *Shubert's algorithm* [174].

Example 8.29 Similarly as in Example 6.12, we will generate a set $\mathcal{A} \subset \Delta = [0, 1]^2 \subset \mathbb{R}^2$ containing 100 random points in the neighborhood of the line $y = \frac{5}{4}(x - 1)$ obtained by adding, to each of 100 uniformly distributed points on the straight line segment between points $A = (1, 0)$ and $B = (9, 10)$, a randomly generated noise from the bivariate normal distribution $\mathcal{N}(0, \sigma^2 I)$ with expectation $0 \in \mathbb{R}^2$ and covariance matrix $\sigma^2 I$, $\sigma = 0.5$, and add 10 points from Δ (10% of $|\mathcal{A}|$). These points will be considered as outliers, and the new set will be denoted by $\mathcal{A}_1$. This simulates the occurrence of errors in an image, something to be always reckoned with in real situations. We will try to reconstruct the original gray line in

Fig. 8.27 by finding the best representative of the set $\mathcal{A}$ solving the GOP (8.58) using $\mathfrak{D}_1$ and $\mathfrak{D}_2$ approaches.

Using the $\mathfrak{D}_1$-approach we obtain $\vartheta^\star = 5.60852$ and $\delta^\star = 0.88171$ as the optimal parameters for the red line in Fig. 8.27a. Using the $\mathfrak{D}_2$-approach we obtain two global minimizers: $\vartheta_1^\star = 2.50614$ and $\vartheta_2^\star = 5.64773$, both defining the same red line shown in Fig. 8.27b. The centroid of our data is $(\bar{x}, \bar{y}) = (4.75054, 5.27095)$ and therefore $\bar{x} \cos \vartheta_1^\star + \bar{y} \sin \vartheta_1^\star = -0.69471$, $\bar{x} \cos \vartheta_2^\star + \bar{y} \sin \vartheta_2^\star = 0.69471$, and $\delta^\star = |\bar{x} \cos \vartheta^\star + \bar{y} \sin \vartheta^\star| = 0.69471$, where $\vartheta^\star = \vartheta_1^\star$ or $\vartheta_2^\star$.

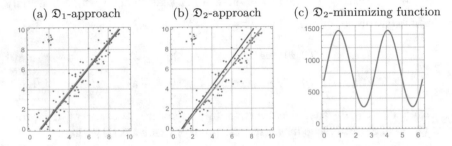

(a) $\mathfrak{D}_1$-approach   (b) $\mathfrak{D}_2$-approach   (c) $\mathfrak{D}_2$-minimizing function

Fig. 8.27 Searching for the best representative as a line using $\mathfrak{D}_1$ and $\mathfrak{D}_2$ approach

Remark 8.30 Solutions $\vartheta_1^\star$ and $\vartheta_2^\star$ of the problem (8.61) in the previous example differ by π, which just means that these two angles determine opposite orientations of the normal vector, which does not affect the red line itself. Figure 8.27c shows the graph of the minimizing function from problem (8.61).

This happens in general, i.e. the problem (8.61) has always two solutions $\vartheta_1^\star$ and $\vartheta_2^\star$ which differ by π. Namely, if $\vartheta_1^\star < \pi$ is a solution to the problem (8.61), then $\vartheta_2^\star = \vartheta_1^\star + \pi$ is also a solution since

$$\big((x_i - \bar{x}) \cos(\vartheta_1^\star + \pi) + (y_i - \bar{y}) \sin(\vartheta_1^\star + \pi)\big)^2 = \big((x_i - \bar{x}) \cos \vartheta_1^\star + (y_i - \bar{y}) \sin \vartheta_1^\star\big)^2,$$

and similarly, if $\vartheta_1^\star > \pi$ is a solution, then so is $\vartheta_2^\star = \vartheta_1^\star - \pi$.

As Fig. 8.27 shows, $\mathfrak{D}_2$-approach is sensitive regarding errors, i.e. outliers, while the $\mathfrak{D}_1$-approach is almost not affected by them. This is something to be taken into account in real-world applications.

Multiple Lines Detection Problem in the Plane

Let $\mathcal{A} = \{a_i = (x_i, y_i) \in \mathbb{R}^2 : i = 1, \ldots, m\} \subset \Delta = [a, b] \times [c, d]$ be a data set originating from unknown lines in the plane, which have to be recognized or reconstructed. The corresponding artificial data set can be constructed similarly as was described for data originating from a single line (see Example 6.12). In addition, assume that data originating from each line satisfy the homogeneity property, i.e. that they are mostly homogeneously scattered around the line (see Definition 8.1).

Searching for k-MAPart $\Pi^* = \{\pi_1^*, \ldots, \pi_k^*\}$, where the cluster centers are lines $\ell_j^*(\vartheta_j^*, \delta_j^*)$, $j = 1, \ldots, k$, boils down to finding optimal parameters $(\vartheta_j^*, \delta_j^*)$, $j = 1, \ldots, k$, which are in accordance with (8.2) the solution to the GOP (cf. (3.40))

$$\underset{\vartheta_j \in [0, 2\pi], \, \delta_j \in [0, M]}{\arg\min} \sum_{i=1}^{m} \min_{1 \le j \le k} \mathfrak{D}(a_i, \ell_j(\vartheta_j, \delta_j)), \tag{8.63}$$

where $M := \underset{i=1,\ldots,m}{\max} \|a_i\|$.

In general, the function in (8.63) is non-convex and non-differentiable, but, similarly as in [162], it can be proved to be Lipschitz-continuous. Therefore, one could try to solve the problem (8.63) using some globally optimizing method, e.g. the DIRECT algorithm, but that would be an extremely demanding task.

The Number of Lines Is Known in Advance

The problem of finding MAPart of the set $\mathcal{A}$ with cluster centers being lines and the number k of clusters is known can be solved using the KCG algorithm (Algorithm 8.2), where the γ-cluster-centers are lines, hence the algorithm will be called the k-*closest lines algorithm*, KCL (vidi [143, 144, 159]).

KCL Algorithm

In Step A of the KCL algorithm the set $\mathcal{A}$ is divided into k clusters using the Minimal Distance Principle. The distance-like function $\mathfrak{D}$ has to be the same as the one which is going to be used for finding the best representative.

In Step B of the KCL algorithm, for each of these k clusters one has to determine its best representative as a line. Earlier, starting on page 199, we discussed how to do that.

The algorithm is started by choosing initial center-lines or by choosing an initial partition. For initial center-lines one can take approximation of the solution to the GOP (8.63) obtained after a few, say 10 iterations of the DIRECT algorithm.

Example 8.31 Figure 8.28 shows data originating from three, five, and seven lines. The initial approximations for the KCL algorithm were obtained by solving the GOP (8.63) using 10 iterations of the DIRECT algorithm (red lines in Fig. 8.28). In case of three and five lines all lines were detected after several iterations of the KCL algorithm (see Fig. 8.29a, b). But in the case of seven lines, the KCL algorithm detected only four lines (red lines in Fig. 8.29c), and only two of them correspond to some of the original lines. Obviously, in this case the initial approximation for the KCL algorithm was not good enough.

(a) k=3, |A| = 621   (b) k=5, |A| = 916   (c) k=7, |A| = 1298

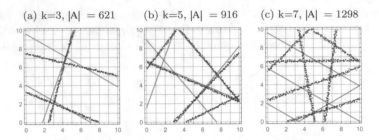

Fig. 8.28 Initial approximations obtained by the DIRECT algorithm

(a) k=3, it(KCL)=3   (b) k=5, it(KCL)=6   (c) k=7, it(KCL)=19

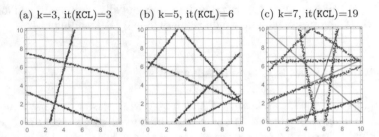

Fig. 8.29 Optimal partitions obtained by the KCL algorithm

The Number of Lines Is Not Known in Advance

If the number of lines from which data originated is not known, we are going to search for optimal partitions with $k = 2, 3, \ldots, K$ clusters by adapting the incremental algorithm for line-centers, and try to determine the MAPart among obtained partitions.

The algorithm starts by choosing an initial line-center in Hesse normal form $\hat{\ell}_1(\hat{\vartheta}_1, \hat{\delta}_1)$. For instance, this could be the best representative of the set $\mathcal{A}$. The approximation of the next line-center $\hat{\ell}_2(\hat{\vartheta}_2, \hat{\delta}_2)$ will be obtained by solving the GOP

$$\underset{\vartheta \in [0, 2\pi], \, \delta \in [0, M]}{\arg \min} \sum_{i=1}^{m} \min\{\mathfrak{D}(a_i, \hat{\ell}_1(\vartheta_1, \delta_1)), \mathfrak{D}(a_i, \ell(\vartheta, \delta))\}, \qquad (8.64)$$

where the distance-like function $\mathfrak{D}$ has to be the same as the one used to find the best representative.

Applying the KCL algorithm to line-centers $\hat{\ell}_1$, $\hat{\ell}_2$ gives locally optimal 2-partition $\Pi^{(2)}$ with line-centers $\ell_1^\star$ and $\ell_2^\star$. Note that for solving the problem (8.64) it suffices to perform just a few, say 10 iterations of the optimization algorithm DIRECT in order to obtain a sufficiently good initial approximation for the KCL algorithm.

In general, knowing $k - 1$ line-centers $\hat{\ell}_1, \ldots, \hat{\ell}_{k-1}$, the next one, $\hat{\ell}_k$, will be found by solving the GOP

$$\underset{\vartheta \in [0,2\pi],\, \delta \in [0,M]}{\arg\min} \sum_{i=1}^{m} \min\{d_{k-1}^{(i)}, \mathfrak{D}(a_i, \ell(\vartheta, \delta))\}, \qquad (8.65)$$

where $d_{k-1}^{(i)} = \min_{1 \leq s \leq k-1} \mathfrak{D}(a_i, \ell(\vartheta_s, \delta_s))$. Then, applying the KCL algorithm to line-centers $\hat{\ell}_1, \ldots, \hat{\ell}_k$ one obtains k-LOPart $\Pi^\star = \{\pi_1^\star, \ldots, \pi_k^\star\}$ with line-centers $\ell_1^\star, \ldots, \ell_k^\star$.

The incremental algorithm can be stopped using the criterion (8.8), but in case of MGOD-problem it is more effective to use criterion (8.11).

Notice that we could have started the algorithm also with several initial line-centers.

The MAPart has to be chosen among the partitions obtained by the described incremental algorithm. To this end one can adapt CHG and DBG indexes given by (8.12), respectively (8.13), or use the GO index (8.17). While determining the value of DBG index it will be necessary to find the distance between two lines. Since relevant are only segments of these lines contained in Δ, one can consider the area between these segments (see [190]). However, it seems that it is best to take the Hausdorff distance between these straight line segments, which is in fact equal to the Hausdorff distance between two-point sets of their endpoints.

Nevertheless, it turned out in numerous examples (see [143, 146, 156, 163]) that for solving the multiple lines detection problem it is best to utilize the specific structure of the problem and use the necessary conditions (8.15) in order that the LOPart be an MAPart. This will also be illustrated in the following examples.

Example 8.32 Let $\mathcal{A}$ be the data set as in Fig. 8.28b. We will look for MAPart using the incremental algorithm.

We start the incremental algorithm with the best representative of the set $\mathcal{A}$, i.e. with the line $\hat{\ell}_1(\hat{\vartheta}_1, \hat{\delta}_1)$ obtained by solving the GOP (8.58) using the $\mathfrak{D}_1$-approach. The course of the incremental algorithm is shown in Fig. 8.30a–e. Since MinPts $= \lfloor \log |\mathcal{A}| \rfloor = 6$ and ϵ-density of the set $\mathcal{A}$ based on 99% quantile is $\epsilon(\mathcal{A}) = 0.225$, the algorithm has to be stopped when QD < 0.225. As it can be seen, this happened after 5 iterations.

As one can see from data in headings of the figures, in this case the DBG index was useless, and the GO index indicates the partition shown in Fig. 8.30d, which is also unacceptable.

(a) It=2 QD=2.33, (b) It=3 QD=1.05, (c) It=4 QD=1.05, (d) It=5 QD=0.08 (e) It=6 QD=0.08
DBG=0.16, GO=79.6   DBG=0.08, GO=9.8   DBG=0.13, GO=61.0   DBG=0.08, GO=60.0   DBG=0.01, GO=65.3

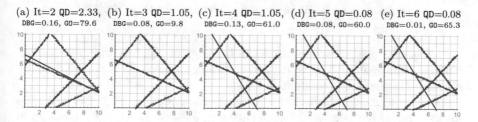

Fig. 8.30 Applying the incremental algorithm to the set from Fig. 8.28b

Let us check the necessary conditions (8.15). In this case the lower bound of local density (8.14) for each cluster is $\frac{\text{MinPts}}{2\epsilon(\mathcal{A})} = 13.33$. Local densities in the sixth iterations are: 20.8, 21.1, 20.7, 1.1, 21.0, and 20.8. This means that, according to (8.15), the red line with local density 1.1 has to be left out, and by the Minimal Distance Principle, its elements should be allotted to other clusters. In this way we obtained MAPart shown in Fig. 8.30e without the isolated red line.

Example 8.33 Let $\mathcal{A}$ be the data set as in Fig. 8.28c. The standard incremental algorithm does not provide an acceptable solution. In this case we will therefore apply an improved version of the incremental algorithm: the *Incremental method for multiple line detection problem—iterative reweighed approach* (see [143]), and search for MAPart in this way.

The incremental algorithm will be started with the best representative of the set $\mathcal{A}$, i.e. with the line $\hat{\ell}_1(\hat{\vartheta}_1, \hat{\delta}_1)$ obtained by solving the GOP (8.58) using $\mathfrak{D}_1$-approach. The course of the incremental algorithm is shown in Fig. 8.31a–g. Since MinPts $= \lfloor \log |\mathcal{A}| \rfloor = 7$ and ϵ-density of the set $\mathcal{A}$ based on 99% quantile is $\epsilon(\mathcal{A}) = 0.292$, the algorithm has to be stopped when QD < 0.292, which occurs after 8 iterations.

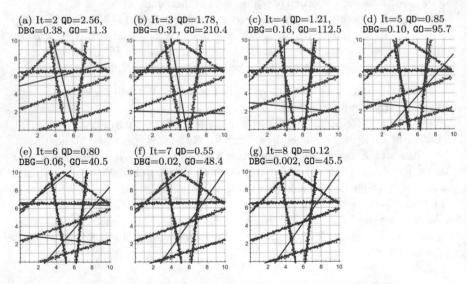

Fig. 8.31 Applying the incremental algorithm to the set from Fig. 8.28c

In this case one can see from data in headings of Fig. 8.31 that DBG and GO indexes indicate Fig. 8.31g as depicting MAPart, which is also confirmed by checking the necessary conditions (8.15).

Let us take a closer look at the necessary conditions (8.15) in this case. The lower bound of local density for each cluster is $\frac{\text{MinPts}}{2\epsilon(\mathcal{A})} = 11.97$. Local densities in the eighth iteration are: 20.3, 21.9, 20.8, 19.9, 1.6, 20.3, 20.8, 20.5. This means that, according to (8.15), the red line with local density 1.6 has to be left out, and by the Minimal Distance Principle, its elements should be allotted to other clusters. In this way we obtain MAPart which matches the original very well.

Example 8.34 Let us illustrate in more details the method proposed in [143], considering data sets of local density $\rho = 21$ originating from 5 or 10 lines[1] contained in the rectangle $\Delta = [0, 10]^2$. On line ℓ of length $|\ell|$ there were generated $\lceil \rho |\ell| \rceil$ uniformly distributed data points, and to each point, noise was added by generating pseudorandom numbers from bivariate normal distribution with mean zero and covariance matrices $\sigma^2 I$, for $\sigma^2 = 0.005$ and $\sigma^2 = 0.01$, where I is the 2×2 identity matrix.

For $n = 5$, the proposed method successfully detected all 5 lines in all examples.

For $n = 10$ and $\sigma^2 = 0.005$, the method successfully detected 10 lines in 96% of cases, and in the rest 4% of cases it detected 9 lines. For $n = 10$ and $\sigma^2 = 0.01$, the method successively detected all 10 lines in 88% of cases, and in the rest 12% of cases it detected 9 lines. From this last group of test-examples we will choose a few typical cases (see Fig. 8.32) on which we will illustrate the proposed method. Incremental iterative process finished according to criterion (8.11) (i.e. A_0 from the necessary conditions (8.15)).

(a) $|A| = 1934$, $\epsilon(A) = 0.358$ (b) $|A| = 1649$, $\epsilon(A) = 0.341$ (c) $|A| = 1704$, $\epsilon(A) = 0.328$ (d) $|A| = 1975$, $\epsilon(A) = 0.321$

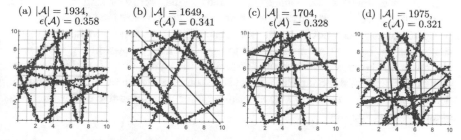

Fig. 8.32 Illustration of the method from [143]

For the data set shown in Fig. 8.32a the method successively detected all 10 lines in 10 iterations.

For data set shown in Fig. 8.32b the lower bound for local density is $\frac{\text{MinPts}}{2\epsilon(A)} = 10.26$. The method detected 11 lines in 11 iterations. Checking the condition A_1 from (8.15), the line with local density 5.87 is dropped and remains MAPart with 10 lines.

For data set shown in Fig. 8.32c the lower bound for local density is $\frac{\text{MinPts}}{2\epsilon(A)} = 10.65$. The method detected 11 lines in 11 iterations. Checking the condition A_1, two lines with local densities 4.58 and 6.85 are dropped. Thus we obtain 9 lines $\tilde{\ell}_s$, $s = 1, \ldots, 9$, to which we associate the sets

---

[1]Recall that *line* means *straight line segment*.

$$\tilde{\pi}_s = \{a \in \mathcal{A} : \mathfrak{D}_1(a, \tilde{\ell}_s) \le \epsilon(\mathcal{A})\}, \quad s = 1, \ldots, 9,$$

and define the residual $\mathcal{R} = \mathcal{A} \setminus \bigcup_{s=1}^{9} \tilde{\pi}_s$. Repeating the method for the set $\mathcal{R}$ detects the remaining line and produces MAPart.

For the data set shown in Fig. 8.32d the lower bound for local density is $\frac{\text{MinPts}}{2\epsilon(\mathcal{A})} = 10.91$. The method detected 13 lines in 13 iterations. Checking the condition $\mathbf{A}_1$ drops four lines with local densities: 3.99, 4.70, 9.39, and 7.00. The remaining line was found similarly as in the previous case giving MAPart.

For detailed description and justification of this method and the corresponding algorithm, see [143].

Real-World Images

Let us consider the problem of recognizing maize rows (Fig. 8.33a) in Precision Agriculture [70, 143, 190, 191]. Figure 8.33b shows the data set $\mathcal{A}$ ($|\mathcal{A}| = 5037$) obtained by using suitable preprocessing. With MinPts $= \lfloor \log |\mathcal{A}| \rfloor = 8$, we obtain $\epsilon(\mathcal{A}) = 4.123$. Notice that in this case data do not satisfy the homogeneity property. If QD(π) is defined as 50% quantile (median) of the corresponding set (8.9), the incremental algorithm produces 6-GOPart with 6 center-lines (see Fig. 8.33c).

Condition $\mathbf{A}_1$ does not eliminate any center-line, and condition $\mathbf{A}_2$ eliminates center-line ℓ_6. Namely, points assigned to the cluster π_6 (shown in Fig. 8.33d) are obviously not homogeneously scattered around the center-line ℓ_6, which was formally confirmed by applying Remark 2.12. All other center-lines satisfy conditions $\mathbf{A}_1$ and $\mathbf{A}_2$. If we perform the KCL algorithm on the remaining five center-lines, we obtain the final solution.

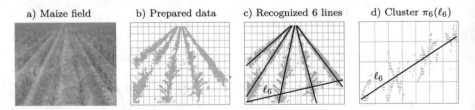

a) Maize field   b) Prepared data   c) Recognized 6 lines   d) Cluster $\pi_6(\ell_6)$

Fig. 8.33 Maize rows detection

8.1.8 Solving MGOD-Problem by Using the RANSAC Method

To solve the general MGOD-problem (8.2) one can apply the RANSAC (RANdom SAmple Consensus) method (see [36, 37, 57, 205]) with already mentioned parameters MinPts and $\epsilon(\mathcal{A})$ of the DBSCAN method and appropriate modifications of the k-means algorithm. We will describe the method in more details for the MCD-problem. Other MGOD-problems are dealt with similarly.

As before, let $\mathcal{A} = \{a_i = (x_i, y_i) : i = 1, \ldots, m\} \subset \Delta = [a, b] \times [c, d] \subset \mathbb{R}^2$ be a data set originating from several unknown circles in the plane, which have to be recognized or reconstructed. We assume that data coming from some circle $K(C, r)$ satisfy the homogeneity property with local density $\rho = \frac{|\pi(K)|}{2r\pi}$ (see Definition 8.1), and that the whole set $\mathcal{A}$ has ϵ-density $\epsilon(\mathcal{A})$ obtained with $\mathsf{MinPts} = \lfloor \log |\mathcal{A}| \rfloor$ based on 99.5% quantile of the set $\{\epsilon_a : a \in \mathcal{A}\}$ (see Sect. 4.5.1).

Any random choice of at least three points $a_1, \ldots, a_t \in \mathcal{A}$, $t \geq 3$, which are not collinear, i.e. $\mathrm{rank}[\tilde{a}_1^T, \ldots, \tilde{a}_t^T] = 3$, where $\tilde{a}_i = (x_i, y_i, 1)$, uniquely determines their circle representative $\hat{K}(\hat{C}, \hat{r})$ (see section "The Best Representative"). Circle $\hat{K}$ obtained in this way will be deemed acceptable for further procedure if

$$\hat{K}(\hat{C}, \hat{r}) \subset \Delta \quad \text{and} \quad \hat{r} > 2\epsilon(\mathcal{A}). \tag{8.66}$$

The set of all points $a \in \mathcal{A}$ contained in the annulus of width $2\epsilon(\mathcal{A})$ around the circle $\hat{K}$ forms the cluster

$$\hat{\pi} = \{a \in \mathcal{A} : \mathfrak{D}(a, \hat{K}) < \epsilon(\mathcal{A})\}. \tag{8.67}$$

The corresponding pair (cluster, circle) will be denoted $(\hat{\pi}, \hat{K})$.

We repeat this procedure, according to Proposition 8.35, N times and keep that pair $(\hat{\pi}_1, \hat{K}_1)$ for which the cluster (8.67) is the most numerous one. The pair $(\hat{\pi}_1, \hat{K}_1)$ can be further improved by solving the optimization problem

$$\underset{p,q,r}{\arg\min} \sum_{a_i \in \hat{\pi}_1} \mathfrak{D}(a_i, K((p, q), r)), \tag{8.68}$$

where $\mathfrak{D}$ is given by either (8.19), (8.20), or (8.21), and for the initial approximation of the center and radius of the circle, one can take $\hat{C}_1$ and $\hat{r}_1$.

Repeat the described procedure on the set $\mathcal{B} := \mathcal{A} \setminus \hat{\pi}_1$ with fewer, but not fewer than $\frac{1}{2} N$ attempts as predicted by Proposition 8.35. This procedure is repeated until the number of points in the set $\mathcal{B}$ drops below some given threshold, say 5–10% of $|\mathcal{A}|$. In this way, left out are only traces of data from previous iteration.

From the set of cluster-circle pairs $\{(\hat{\pi}_j, \hat{K}_j) : j = 1, \ldots, s\}$ obtained in this way, leave out those which do not satisfy the condition

$$\mathbf{A}_1 : \quad \rho(\pi_j) \geq \frac{\mathsf{MinPts}}{2\epsilon(\mathcal{A})}.$$

Since, in general, the remaining clusters do not include the whole data set $\mathcal{A}$, and the remaining circles are not in optimal positions, we will use these remaining circles as initial circles for the KCC algorithm applied to $\mathcal{A}$. In this way we obtain the reconstructed circles $K_j^\star$, $j = 1, \ldots, k$.

The following simple proposition [156] indicates the necessary number of attempts to choose t points from $\mathcal{A}$ in order to expect at least one of the acceptable circles to be close to an original one.

Proposition 8.35 *Let* $\Pi = \{\pi_1, \ldots, \pi_k\}$ *be a k-partition of the set* $\mathcal{A}$. *The probability that all t randomly chosen elements from the set* $\mathcal{A}$ *belong to a single cluster equals*

$$P(t) = \frac{\binom{|\pi_1|}{t} + \cdots + \binom{|\pi_k|}{t}}{\binom{|\mathcal{A}|}{t}}. \qquad (8.69)$$

In addition, $P(j + 1) < P(j)$ *for every* $j = 3, \ldots, K$, $K = \min\limits_{1 \leq j \leq k} |\pi_j|$, *and the sequence* $\big(P(j)\big)$ *decreases rapidly.*

The proposition suggests that out of $N = \lceil \frac{1}{P(t)} \rceil$ random choices of t points from the set $\mathcal{A}$, we might expect to get at least one set of t points from the same cluster.

Example 8.36 We will illustrate the described procedure applied to the data set $\mathcal{A}$ shown in Fig. 8.34a. In this example MinPts $= 6$, $\epsilon(\mathcal{A}) = 0.316$, $\frac{\text{MinPts}}{2\epsilon(\mathcal{A})} = 12.67$, and for $t = 3$, $N = 18$ in accordance with Proposition 8.35. The course of the algorithm is shown in Fig. 8.34.

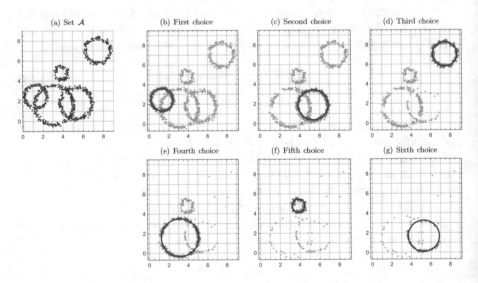

Fig. 8.34 Solving an MCD-problem using the RANSAC method

In this way we found 6 cluster-circle pairs shown in Fig. 8.35a. The obtained clusters have local densities

$$28.012, \ 22.151, \ 22.949, \ 16.93, \ 23.433, \ 4.493.$$

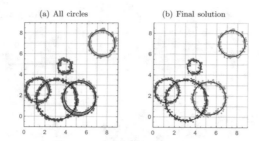

(a) All circles   (b) Final solution

Fig. 8.35 Final steps in applying the RANSAC method

This means that, according to condition $\mathbf{A_1}$, only the circle whose cluster has local density 4.493 needs to be dropped out. The final result obtained by the KCC algorithm is shown in Fig 8.35b.

One should emphasize the high efficiency of the described procedure. The CPU-time needed to recognize 5 circles runs between 0.5 and 1.5 s, with less than half the time used by the KCC algorithm.

8.2 Determining Seismic Zones in an Area

Let us consider a data set $\mathcal{A} = \{a_i = (\lambda_i, \varphi_i) : i = 1, \ldots, m\} \subset \mathbb{R}^2$, where $\lambda_i \in [-180°, 180°]$ and $\varphi_i \in [-90°, 90°]$ are the longitude and latitude of the geographical position a_i where an earthquake occurred in some area of interest. Associated with each datum a_i is the corresponding weight $M_i > 0$—the magnitude of the earthquake which occurred at some depth below the point a_i. In addition, to each datum a_i we will later associate also the information about the moment at which the earthquake occurred. Examples will be taken from geophysical data about earthquakes which occurred in the wider area of the Iberian Peninsula and in the wider area of the Republic of Croatia (see Fig. 8.36).

The first set consists of points with longitudes between 12°W (i.e. $-12°$) and 5°E, and latitudes between 34°N and 44°N, which determine the geographical positions at which an earthquake of magnitude $\geq$3.5 [118] occurred in the period from June 24, 1910 until October 25, 2019 (based on the catalogue of the National Geographic Institute of Spain (NGIS: www.ign.es)). There were 5 618 earthquakes of magnitude $M \geq 3.5$, including 3 798 of magnitude $M \geq 4$, 1 694 of magnitude $M \geq 5$, and 126 of magnitude $M \geq 6$, (see Fig. 8.36a).

The second set consists of points with longitudes between 13°E and 19.5°E and the latitudes between 42°N and 46.5°N, which determine the geographical positions at which an earthquake of magnitude $\geq$ 3.5 occurred in the period from December 9, 1880 until April 24, 2020 (public, available at: https://earthquake.usgs.gov/earthquakes/search/). There were 2 795 earthquakes of magnitude $M \geq 3.5$,

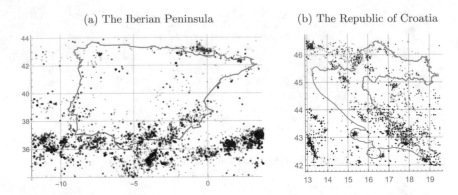

(a) The Iberian Peninsula   (b) The Republic of Croatia

Fig. 8.36 Earthquakes in the wider area of the Iberian Peninsula and in the wider area of the Republic of Croatia

including 1483 of magnitude $M \geq 4$, 1141 of magnitude $M \geq 5$, and 171 of magnitude $M \geq 6$, (see Fig. 8.36b).

8.2.1 Searching for Seismic Zones

For data set $\mathcal{A}$ with data weights $M_i \geq 3.5$ we will find MAPart with clusters representing seismic zones in the considered geographical region. Further detailed research would represent useful information for civil engineering.

One possibility to determine seismic zones could be center-based clustering with spherical clusters (see Chap. 3). First we need to implement the Incremental Algorithm 2 in order to reach the highest number of LOParts. Then we use indexes, introduced in Chap. 5, to determine MAPart.

Another, and a better option, is to look for MAPart with ellipsoidal clusters. In order to reach the highest number of LOParts, we implement the Mahalanobis incremental algorithm from Sect. 6.4.2, and then use Mahalanobis indexes from Sect. 6.5 to determine MAPart.

The described procedure has been implemented in [5, 166] for data from the wider area of the Republic of Croatia, and in [116, 117] the same was done for data from the wider area of the Iberian Peninsula.

8.2.2 The Absolute Time of an Event

In order to carry out the analysis of earthquake activity in clusters from MAPart, it is important to properly define the time (moment) of earthquake occurrence. Based on the time data available for each earthquake, i.e. Year, Month, Day, Hour, Minutes,

Seconds, which are not easily manageable in numerical calculations, we will devise a simple scale for timing events.

Using inputted time data, the *Mathematica*-module `AbsoluteTime[]` gives the number of seconds elapsed from, or if negative, preceding, January 1, 1990 at 00:00:00. For instance, for April 15, 1979 at 6:19:44 AM, `AbsoluteTime[{1979,4,15,6,19,44}]` gives 2501993984, and for January 27, 1880 at 3:30:00 PM, i.e. at 15:30:00, gives −628849800. Such numbers are suitable for numerical calculations, but are not very *user friendly*. To make them more *human-readable* we will divide them by $\kappa = 31536000$—the number of seconds in an average year, and for our purposes, call this number the *absolute time*. So for April 15, 1979 at 6:19:44 a.m. we get 79.3377 and for January 27, 1880 at 3:30:00 p.m. we obtain −19.9407. The largest integer less than or equal[2] these numbers will be called the *absolute year* of the respective event, and will be denoted by $Y^{(x)}$. So April 15, 1979 at 6:19:44 a.m. belongs to the absolute year $Y^{(79)}$, and January 27, 1880 at 3:30:00 p.m. belongs to the absolute year $Y^{(-20)}$. Several more examples are shown in Table 8.1.

Note that the minus first absolute year $Y^{(-1)}$ begins on January 1, 1899 at 00:00:00, and ends on January 1, 1900 at 00:00:00; the zeroth absolute year $Y^{(0)}$ begins on January 1, 1900 at 00:00:00, and ends on January 1, 1901 at 00:00:00; the first absolute year $Y^{(1)}$ begins on January 1, 1901 at 00:00:00, and ends on January 1, 1902 at 00:00:00, and so forth (see Fig. 8.37).

In this way, we will be able to observe any interval (in years) that can begin in any moment defined as the absolute time.

Table 8.1 Absolute years for some events

	Event (Y M D min sec)	absolute time	$Y^{(x)}$
1	{1880 1 27 15 30 0}	−19.94	$Y^{(-20)}$
2	{1899 1 27 15 30 0}	−0.93	$Y^{(-1)}$
3	{1900 1 27 15 30 10}	0.07	$Y^{(0)}$
4	{1901 1 27 15 30 10}	1.07	$Y^{(1)}$
5	{1979 4 15 6 19 44}	79.34	$Y^{(79)}$
6	{2019 11 27 14 45 24}	119.99	$Y^{(119)}$

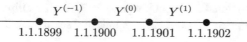

$$Y^{(-1)} \qquad Y^{(0)} \qquad Y^{(1)}$$

1.1.1899 1.1.1900 1.1.1901 1.1.1902

Fig. 8.37 Absolute years

---

[2] See the footnote on page 46.

8.2.3   The Analysis of Earthquakes in One Zone

Each cluster $\pi_j^\star$ from MAPart $\Pi^\star$ contains earthquake data with known geographical positions $a_i = (\lambda_i, \varphi_i)$, absolute times τ_i and magnitudes $M_i > 0$. We will study incidents of earthquake events in cluster $\pi_j^\star$ of magnitude $4 \leq M < 5$, and separately of magnitude $M \geq 5$.

(i) Let $Y^{(j_4)}$ (resp. $Y^{(j_5)}$) be the first absolute year in which an earthquake of magnitude $4 \leq M < 5$ (resp. $M \geq 5$) occurred in the cluster $\pi_j^\star$;

(ii) Let $r \geq 1$ be an integer and consider the sequences of the following $r + 1$ absolute years:

$$G_r^{(s)} = \left(Y^{(s)}, Y^{(s+1)}, \ldots, Y^{(s+r)}\right), \quad s + r \leq 120, \tag{8.70}$$

for $s = j_4, j_4 + 1, \ldots$ in case $4 \leq M < 5$, and $s = j_5, j_5 + 1, \ldots$ in case $M \geq 5$;

(iii) Let $\hat{r}^{(j_4)} \geq 1$ be the smallest integer such that in each sequence $G_r^{(s)}$, $s = j_4, j_4 + 1, \ldots, 120 - \hat{r}^{(j_4)}$ at least one earthquake of magnitude $4 \leq M < 5$ occurred in the cluster $\pi_j^\star$;

Similarly, let $\hat{r}^{(j_5)} \geq 1$ be the smallest integer such that in each sequence $G_r^{(s)}$, $s = j_5, j_5 + 1, \ldots, 120 - \hat{r}^{(j_5)}$ at least one earthquake of magnitude $M \geq 5$ occurred in the cluster $\pi_j^\star$;

(iv) The number $\hat{r}^{(j_4)}$ represents the incidence with which earthquakes of magnitude $4 \leq M < 5$ occur in the cluster $\pi_j^\star$. For example, $\hat{r}^{(j_4)} = 10$ means that every 10 years one can expect at least one earthquake of magnitude $4 \leq M < 5$ in this cluster, and similarly for $\hat{r}^{(j_5)}$.

Note that, by using this approach, we did not have to consider fore and aftershocks [112].

8.2.4   The Wider Area of the Iberian Peninsula

Implementing the Mahalanobis incremental algorithm (Sect. 6.4.2) for data from the wider area of the Iberian Peninsula in the period from June 24, 1910 until October 25, 2019, we obtained LOParts with $k = 3, \ldots, 22$ ellipsoidal clusters. The values of MCH, MDB, MSSWC, and MArea indexes (see Sect. 6.5) are shown in Fig. 8.38.

It is evident that the MAPart has 4 clusters, as depicted in Fig. 8.39a: $\pi_1^\star$ (South-west of the Iberian Peninsula), $\pi_2^\star$ (South-east of the Iberian Peninsula), $\pi_3^\star$ (area to the east of the Iberian Peninsula), and $\pi_4^\star$ (North of Spain). Note that ellipses in both

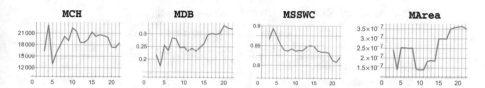

Fig. 8.38 MCH, MDB, MSSWC, and MArea indexes for the Iberian Peninsula

(a) MAPart $\Pi_4^\star$ with 4 clusters (b) MAPart $\Pi_{11}^\star$ with 11 clusters

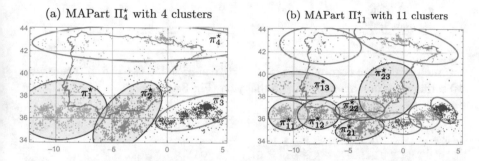

Fig. 8.39 MAPart $\Pi_4^\star$ with 4 and MAPart $\Pi_{11}^\star$ with 11 ellipsoidal clusters in the region of the Iberian Peninsula. Light blue clusters $\pi_{11}^\star, \pi_{12}^\star, \pi_{13}^\star \in \Pi_{11}^\star$ stem from the cluster $\pi_1^\star \in \Pi_4^\star$. Brown clusters $\pi_{21}^\star, \pi_{22}^\star, \pi_{23}^\star \in \Pi_{11}^\star$ stem from the cluster $\pi_2^\star \in \Pi_4^\star$

figures include 95 % of data points $a \in \pi_j$ in the d_M-neighborhood[3] of the cluster's centroid c_j, i.e. such that $d_M(a, c_j) \leq r_j$, where r_j is the 95 % quantile of the set $\{\|a - c_j\|_{\Sigma_j} : a \in \pi_j\}$.

Also, based on indexes mentioned earlier (see Fig. 8.38), the partition $\Pi_{11}^\star$ with 11 clusters (see Fig. 8.39b) can be considered as a sufficiently appropriate partition. In this case, 6 clusters which stem from clusters $\pi_1^\star$ and $\pi_2^\star$ are especially important:

(a) clusters that stem from $\pi_1^\star$ (SW IP) are denoted blue: $\pi_{11}^\star$ (Western Azores–Gibraltar fault), $\pi_{12}^\star$ (Eastern Azores–Gibraltar fault), and $\pi_{13}^\star$ (Central Portugal);

(b) clusters that stem from $\pi_2^\star$ (SE IP) are denoted brown: $\pi_{21}^\star$ (Alboran Sea), $\pi_{22}^\star$ (South-eastern Spain), and $\pi_{23}^\star$ (East of Spain).

The results obtained for these areas are summarized in Table 8.2.

We will further analyze only the cluster $\pi_2^\star \in \Pi_4^\star$. The analysis of the cluster $\pi_1^\star$ and their sub-clusters can be found in [117].

In the cluster $\pi_2^\star$ (see Fig. 8.39a) 643 earthquakes occurred with magnitude $4 \leq M < 5$ and with occurrence incidence of 2 years—the last one occurred on October 18, 2019 at 15:54:11 ($M = 4.5$). Also, in this cluster 49 earthquakes occurred with magnitude $M \geq 5$ and with occurrence incidence of 10 years—the

---

[3]d_M is the normalized Mahalanobis distance-like function, see (6.18).

Table 8.2 Summary of the results selected from partitions $\Pi_4^\star$ and $\Pi_{11}^\star$

Area	The number of earthquakes			Occurrence incidents	
	$M \geq 3.5$	$4 \leq M < 5$	$M \geq 5$	$4 \leq M < 5$	$M \geq 5$
π_{11} (W Azores–Gibraltar)	605	271	22	10	32
π_{12} (E Azores–Gibraltar)	361	171	10	5	15
π_{13} (Central Portugal)	142	57	3	10	–
π_{21} (Alboran Sea)	1111	353	30	5	20
π_{22} (SE Spain)	610	209	14	5	20
π_{23} (East Spain)	286	81	5	5	–

last such earthquake occurred on May 28, 2016 at 23:54:53 ($M = 5.4$). There were 10 earthquakes of magnitude $M \geq 5$ that year (Fig. 8.40c).

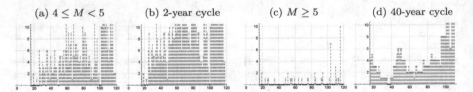

(a) $4 \leq M < 5$ (b) 2-year cycle   (c) $M \geq 5$ (d) 40-year cycle

Fig. 8.40 Incidents of earthquakes of magnitudes $4 \leq M < 5$ and $M \geq 5$ in the cluster $\pi_2^\star$

The cluster $\pi_2^\star$ consists of sub-clusters: $\pi_{21}^\star$ (Alboran Sea), $\pi_{22}^\star$ (South-east of Spain), and $\pi_{23}^\star$ (East of Spain), which appear in the partition $\Pi_{11}^\star$ (see Fig. 8.39b).

In the cluster $\pi_{21}^\star$, 1111 earthquakes of magnitude $M \geq 3.5$ occurred in the period 1915–2020, including 353 of magnitude $4 \leq M < 5$, with at least one earthquake of magnitude $4 \leq M < 5$ occurring every 5 years—the last such earthquake occurred on February 16, 2019 at 4:35:35 ($M = 4.3$). Also, during that period, 30 earthquakes of magnitude $M \geq 5$ occurred in this cluster, with at least one earthquake of such magnitude occurring every 35 years—the last such earthquake occurred on March 15, 2016 at 4:40:39 ($M = 5.2$).

In the cluster $\pi_{22}^\star$, 610 earthquakes of magnitude $M \geq 3.5$ occurred in the period 1915—2020, including 209 of magnitude $4 \leq M < 5$, with at least one earthquake of magnitude $4 \leq M < 5$ occurring every 5 years—the last such earthquake occurred on October 25, 2019 at 9:35:48 ($M = 4.4$). Also, during that period, 14 earthquakes of magnitude $M \geq 5$ occurred in this cluster, with at least one earthquake of magnitude $M \geq 5$ occurring every 20 years—the last such earthquake occurred on March 19, 2013 at 3:11:31 ($M = 5.5$).

In the cluster $\pi_{23}^\star$, 286 earthquakes of magnitude $M \geq 3.5$ occurred in the period 1915—2020, including 81 of magnitude $4 \leq M < 5$, with at least one earthquake of magnitude $4 \leq M < 5$ occurring every 5 years—the last such earthquake occurred on August 13, 2018 at 14:40:4 ($M = 4.0$). Also, during that period, 5 earthquakes of magnitude $M \geq 5$ occurred in this cluster, with at least one earthquake of magnitude

$M \geq 5$ occurring every 55 years—the last such earthquake occurred on May 11, 2011 at 16:47:26 ($M = 5.1$).

8.2.5   The Wider Area of the Republic of Croatia

Implementing the Mahalanobis incremental algorithm (Sect. 6.4.2) for data from the wider area of the Republic of Croatia in the period from December 9, 1880 until April 24, 2020, we obtained LOParts with $k = 3, \ldots, 17$ ellipsoidal clusters. The values of MCH, MDB, MSSWC, and MArea indexes (see Sect. 6.5) are shown in Fig. 8.41.

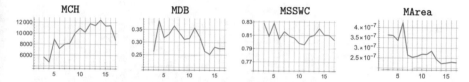

Fig. 8.41   MCH, MDB, MSSWC, and MArea indexes for Croatia

Based on these indexes, among obtained partitions we choose the one with 14 clusters as the MAPart (see Fig. 8.41). The following clusters refer specifically to Croatia: $\pi_1^\star$ (Dubrovnik area), $\pi_2^\star$ (Ston–Metković area), $\pi_3^\star$ (Šibenik–Split area), $\pi_4^\star$ (Gospić–Ogulin area), and $\pi_5^\star$ (Zagreb area) (see Fig. 8.42).

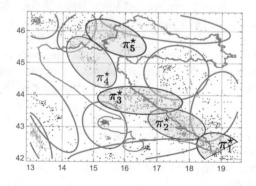

Fig. 8.42   Data points and the MAPart with 14 ellipsoidal clusters in the region of Croatia—five clusters are highlighted

The results obtained for these areas are summarized in Table 8.3.

We will further analyze only the cluster $\pi_1^\star$ (Dubrovnik area). The analysis of other clusters can be found in [117].

In the cluster $\pi_1^\star$ (Dubrovnik area) 231 earthquakes occurred of magnitude $M \geq 3.5$, including 123 of magnitude $4 \leq M < 5$ (see Fig. 8.43a). At least one earthquake of magnitude $4 \leq M < 5$ occurred every 7 years. The last such earthquake occurred on November 26, 2018 at 14:17:38 ($M = 4.2$).

Table 8.3 Summary of the results for the Croatia region

Area	Number of earthquakes			Occurrence incidents	
	$M \geq 3.5$	$4 \leq M < 5$	$M \geq 5$	$4 \leq M < 5$	$M \geq 5$
Dubrovnik	231	123	18	7	40
Ston–Metković	529	263	32	5	22
Šibenik–Split	279	150	18	7	30
Gospić–Ogulin	125	52	17	12	45
Zagreb	199	83	16	10	50

(a) $4 \leq M < 5$ (b) 7-year cycle   (c) $M \geq 5$ (d) 40-year cycle

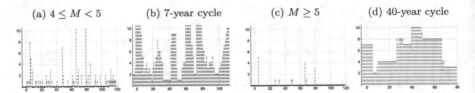

Fig. 8.43 Incidents of earthquakes of magnitudes $4 \leq M < 5$ and $M \geq 5$ in the cluster $\pi_1^\star$ (Dubrovnik)

Also, 18 earthquakes of magnitude $M \geq 5$ occurred in this cluster (see Fig. 8.43c). At least one such earthquake occurred every 40 years (see Fig. 8.43d)— the last three such earthquakes happened in 1979: on April 15, 1979 at 6:19:46 ($M = 6.8$), on April 15, 1979 at 14:43:6 ($M = 5.8$), and on May 24, 1979 at 17:23:18 ($M = 6.1$).

To determine the zones of seismic activity one can also apply the DBSCAN method. In [169], this method is applied to the problem of earthquake data zoning in the wider area of Croatia. Using density-based clustering for earthquake zoning, it is possible to recognize non-convex shapes, which gives much more realistic results. Besides, MAPart is obtained automatically without using indexes. The size of the DBSCAN-parameter ϵ (see Sect. 4.5.1) significantly influences the recognition of the number and the configuration of earthquake zones. Therefore, special attention in the algorithm is given to the problem of determining the corresponding value of the DBSCAN-parameter ϵ.

8.3   Temperature Fluctuations

Analyzing climate changes is a very elaborate and demanding task, and we do not intend to dwell on it. For example, the air temperature at a certain location depends on numerous climate factors, including incoming solar radiation, humidity, altitude,

etc. We are going to look only at fluctuations of mean daily temperatures[4] in the city
of Osijek (Croatia) from 1918 to 2018[5], and illustrate the use of cluster analysis to
examine such data.

We will use data of the form $(d, \hat{T})$, where d is the day in a particular year
determined by the ordered triple $(yyyy, mm, dd)$: year–month–day, and $\hat{T}$ is the
mean temperature in °C on that day.

To make things simpler, we will assume that a year has 365 days, and, similarly
as in Sect. 2.4.1, we assign to the day d the real number

$$t = yyyy + (mm - 1)/12 + (dd - 1)/365.$$

Important in our discussion is only the moment of the year, and not the particular
year in which the event occurred. Because the mean daily temperatures are periodic
data with fundamental period 365, we will regard the day d as the point $(\cos \tau, \sin \tau)$
on the unit circle, where

$$\tau = 2\pi(t - yyyy) \in [0, 2\pi]. \tag{8.71}$$

The number $\tau \in [0, 2\pi]$ represents the moment which is $\tau/2\pi$ -th part of a year
apart from January 1.

To moments (days) defined in this way we assign the corresponding mean
daily temperatures. Since the temperatures can be negative[6], in order to be able to
better graphically represent temperature values, the measure for temperature will be
defined as

$$T = \hat{T} + \Delta, \quad \Delta = 30. \tag{8.72}$$

After completing the calculations, the scale, as well as the temperature data in
graphical representations, will be reversed back to the original scale by subtract-
ing Δ.

Using (8.71) and (8.72), our set of data $(d_i, \hat{T}_i), i = 1, \ldots, m$, is the set

$$\mathcal{A} = \{a_i = (T_i \cos \tau_i, T_i \sin \tau_i) : i = 1, \ldots, m\} \subseteq \mathbb{R}^2. \tag{8.73}$$

We will depict such data sets using Burn diagrams (see Sect. 2.4.2). Each point
$a_i \in \mathcal{A}$ represents the mean daily temperature at some moment (day) in the year:
the temperature T_i is represented by the distance from a_i to the origin O, and the

---

[4]The mean daily temperature is calculated by the Kämtz' formula dating back to 1831: $T_{mean} = (T_7 + T_{14} + 2T_{21})/4$, where T_7, T_{14}, and T_{21} are temperatures at 7:00, 14:00, and 21:00—the three
standard observation times.

[5]Source: Croatian Meteorological and Hydrological Service (DHMZ).

[6]In our case, the minimal mean temperature, -20.6 °C, occurred on January 23, 1942.

day to which this temperature is assigned is represented by the number τ_i, i.e. the polar angle of the point a_i (see e.g. Fig. 8.44).

8.3.1   Identifying Temperature Seasons

Based on the data set $\mathcal{A}$ of mean daily temperatures during a period of one or several years, we will try to estimate *temperature seasons* (TS), i.e. *seasons* which will be determined as clusters in 4-GOPart of the set $\mathcal{A}$. Then, using appropriate cluster analysis, we will ponder over *seasons'* characteristics during that period. We will name the TS by standard names: Spring, Summer, Autumn, and Winter.

Let $\mathcal{A}$ be the data set of mean daily temperatures between years n_1 and n_2. A globally optimal k-partition of the set $\mathcal{A}$ given by (8.73) can be found by solving the GOP (3.40), where d is some distance-like function. We will use the LS distance-like function.

The GOP (3.40) can be solved by applying the standard k-means Algorithm 4.7 using randomly chosen initial centers, but doing so we can expect to get only a k-LOPart.

Looking for a k-GOPart can be done like this [161, 162]: first, using only a few, say 30 iteration of the global optimization algorithm DIRECT, find a favorable initial approximation, and then improve the obtained solution by k-means algorithm.

Example 8.37 We will illustrate the proposed method to find TS as clusters of a 4-GOPart by the following example. Consider the data sets of average daily temperatures in Osijek in the 7-year periods: 1939–1945 and 2011–2017.

Using the previously described procedure we obtained 4-GOPart shown in Fig. 8.44 and in Table 8.4. Figure 8.44 shows mean daily temperatures and marks for the beginning and end of each TS (season), and Table 8.4 provides duration of each TS and the maximal, minimal, and average of the mean daily temperatures in those TS.

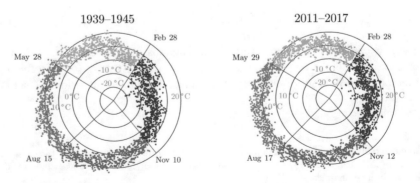

Fig. 8.44 Temperatures according to TS in Osijek in the chosen 7-year periods

Table 8.4 Durations, temperatures and average mean daily temperatures of TS in the chosen 7-year periods

	1939–1945			2011–2017		
	Days	Temperatures	Average	Days	Temperatures	Average
Spring	89	[−4.7, 25.7]	10.7	90	[−2.7, 23.8]	12.2
Summer	79	[12.5, 30.1]	20.8	80	[12.4, 30.6]	22.4
Autumn	87	[3.1, 28.7]	14.9	87	[2.6, 30.2]	15.4
Winter	110	[−20.6, 14.8]	0.4	108	[−17.7, 12.5]	2.5

As one can see, there are no substantial differences in beginnings and durations of TS, but already this example shows a notable raise of temperature in every TS, although more detailed analysis is required to make definite conclusions.

Just for the record, let us mention that the initial approximation of 4-GOPart was obtained using 30 iterations of the DIRECT algorithm which required 5 s of the CPU time. Then, the final 4-GOPart itself was obtained using the k-means algorithm which required additional 0.1 s.

Example 8.38 Let us consider the data sets of mean daily temperatures in Osijek from 1918 to 2018.[7]

Following (8.73), for each year we defined the corresponding data set A, and performed the previously described procedure to obtain the 4-GOPart. This produced the beginning and duration of each TS (temperature season), the minimal, maximal, and average mean daily temperature in each TS, and the CPU-time for DIRECT and k-means algorithms.

The following characteristics of the average TS (temperature seasons) in Osijek during this 100-year period, are shown in Table 8.5:

- the average dates of beginnings and average durations of every TS with the corresponding standard deviations expressed in days (see also Fig. 8.45). For instance, the average Spring started on March 1 with standard deviation of 8 days, and lasted for 90 days, with standard deviation of 2.9 days;
- the average mean daily temperature in every TS with the corresponding standard deviation expressed in °C. For instance, the average mean daily temperature in an average Spring was 11.3 °C with standard deviation of 1.5 °C. In addition, the mean daily temperatures in Spring were between −11.7 °C and 28.5 °C (see also Fig. 8.46).

The average CPU-time for each year was 1.20 s for the DIRECT algorithm and 0.02 s for the k-means algorithm.

---

[7]Data for 1991 and 1992 were left out since these data are incomplete due to the war.

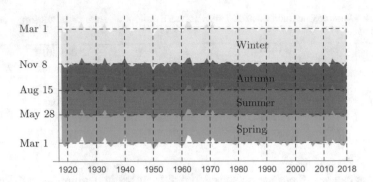

Fig. 8.45 TS (temperature seasons) in Osijek

Table 8.5 TS-clusters, their durations and average mean daily temperatures in Osijek in the period from 1918 to 2018

	Beginning of TS		Duration		Mean daily temperatures		
	Date	Stdev	Days	Stdev	Temp	Stdev	Interval (range)
Spring	Mar 01	8	90	3.1	11.3	1.5	[−11.7, 28.5]
Summer	May 28	6	77	1.7	21.0	1.1	[8.4, 31.6]
Autumn	Aug 15	6	85	2.9	14.7	1.5	[−1.8, 30.2]
Winter	Nov 8	8	113	3.5	1.4	1.5	[−20.6, 17.5]

Using linear regression for the beginnings of TS in the above 100-year period, shows that one can expect that all TS will start a few days earlier: Spring about 5 days, Summer about 4 days, Autumn about 3 days, and Winter about 4 days.

The fluctuation of intervals (ranges) of mean daily temperatures in each TS is represented in Fig. 8.46 by light green, red, brown, and blue areas, respectively, and the fluctuation of average mean daily temperatures in each TS is depicted by the wiggly red line. For further analysis, based on these data, for each TS we will determine the model-function of the form

$$f(t; \alpha, \beta, a) = \alpha + \beta t + \sum_{j=1}^{n_T} A_j \cos(\omega_j t + \varphi_j), \qquad (8.74)$$

with $a = (A_1, \omega_1, \varphi_1, \ldots, A_{n_T}, \omega_{n_T}, \varphi_{n_T})$,

where $t \mapsto \alpha + \beta t$ is the linear part in the model, and $t \mapsto A_j \sin(\omega_j t + \varphi_j)$ describes the j-th periodic influence with amplitude A_j, basic period $\tau_j = \frac{2\pi}{\omega_j}$ and delay $\frac{\varphi_j}{\omega_j}$.

For example, for data related to Spring in the period from 1918 to 2018 (without 1991 and 1992)

$$\left(t_i, \hat{T}_{\mathrm{Sp}}^{(i)}\right), \quad i \in J,$$

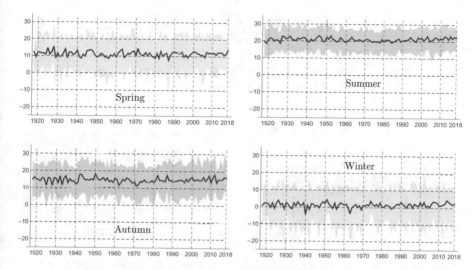

Fig. 8.46 Temperature intervals (ranges) and average mean daily temperatures (wiggly red line) in four TS in Osijek

where $J = \{1,\dots,73\} \cup \{76,\dots,101\}$ and $\hat{T}_{Sp}^{(i)}$ is the average mean daily temperature in Spring of the year $t_i = 1918 + i - 1$, $i \in J$, the corresponding optimal parameters of the model-function (8.74) for $n_T = 3$ were found by solving the GOP

$$\underset{\alpha,\beta\in\mathbb{R},\, a\in\mathbb{R}^{3n}}{\arg\min} F(\alpha,\beta,a), \quad F(\alpha,\beta,a) = \sum_{i\in J}\left(\hat{T}_{Sp}^{(i)} - f(t_i;\alpha,\beta,a)\right)^2. \quad (8.75)$$

For this problem we used the *Mathematica*-module `NonlinearModelFit[]` [195], with the initial approximation $(\hat{\alpha}, \hat{\beta}, \hat{a})$ found (see also [157]) by Algorithm 8.39 (to simplify the notation we use the substitution $y_i := \hat{T}_{Sp}^{(i)}$). In order to find an appropriate initial approximation, the nonlinear GOP (8.77) was solved using the `DIRECT` algorithm, which required between 4 and 6 s. The solution to GOP (8.77) obtained using the *Mathematica*-module `NonlinearModelFit[]` with such a choice of initial approximation is shown in Fig. 8.47.

Algorithm 8.39 (Initial Approximation)

Step 0:   Input $99 \gg n_T \geq 1$, $\quad J = \{1,\dots,73\} \cup \{76,\dots,101\}$, (t_i, y_i), $i \in J$;

Step 1:   Determine $(\hat{\alpha}, \hat{\beta}) \in \mathbb{R}^2$ by solving the simple linear least squares problem

$$(\hat{\alpha}, \hat{\beta}) = \underset{\alpha,\beta\in\mathbb{R}}{\arg\min} \sum_{i\in J}(y_i - \alpha - \beta t_i)^2; \quad (8.76)$$

Set $\hat{y}_i = y_i - \hat{\alpha} - \hat{\beta} t_i$, $i \in J$, and set $j = 1$;

Step 2:   Solve the nonlinear least squares problem

$$(\hat{A}_j, \hat{\omega}_j, \hat{\varphi}_j) = \underset{A,\omega,\varphi \in \mathbb{R}}{\arg\min} \sum_{i \in J} (\hat{y}_i - A\cos(\omega t_i + \varphi))^2; \qquad (8.77)$$

Step 3:   If $j < n_T$, set $\hat{y}_i = \hat{y}_i - \hat{A}_j \cos(\hat{\omega}_j t_i + \hat{\varphi}_j), i \in J$;

Step 4:   Set $j = j + 1$ and go to *Step 2*;   Else STOP.

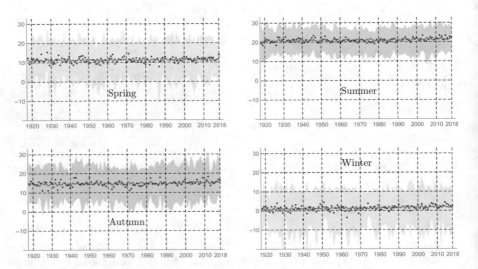

Fig. 8.47 Intervals (ranges) of mean daily temperatures and average mean daily temperatures (red points) with graph of the corresponding model-function (8.74) for TS from 1918 until 2018 in Osijek

The temperature characteristics of TS Spring in the period 1918–2018 are the following: the average mean daily temperature was 11.13 °C, the raise of temperatures in those 100 years was 0.25 °C.

The temperature characteristics of TS Summer in the period 1918–2018 are the following: the average mean daily temperature was 20.52 °C, the raise of temperatures in those 100 years was 0.9 °C.

The temperature characteristics of TS Autumn in the period 1918–2018 are the following: the average mean daily temperature was 14.34 °C, the raise of temperatures in those 100 years was 0.77 °C.

The temperature characteristics of TS Winter in the period 1918–2018 are the following: the average mean daily temperature was 0.76 °C, the raise of temperatures in those 100 years was 1.22 °C.

Yearly periodic influences determined by the amplitudes, fundamental periods and delays for all temperature seasons are listed in Table 8.6.

Table 8.6 Yearly periodic influences are determined by amplitudes A_j, fundamental periods τ_j, and delays $\frac{\varphi_j}{\omega_j}$ of the trigonometric function $t \mapsto \sum_{j=1}^{3} A_j \cos(\omega_j t + \varphi_j)$ from (8.74)

TS	Amplitudes A_j			Periods $\tau_j = \frac{2\pi}{\omega_j}$			Delays		
	$j=1$	$j=2$	$j=3$	τ_1	τ_2	τ_3	$\frac{\varphi_1}{\omega_1}$	$\frac{\varphi_2}{\omega_2}$	$\frac{\varphi_3}{\omega_3}$
Spring	0.42	2.43	6.95	0.44	1.87	4.73	0.33	5.62	16.98
Summer	0.25	2.43	6.84	0.20	1.85	5.79	0.44	6.02	16.44
Autumn	0.43	2.41	7.49	0.17	1.86	4.54	0.41	5.72	16.92
Winter	0.36	2.44	6.62	0.41	1.89	4.54	0.18	5.19	15.72

8.4 Mathematics and Politics: How to Determine Optimal Constituencies?

Determining constituencies is a difficult optimization problem, in particular in states with many territorial units. For that reason, this problem is still popular in scientific and professional literature where one can find algorithms for generating constituencies based on various heuristic approaches [27, 41, 74, 137, 138].

History is full of examples where political parties showed strong partiality generating constituencies according to their political interests. The best known such example occurred in 1812 when the state of Massachusetts was divided, showing bias, into unnatural constituencies. Such a redrawing of constituencies is known as *Gerrymandering* after Elbridge Gerry, the then governor of Massachusetts (see e.g. [27, 41]).

We will present here a modified form of a mathematical model which allows for creation of optimal constituencies with evenly distributed number of voters (see also [79, 110, 147]). The model is set up so that no favor is given to any political option, and is based on applying cluster analysis respecting the requirement that all constituencies have similar number of voters.

The method will be illustrated in case of the Republic of Croatia, based on data from 2020. The problem to determine constituencies depends on numerous conditions depending on time, e.g. migration of the population. For this reason the borders of constituencies should be periodically revised, e.g. after every census.

Defining the Problem

We are looking at the territory of a state organized into m territorial units (in Croatia these are called cities and municipalities), which we want to group into k, $1 < k < m$, constituencies, such that:

(i) the constituencies consist of territorial units which are in some distance-like sense mutually close;

(ii) the number of voters in constituencies does not differ by more than $\pm p\%$ from the average number of voters in constituencies.

The requirement that the territorial units in a constituency be spatially close can be ensured by applying some distance-like function $d : \mathbb{R}^2 \times \mathbb{R}^2 \rightarrow [0, +\infty)$. In what follows, we will use the LS distance-like function because of its simplicity and naturalness.

The mathematical model which we consider here is based only on criteria (i) and (ii). In order to improve the model, one can take into account also some other criteria to ensure additional evenness, like:

- *socio-economic homogeneity* to ensure that the constituencies have approximately equal total income [26, 27];
- *similarity with existing constituencies* to ensure that revising constituencies does not result in big changes with respect to existing ones [27];
- *areal similarity* to ensure that the constituencies are of similar area [27];
- *preserving larger territorial units* to ensure that the borders of constituencies coincide with borders of larger territorial units as much as possible.

8.4.1 Mathematical Model and the Algorithm

Assume that the state area is partitioned into m territorial units determined by their geographical position in Gauss–Krüger coordinate system by points $a_i = (x_i, y_i)$, $i = 1, \ldots, m$, that there are q_i voters in the unit a_i, and let $Q = \sum_{i=1}^{m} q_i$ denote the total number of voters. We want to divide the state area into k, $1 < k < m$, constituencies $\pi_1, \ldots, \pi_k$ with corresponding numbers $Q_1, \ldots, Q_k$ of voters, taking account of requirements (i) and (ii) on page 223. In addition we do not want that the number of voters in any constituency differs by more than $p\%$ of Q/k—the average number of voters in constituencies, i.e. that the following holds:

$$\left(1 - \frac{p}{100}\right) \frac{Q}{k} \leq Q_j \leq \left(1 + \frac{p}{100}\right) \frac{Q}{k}, \quad j = 1, \ldots, k.$$

We are going to describe two approaches to this problem—the ***Integer Approach*** and the ***Linear Relaxation Approach***.

Integer Approach

To each constituency π_j assign the corresponding center c_j. Using the LS distance-like function, the center c_j is obtained as the weighted arithmetical mean of territorial units a_i belonging to the constituency π_j, and the weights are taken to be the numbers of voters in those territorial units. The problem of determining optimal constituencies boils down to the following center-based clustering problem (see Sect. 3.4):

$$\min_{c_1,\dots,c_k \in \mathbb{R}^2} \sum_{i=1}^{m} q_i \min\left\{\|a_i - c_1\|^2, \dots, \|a_i - c_k\|^2\right\}. \tag{8.78}$$

Using the integer approach, we assume that each territorial unit a_i completely belongs to a single constituency, although in general, parts of a territorial unit a_i can be partially contained in several constituencies.

Defining the membership matrix

$$W = (w_{ij}) \in \mathbb{R}^{m \times k}, \qquad w_{ij} = \begin{cases} 1 & a_i \in \pi_j \\ 0 & a_i \notin \pi_j \end{cases} \tag{8.79}$$

and taking account of conditions (i) and (ii), all specifications can be stated as follows:

$$\sum_{j=1}^{k} w_{ij} = 1, \quad i = 1, \dots, m \tag{8.80}$$

$$\sum_{i=1}^{m} \sum_{j=1}^{k} w_{ij} q_i = Q \tag{8.81}$$

$$(1 - \tfrac{p}{100})\tfrac{Q}{k} \le \sum_{i=1}^{m} w_{ij} q_i \le (1 + \tfrac{p}{100})\tfrac{Q}{k} \tag{8.82}$$

$$w_{ij} \in \{0, 1\}, \quad i = 1, \dots, m, \quad j = 1, \dots, k. \tag{8.83}$$

The condition (8.80) ensures that every territorial unit a_i belongs to some constituency π_j, (8.81) ensures that every voter in every territorial unit will be included in some constituency, (8.82) ensures evenness of the number of voters in constituencies of up to $p\%$, and (8.83) ensures that every territorial unit a_i entirely belongs to exactly one constituency.

Following [74], the minimizing objective function (8.78) can be written as

$$\min_{w_{ij}} \sum_{i=1}^{m} \sum_{j=1}^{k} w_{ij} q_i \left\| a_i - \frac{\sum_{s=1}^{m} w_{sj} q_s a_s}{\sum_{s=1}^{m} w_{sj} q_s} \right\|^2. \tag{8.84}$$

If among territorial units there is one with the number q_{i_0} of voters more than $p\%$ above the average, i.e. such that

$$q_{i_0} \ge (1 + \tfrac{p}{100})\tfrac{Q}{k}, \tag{8.85}$$

then the optimization problem (8.84) under the conditions (8.80)–(8.83) will not have a solution. In this case it is necessary to allow that such territorial unit be

associated with more than one constituency. Such a situation can occur in states with
a markedly large city, as, for instance, Zagreb in Croatia or Budapest in Hungary.

For the mathematical model this means that such territorial units should be
allowed to split among several constituencies. This will be achieved by introducing
the set I_0 of indexes of those territorial units which satisfy (8.85), and instead of
condition (8.83) require

$$w_{ij} \in \begin{cases} [0, 1], & i \in I_0 \\ \{0, 1\}, & i \in \{1, \ldots, m\} \setminus I_0 \end{cases}, \quad j = 1, \ldots, k. \quad (8.86)$$

Under conditions (8.80)–(8.82) and (8.86), the optimization problem (8.84) is a
problem of nonlinear global optimization with constraints [60, 122], with immense
number of variables and many potential local solutions. For example, determining
10 constituencies in Croatia, involves 5560 variables. Solving this problem directly
is, from the numerical aspect, extremely demanding.

Motivated by [123], instead of solving the above mentioned problem of global
optimization, we will construct an algorithm which, with good initial approximation, gives a solution for which we can quite confidently claim to be optimal.

Algorithm 8.40

Step 0:   Input $1 \le k \le m$; $I = \{1, \ldots, m\}$; $J = \{1, \ldots, k\}$; $A = \{a_i \in \mathbb{R}^2 : i \in I\}$.
           Choose distinct centers $c_1 \ldots, c_k$ which are assumed to be heuristically
           close to constituencies' centers;
Step 1:   Solve the integer programming problem [175]

$$\min_{w_{ij}} \sum_{i=1}^{m} \sum_{j=1}^{k} w_{ij} q_i \left\| a_i - c_j \right\|^2,$$

           conforming to conditions (8.80)–(8.82) and (8.86). The solution is the
           matrix denoted by $\hat{W} = (\hat{w}_{ij})$;
Step 2:   Calculate the new centers as weighted arithmetical means

$$\hat{c}_j := \frac{\sum_{s=1}^{m} \hat{w}_{sj} q_s a_s}{\sum_{s=1}^{m} \hat{w}_{sj} q_s}, \quad j = 1, \ldots, k$$

Step 3:   For all $j \in J$ such that $c_j \ne \hat{c}_j$, if such does exist, set $c_j = \hat{c}_j$ and go to
           Step 1;
           Else set $W^\star = \hat{W}, c_j^\star = \hat{c}_j$ for every $j \in J$, and stop the algorithm.

As one can see, the Algorithm 8.40 successively solves the optimization problem (8.78) with known centers complying with (8.80)–(8.82) and (8.86) (Step 1)
and finds new cluster centers as weighted arithmetic means of data a_i with weights
obtained in Step 1 (Step 2). Provided that the algorithm started with good initial
approximation of centers $c_1, \ldots, c_k$, we can confidently assert that the result are
optimal centers $c_1^\star, \ldots, c_k^\star$ of constituencies, and constituencies themselves are
defined as

$$\pi_j^\star = \{a_i \in \mathcal{A} : w_{ij}^\star \neq 0\}, \quad j = 1, \ldots, k.$$

In so doing, one has to bear in mind that two possibilities may occur:

- if $w_{ij}^\star = 1$ then the entire territorial unit a_i is contained in the constituency $\pi_j^\star$;
- if $w_{ij}^\star \in [0, 1]$ then the constituency $\pi_j^\star$ contains the $w_{ij}^\star$-th part of the
  territorial unit a_i (this will happen only when $i \in I_0$ (see the definition of I_0
  preceding (8.86)).

The method for finding a good initial approximation will be described at the very
end of the current subsection, on page 228.

Linear Relaxation Approach

The condition (8.83) guarantees that borders of resulting constituencies do not
split territorial units. But sometimes it may make sense to relax this condition and
allow some splitting. In practice this means that some numbers w_{ij} assigned to the
territorial unit a_i will be neither 0 nor 1 as in (8.79), but some real number from the
interval [0, 1]

$$w_{ij} \in [0, 1], \quad j = 1, \ldots, k, \tag{8.87}$$

also satisfying (8.80), i.e.

$$\sum_{j=1}^{k} w_{ij} = 1, \quad i = 1, \ldots, m.$$

As a consequence, voters in some territorial units may be separated into different
constituencies, while still satisfying other conditions in the model. Of course, in this
case the condition (8.82) has to be modified as follows:

$$\left\lfloor (1 - \tfrac{P}{100}) \tfrac{Q}{k} \right\rfloor + 1 \leq \sum_{i=1}^{m} w_{ij} q_i \leq \left\lfloor (1 + \tfrac{P}{100}) \tfrac{Q}{k} \right\rfloor - 1, \tag{8.88}$$

in order that the numbers of voters in those constituencies be integers.

With such relaxation, the global optimization problem (8.78) satisfying conditions (8.80), (8.81), (8.87), and (8.88), becomes a bit simpler than the original

problem, but to solve it, we will again, instead of global optimization use a modification of the Algorithm 8.40 based on the approach from [123]. In contrast to the original integer programming problem in Step 1, in this case it is necessary to solve a considerably simpler linear programming problem [175]. Therefore, rather than search for a good initial approximation, following [100] we will start the iterative process with, say 1000 randomly generated initial centers, and for the solution take the one which gives the smallest value of the objective function (8.78).

The solution obtained by this linear relaxation approach can be considered to be the final solution (observe that it might happen that the borders of constituencies, besides large cities, split also some other territorial units), or as a good initial approximation for the integer approach.

8.4.2 Defining Constituencies in the Republic of Croatia

The territory of Croatia is organized into $m = 556$ territorial units (cities and municipalities). First we will try to determine $k = 10$ constituencies by applying the linear relaxation approach and the integer approach, based on the number of voters in 2020, and requiring that the number of voters in no constituency differs by more than 5% from the average. After that, we will analyze options with $k = 2, \ldots, 10$ constituencies and try to suggest the most appropriate number of them.

Applying the Linear Relaxation Approach to the Model with 10 Constituencies

Applying the linear relaxation approach to the model with 10 constituencies, we obtain constituencies π_j, $j = \mathrm{I}, \ldots, \mathrm{X}$,[8] with Q_j voters and relative deviation δ_j from the average $\overline{Q} = 363\,430$ (see Table 8.7)

$$\delta_j = \frac{Q_j - \overline{Q}}{\overline{Q}} \cdot 100 \,. \tag{8.89}$$

Figure 8.48 shows the smallest convex sets containing those territorial units which belong to the same constituencies. Black dots denote geographical positions of centers of each cluster, i.e. of each constituency. As expected, some territorial units are split to several constituencies.

As was already mentioned, the obtained solution can be considered as being final, or as a good initial approximation for the integer approach.

---

[8]Numbering constituencies in Croatia is done exclusively using Roman numerals.

Applying the Integer Approach to the Model with 10 Constituencies

Applying the Algorithm 8.40 with initial approximation obtained by the linear relaxation approach, we get 10 constituencies with number of voters shown in Table 8.7. Note that these numbers comply with the condition that the number of voters in constituencies should be approximately even.

Table 8.7 Number of voters per constituencies obtained by applying the linear relaxation approach and the integer approach

Constituency	Relaxation approach		Integer approach	
	Q_j	δ_j	Q_j	δ_j
I	372 302	2.44	372 294	2.44
II	372 302	2.44	371 927	2.34
III	350 123	−3.66	361 953	−0.41
IV	350 123	−3.66	355 017	−2.31
V	350 123	−3.66	354 655	−2.41
VI	350 123	−3.66	355 153	−2.28
VII	372 302	2.44	354 567	−2.44
VIII	372 302	2.44	371 533	2.23
IX	372 302	2.44	365 066	0.45
X	372 302	2.44	372 140	2.40

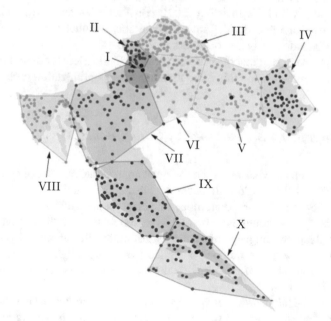

Fig. 8.48 Division into $k = 10$ constituencies according to linear relaxation approach

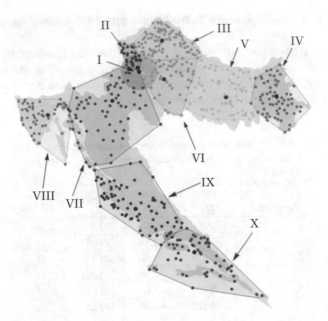

Fig. 8.49 Division into $k = 10$ constituencies according to integer approach

In this case every territorial unit belongs to exactly one constituency, except for the city of Zagreb which has to be divided into three constituencies: I (56.3%), II (42.7%), and VII (1%). Insisting on 10 constituencies resulted in an abnormal situation where Zagreb is divided into considerably unequal parts (Fig. 8.49), with the smallest including only 1% of voters in Zagreb.

It should be perceived that the problem how to divide a territorial unit into several constituencies, remains, and this issue presents a new optimization problem.

8.4.3   Optimizing the Number of Constituencies

A natural question is "what would be the most appropriate number of constituencies in Croatia?" In other words, in order to ensure the areal connectedness while preserving the evenness of the number of voters as specified for the model, how many constituencies should there be and what should their configuration look like?

Translated into the language of mathematical model, this amounts to determining optimal partition with the number of clusters ensuring as good as possible compactness and separation of clusters. To this end one can use some of the indexes introduced in Chap. 5.

The Calinski–Harabasz index applied to our data set, indicates that the appropriate number of constituencies is 6, while the Davies–Bouldin index suggests that this number is 5 (see Fig. 8.50).

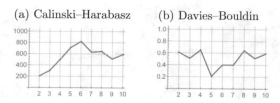

(a) Calinski–Harabasz (b) Davies–Bouldin

Fig. 8.50 Values of Calinski–Harabasz and Davies–Bouldin indexes depending on the number of constituencies

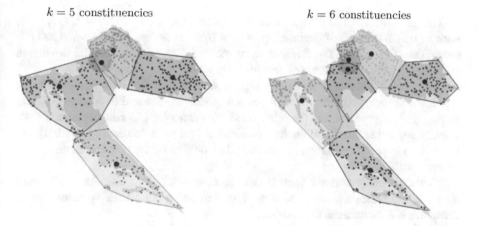

$k = 5$ constituencies

$k = 6$ constituencies

Fig. 8.51 Division into 5 and 6 constituencies according to integer approach

Figure 8.51 shows a possible partition of Croatia into 5, respectively 6, constituencies. Black dots denote the geographical positions of centers of individual clusters, i.e. constituencies.

Another possibility would be to give up on the requirement that the number of voters be more or less even throughout constituencies, and require instead that the number of voters per elected MP be up to $p\%$ even (see also [110]). In this case it would be necessary to determine the most acceptable number of constituencies with possibly various numbers of MPs.

8.5 Iris

The *IRIS data set*[9] was introduced by the British statistician and biologist Ronald Fisher (see [58]).

---

[9]https://archive.ics.uci.edu/ml/datasets/iris.

Fig. 8.52 Iris setosa, Iris virginica, and Iris versicolor

The set consist of 150 four-dimensional points $\mathcal{A} = \{a_i \in \mathbb{R}^4 : i = 1, \ldots, 150\}$, with 50 points for each of the three species of Iris: *Iris setosa, Iris virginica* and *Iris versicolor* (Fig. 8.52). Four features were measured from each sample: the length and the width (in centimeters) of sepals and petals.

One species, Iris setosa, is *linearly separable* from the other two. This means that one can draw a hyperplane between Iris setosa sample data and data corresponding to the other two species. On the other hand Iris versicolor and Iris virginica are not linearly separable from one another (their data sets have substantial overlap), i.e. there is no hyperplane that separates their data on any subset of features one wants to work with.

Therefore many authors assume that there are only two clusters in the IRIS data set, and hence a good clustering algorithm and a good validity index should recognize a 2-partition as the MAPart.

We are going to test four different approaches on this data set: spherical clustering, Mahalanobis clustering, fuzzy spherical clustering and fuzzy Mahalanobis clustering. For each of these four approaches, we will perform the appropriate k-means algorithm (WKMeans[], MWKMeans[], cmeans[], GKcmeans[]) with initial centers mean($Iris\,setosa$), mean($Iris\,versicolor$), and mean($Iris\,virginica$).

Table 8.8 Confusion matrices

WKMeans[]	MWKMeans[]	cmeans[]	GKcmeans[]
$\begin{bmatrix} 50 & 0 & 0 \\ 0 & 46 & 4 \\ 0 & 7 & 43 \end{bmatrix}$	$\begin{bmatrix} 50 & 0 & 0 \\ 0 & 50 & 0 \\ 0 & 7 & 43 \end{bmatrix}$	$\begin{bmatrix} 46.9 & 3.2 & 1.6 \\ 3.4 & 37.3 & 10.1 \\ 1.7 & 13.5 & 32.4 \end{bmatrix}$	$\begin{bmatrix} 46.1 & 3.8 & 1.8 \\ 2.3 & 34.1 & 14.4 \\ 1.3 & 16.3 & 30.0 \end{bmatrix}$

Table 8.9 Number of clusters according to spherical and Mahalanobis indexes

	CH	DB	SSWC	Dunn	Area	Rand	Jaccard	Hausdorff dist.	CPU
Inc[]	4	2	2	3	–	3	3	3	0.06
MInc[]	2	2	2	–	2	3	3	3	0.19

Table 8.10 Number of clusters according to fuzzy and Mahalanobis fuzzy indexes

	CH	DB	XB	Area	Rand	Jaccard	Hausdorff dist.	CPU
FInc[]	3	2	2	–	3	3	3	8.59
MFInc[]	3	2	2	3	3	3	3	5.20

Then, for each of these four approaches we will carry out the corresponding incremental algorithm and calculate the appropriate indexes. Table 8.9 lists the number of clusters in the corresponding MAPart obtained using Inc[] and MTnc[]. In most cases using spherical and Mahalanobis clustering, points to the 2-partition as being the MAPart. Comparing the original partition and those obtained using appropriate incremental algorithms, the Rand and Jaccard index attain largest values for the 3-partition. Also, the smallest value of the Hausdorff distance between sets of the original and of the cluster centers obtained by incremental algorithms, is for the 3-partition. One obtains similar results also in case of fuzzy clustering (see Table 8.10).

8.6 Reproduction of *Escherichia coli*

Escherichia coli is the most widely studied prokaryotic model organism, and an important species in the fields of biotechnology and microbiology, where it has served as the host organism for the majority of work with recombinant DNA. Under favorable conditions, it takes as little as 20 min to reproduce.[10]

Since *E. coli* is a rod-shaped bacterium, it is reasonable, while studying its reproduction, to apply the k-means algorithm modified for multiple ovals. The cells with the normal rod shape grow and divide, after which the two daughter cells slip and grow along one another's sides (Fig. 8.53a; the photos were taken at 10 min intervals).

The boundary of bacteria was determined using *Mathematica*-module EdgeDetect[], and the corresponding clusters were identified using the DBSCAN algorithm (Fig. 8.53b).

An appropriate initial approximation for each cluster was found using formulas (8.50)–(8.52) (Fig. 8.53c).

Ovals which define the boundary of each bacterium were obtained using the k-means algorithm adapted for multiple ovals (Fig. 8.53d), and the perfect fit can be seen in Fig. 8.53e.

This facilitates the determination of the precise position and size of each bacterium at every stage of division.

---

[10]https://en.wikipedia.org/wiki/Escherichia_coli.

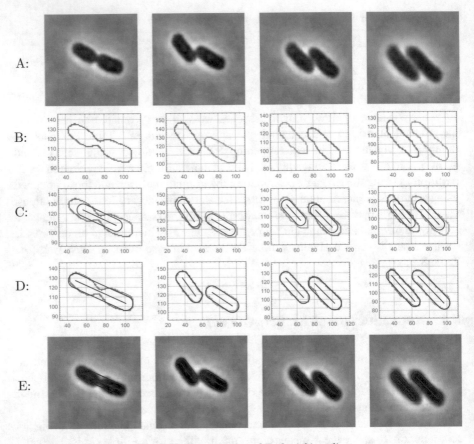

Fig. 8.53 Reproduction of *Escherichia coli*

The complete *Mathematica*-program OvalRecognizing[] used for all calculations and for producing related graphics, uses the following modules: DBSCAN[], DBSCAN1[], EPSILON[], Neps[], CorePoint[], dist[], Dmn[], DGcir[], WGLS[], WKGC[], and DBC[]. The program OvalRecognizing[] is described in Sect. 9.2, and the link to appropriate *Mathematica*-code is supplied. At the same time, this is an example showing how one should use all these modules.

Chapter 9
Modules and the Data Sets

In this chapter, we are going to describe functions, algorithms, and data sets used in this textbook, coded as *Mathematica*-modules. The respective links to all freely available *Mathematica*-modules are provided. If the data a_i are weighted with weights $w_i > 0$, then they are denoted as $\{w_i, a_i\}$. If an algorithm requires other modules, these are listed under *Submodules*. The algorithms can be tested on test-examples or the data sets, also available from the corresponding url-addresses.

9.1 Functions

d[]	p-distance like function, $p = 1$ or 2 (Chap. 2);
Input:	data points $x, y \in \mathbb{R}^n$, parameter $p = 1$ or 2;
Output:	p-distance between the points $x, y \in \mathbb{R}^n$;
Submodules:	none needed
url:	http://clusters.mathos.unios.hr/modules/d.nb

dm[], dM[]	Mahalanobis distance-like function (Sect. 6.2) and normalized Mahalanobis distance-like function (Sect. 6.4);
Input:	data points $x, y \in \mathbb{R}^n$, symmetric positive definite matrix $\Sigma \in \mathbb{R}^{n \times n}$;
Output:	Mahalanobis distance between the points $x, y \in \mathbb{R}^n$;
Submodules:	none needed
url:	http://clusters.mathos.unios.hr/modules/dm.nb

`FF[], Fmin[], G[]`	objective functions (Sects. 3.1, 3.4, and 3.3.2);
Input:	partition $\Pi = \{\pi_1, \ldots, \pi_k\}$ of the set $\mathcal{A} \subset \mathbb{R}^n$ (in case of `Fmin[]` only the set $\mathcal{A}$), centers $c_1, \ldots, c_k \in \mathbb{R}^n$, parameter $p = 1$ or 2;
Output:	objective function value $\mathcal{F}(\Pi)$ given by (3.5) or $F(c)$ given by (3.40), or $\mathcal{G}(\Pi)$ given by (3.33);
Submodules:	`d[]`
url:	http://clusters.mathos.unios.hr/modules/FF.nb

`WF[], WFmin[], WG[]`	weighted objective functions (Sects. 3.2, 3.4, and 3.3.2);
Input:	partition $\Pi = \{\pi_1, \ldots, \pi_k\}$ of the set $\mathcal{A} \subset \mathbb{R}^n$ (in case of `WFmin[]` only the set $\mathcal{A}$), centers $c_1, \ldots, c_k \in \mathbb{R}^n$;
Output:	weighted objective function values;
Submodules:	none needed;
url:	http://clusters.mathos.unios.hr/modules/WF.nb

`MWF[], MWFmin[]`	Mahalanobis objective functions (Sect. 6.4);
Input:	partition $\Pi = \{\pi_1, \ldots, \pi_k\}$ of the weighted set $\mathcal{A} \subset \mathbb{R}^n$, centers $c_1, \ldots, c_k \in \mathbb{R}^n$, covariance matrices $\Sigma_j \in \mathbb{R}^{n \times n}$, $j = 1, \ldots, k$;
Output:	Mahalanobis objective function value $\mathcal{F}_M(\Pi)$ given by (6.19), or $F_M(\Pi)$ given by (6.20);
Submodules:	`dM[]`
url:	http://clusters.mathos.unios.hr/modules/MWF.nb

`mu[]`	membership function for spherical fuzzy clustering, Eq. (7.9);
Input:	data set $\mathcal{A} \subset \mathbb{R}^n$, centers $c_1, \ldots, c_k \in \mathbb{R}^n$, indexes i, j;
Output:	membership function u_{ij};
Submodules:	none needed
url:	http://clusters.mathos.unios.hr/modules/mu.nb

`muGK[]`	membership function for Mahalanobis fuzzy clustering, Eq. (7.28);
Input:	data set $\mathcal{A} \subset \mathbb{R}^n$, centers $c_1, \ldots, c_k \in \mathbb{R}^n$, covariance matrices $\Sigma_t \in \mathbb{R}^{n \times n}$, $t = 1, \ldots, k$, indexes i, j;
Output:	membership function u_{ij};
Submodules:	`dM[]`
url:	http://clusters.mathos.unios.hr/modules/muGK.nb

Fz[]	fuzzy objective function, Eq. (7.1);
Input:	data set $\mathcal{A} \subset \mathbb{R}^n$, centers $c_1, \ldots, c_k \in \mathbb{R}^n$, membership matrix U;
Output:	objective function value $\Phi(c, U)$ given by (7.1);
Submodules:	mu[]
url:	http://clusters.mathos.unios.hr/modules/Fz.nb

MFz[]	Mahalanobis fuzzy objective function, Eq. (7.25);
Input:	data set $\mathcal{A} \subset \mathbb{R}^n$, centers $c_1, \ldots, c_k \in \mathbb{R}^n$, membership matrix U;
Output:	objective function value $F_M(c, U)$ given by (7.25);
Submodules:	dM, muGK[]
url:	http://clusters.mathos.unios.hr/modules/MFz.nb

9.2 Algorithms

DIRECT[]	DIRECT algorithm for global optimization (see Sect. 4.1 and references [55, 56, 64, 65, 69, 88, 89, 131, 154, 170]).
Input:	objective function $f : \mathrm{dom} \to \mathbb{R}$, $\mathrm{dom} \subset \mathbb{R}^n$, domain dom of the function f, additional options: $DMin \to 10^{-3}$ (minimal length of rectangle sides), $MaxIter \to 50$ (maximal number of iteration), $Eps \to 10^{-4}$ (parameter ϵ from the method);
Output:	$\{x^\star, f(x^\star)\}$, where $x^\star \in \arg\min_{x \in \mathrm{dom}} f(x)$;
Submodules:	none needed
url:	http://clusters.mathos.unios.hr/modules/DIRECT.nb

TLSline[]	determining the best total least squares normalized line $\alpha x + \beta y + \gamma = 0$, $\alpha^2 + \beta^2 = 1$, in the plane (Sect. 6.1);
Input:	data set $\mathcal{A} = \{a_i = (x_i, y_i) : i = 1, \ldots, m\} \subset \mathbb{R}^2$ with weights $w = (w_1, \ldots, w_m)$;
Output:	parameters $\{\alpha, \beta, \gamma\}$;
Submodules:	none needed
url:	http://clusters.mathos.unios.hr/modules/TLSline.nb

`Proj[]`	projection of the point $T_0 = (x_0, y_0)$ to the normalized line $\alpha x + \beta y + \gamma = 0$, $\alpha^2 + \beta^2 = 1$, in the plane (Sect. 6.1);
Input:	parameters $\alpha, \beta, \gamma, \alpha^2 + \beta^2 = 1$, of the normalized line, point $T_0 = (x_0, y_0)$;
Output:	distance from the point T_0 to the line; projection of the point T_0 to the line;
Submodules:	none needed
url:	http://clusters.mathos.unios.hr/modules/Proj.nb

`Hdist[]`	Hausdorff distance between two sets (Sect. 4.4);
Input:	finite sets $A, B \subset \mathbb{R}^n$;
Output:	Hausdorff distance;
Submodules:	none needed
url:	http://clusters.mathos.unios.hr/modules/Hdist.nb

`Rand[]`	Rand and Jaccard index and the confusion matrix (Sect. 5.2.1);
Input:	two partitions $\Pi^{(1)}$ and $\Pi^{(2)}$ of the set $\mathcal{A} \subset \mathbb{R}^n$;
Output:	Rand index R, Jaccard index J, confusion matrix S, CPU-time;
Submodules:	none needed
url:	http://clusters.mathos.unios.hr/modules/Rand.nb

`RandCompare[]`	Rand and Jaccard index and the confusion matrix according to Definition 5.12 and Remark 5.13 (Sect. 5.2.1);
Input:	two partitions $\Pi^{(1)}$ and $\Pi^{(2)}$ of the set $\mathcal{A} \subset \mathbb{R}^n$;
Output:	sets $\mathcal{C}_1, \mathcal{C}_2, \mathcal{C}_3, \mathcal{C}_4$, Rand index R, Jaccard index J, confusion matrix S, CPU-time;
Submodules:	none needed
Warning:	For large number of data, say 1000, the CPU-time can be rather long.
url:	http://clusters.mathos.unios.hr/modules/RandCompare.nb

`RandFrigue[]`	fuzzy Rand and Jaccard indexes (Sect. 7.4);
Input:	membership matrices of partitions $\Pi^{(1)}$ and $\Pi^{(2)}$;
Output:	fuzzy Rand FR and fuzzy Jaccard FJ indexes;
Submodules:	none needed
url:	http://clusters.mathos.unios.hr/modules/RandFrigue.nb

EPSILON[]	ϵ-density $\epsilon(\mathcal{A})$ of the set $\mathcal{A}$ (Sect. 4.5.1);
Input:	data set $\mathcal{A} \subset \mathbb{R}^n$, MinPts ≥ 2;
Output:	ϵ-density $\epsilon(\mathcal{A})$;
Submodules:	none needed
url:	http://clusters.mathos.unios.hr/modules/EPSILON.nb

Neps[]	set $N_{\mathcal{A}}(p, \epsilon) = \{q \in \mathcal{A} : d(q, p) < \epsilon\} = B(p, \epsilon) \cap \mathcal{A}$, where $B(p, \epsilon) \subset \mathbb{R}^n$ is the open ϵ-ball around p (Sect. 4.5.1);
Input:	data set $\mathcal{A} \subset \mathbb{R}^n$, $p \in \mathcal{A}$, $\epsilon > 0$;
Output:	$N_{\mathcal{A}}(p, \epsilon)$;
Submodules:	none needed
url:	http://clusters.mathos.unios.hr/modules/Neps.nb

CorePoint[]	choice of the core point in the set $\mathcal{A}$, based on It_0 random choices (Sect. 4.5.2);
Input:	data set $\mathcal{A} \subset \mathbb{R}^n$, $\epsilon > 0$, number of random choices It_0;
Output:	core point c_p, number $|C_p|$ of points in ϵ-neighborhood of the core point c_p;
Submodules:	none needed
url:	http://clusters.mathos.unios.hr/modules/CorePoint.nb

DBSCAN1[]	a single step of DBSCAN algorithm (Sect. 4.5.2);
Input:	data set $\mathcal{A} \subset \mathbb{R}^n$, core point c_p, MinPts, ϵ-density $\epsilon(\mathcal{A})$;
Output:	cluster π, border points SSE, outliers SSN;
Submodules:	Neps[], EPSILON[]
url:	http://clusters.mathos.unios.hr/modules/DBSCAN1.nb

DBSCAN[]	DBSCAN algorithm (Sect. 4.5.2);
Input:	data set $\mathcal{A} \subset \mathbb{R}^n$, MinPts, ϵ-density $\epsilon(\mathcal{A})$. The algorithm can use data sets from DataDiscs.txt, DataGenCirc.txt, DataEllipseGarland.txt, Koralj.txt, constructed using modules DataCircle[], DataOval[], DataCAoval[], DataEllipseGarland[];
Output:	partition Π of the set $\mathcal{A}$ with marked border points SSE and outliers SSN;
Submodules:	Neps[], EPSILON[], CorePoint[], DBSCAN1[];
url:	http://clusters.mathos.unios.hr/modules/DBSCAN.nb

`KMeansPart[]`	k-means algorithm for the data set $\mathcal{A} \subset \mathbb{R}$ or $\mathcal{A} \subset \mathbb{R}^2$ (Sect. 4.2) started with an initial partition. One can choose the p-distance like function for $p = 1$ or 2. Options are provided for printout and graphical presentation of intermediate results, as well as for printing the results as fractions or as decimal numbers;
`Input:`	initial partition $\Pi^{(0)}$, parameter $p = 1$ or 2. Examples from Test-example 9.1 can also be used;
`Output:`	partition Π, centers $c_1, \ldots, c_k$, objective function values $\mathcal{F}(\Pi)$, $F(\Pi)$, $\mathcal{G}(\Pi)$, values of CH and DB indexes;
Submodules:	`d[]`, `FF[]`, `Fmin[]`, `Figure1[]`, `Figure2[]`
`url:`	http://clusters.mathos.unios.hr/modules/KMeansPart.nb

`KMeansCen[]`	k-means algorithm for the data set $\mathcal{A} \subset \mathbb{R}$ or $\mathcal{A} \subset \mathbb{R}^2$ (Sect. 4.2) started with an initial center. One can choose the p-distance like function for $p = 1$ or 2. Options are provided for printout and graphical presentation of intermediate results, as well as for printing the results as fractions or as decimal numbers;
`Input:`	data set $\mathcal{A}$, initial centers $z_1, \ldots, z_k$, parameter $p = 1$ or 2. Examples from Test-example 9.1 can also be used;
`Output:`	partition Π, centers $c_1, \ldots, c_k$, objective function values $\mathcal{F}(\Pi)$, $F(\Pi)$, $\mathcal{G}(\Pi)$, values of CH and DB indexes;
Submodules:	`d[]`, `FF[]`, `Fmin[]`, `Figure1[]`, `Figure2[]`
`url:`	http://clusters.mathos.unios.hr/modules/KMeansCen.nb

`Figure1[]`	graphical presentation of the partition of the data set $\mathcal{A} \subset \mathbb{R}$;
`Input:`	partition Π;
`Output:`	graphical presentation;
Submodules:	none needed
`url:`	http://clusters.mathos.unios.hr/modules/Figure1.nb

`Figure2[]`	graphical presentation of the partition of the data set $\mathcal{A} \subset \mathbb{R}^2$;
`Input:`	partition Π;
`Output:`	graphical presentation;
Submodules:	`Needs["ComputationalGeometry`"];`
`url:`	http://clusters.mathos.unios.hr/modules/Figure2.nb

WKMeans[]	weighted k-means algorithm (Sect. 4.2);
Input:	data set $\mathcal{A} \subset \mathbb{R}^n$, data weights w, initial centers $z_1, \ldots, z_k$. Test-examples 9.2–9.4 can be used;
Output:	partition $\Pi = \{\pi_1, \ldots, \pi_k\}$ of the set $\mathcal{A} \subset \mathbb{R}^n$ (associated with each element of the cluster $\pi_j \in \Pi$ is the corresponding weight), centers $c_1, \ldots, c_k$, number of iterations It;
Submodules:	DIRECT[], Hdist[];
url:	http://clusters.mathos.unios.hr/modules/WKMeans.nb

Inc[]	weighted incremental algorithm (Sect. 4.3);
Input:	data set $\mathcal{A} \subset \mathbb{R}^n$, data weights $w \in \mathbb{R}^n$, initial center(s) cen, number of iterations K. Each additional center is obtained by solving a GOP using Compile. In case when $n > 2$, one has to adapt the minimizing function. Test-examples 9.2–9.4 can be used;
Output:	set of partitions (each partition consists of clusters with corresponding weights associated with its elements), CH, DB, SSWC, and Dunn indexes. Graphical representation of each iteration is presented only for $n = 2$;
Submodules:	CH[], DB[], SSWC[], Dunn[], Hdist[], Rand[], DIRECT[], WKMeans[]
url:	http://clusters.mathos.unios.hr/modules/Inc.nb

CH[]	Calinski–Harabasz index for optimal LS partition (Sect. 5.1.1);
Input:	partition Π of the set $\mathcal{A} \subset \mathbb{R}^n$;
Output:	CH index of the partition Π;
Submodules:	WF[], WG[]
url:	http://clusters.mathos.unios.hr/modules/CH.nb

DB[]	Davies–Bouldin index for optimal LS partition (Sect. 5.1.2);
Input:	partition Π of the set $\mathcal{A} \subset \mathbb{R}^n$;
Output:	DB index of the partition Π;
Submodules:	none needed
url:	http://clusters.mathos.unios.hr/modules/DB.nb

`SSWC[]`	Simplified silhouette width criterion for optimal LS partition (Sect. 5.1.3);
`Input:`	partition Π of the set $\mathcal{A} \subset \mathbb{R}^n$;
`Output:`	SSWC index of the partition Π;
`Submodules:`	none needed
`url:`	http://clusters.mathos.unios.hr/modules/SSWC.nb

`Dunn[]`	Dunn index for optimal LS partition (Sect. 5.1.4);
`Input:`	partition Π of the set $\mathcal{A} \subset \mathbb{R}^n$;
`Output:`	Dunn index of the partition Π;
`Submodules:`	none needed
`url:`	http://clusters.mathos.unios.hr/modules/Dunn.nb

`AgglNest[]`	Agglomerative nesting algorithm for data set $\mathcal{A} \subset \mathbb{R}^2$ (Sect. 4.4). The module can easily be adapted for data set $\mathcal{A} \subset \mathbb{R}^n$;
`Input:`	data set $\mathcal{A} \subset \mathbb{R}^2$, parameter $p = 1$ or 2 specifying whether to use ℓ_1-norm or LS distance-like function, parameter `DD` specifying the distance function between sets. Test-examples 9.5 can be used;
`Output:`	all partitions (cluster centers of all partitions, objective function values, and CH and DB indexes);
`Submodules:`	`d[]`, `FF[]`, `Fmin[]`, `G[]`, `MinEl[]`, `DC1[]`, `DC2[]`, `Dmin1[]`, `Dmin2[]`, `HD1[]`, `HD2[]`
`url:`	http://clusters.mathos.unios.hr/modules/AgglNest.nb

`MinEl[]`	position of the minimal element of an upper triangular matrix (Sect. 4.4);
`Input:`	upper triangular matrix;
`Output:`	position of the minimal element;
`Submodules:`	none needed
`url:`	http://clusters.mathos.unios.hr/modules/MinEl.nb

`DC1[]`, `DC2[]`	ℓ_1 and LS distance between centers of two sets A, B (see (4.16));
`Input:`	sets A, B;
`Output:`	distance between sets A, B;
`Submodules:`	none needed
`url:`	http://clusters.mathos.unios.hr/modules/DC1.nb

Dmin1[], Dmin2[]	ℓ_1 and LS minimal distance between two sets A, B (see (4.17));
Input:	sets A, B;
Output:	minimal distance between sets A, B;
Submodules:	none needed
url:	http://clusters.mathos.unios.hr/modules/Dmin1.nb

Indexes[]	CH and DB indexes for LS distance-like function (Sect. 5.1.1 and 5.1.2);
Input:	data set $\mathcal{A} \subset \mathbb{R}^n$, partition Π, centers cen;
Output:	CH and DB indexes;
Submodules:	Fmin[]
url:	http://clusters.mathos.unios.hr/modules/Indexes.nb

HD1[], HD2[]	Hausdorff ℓ_1 and LS distance between two sets A, B (see (4.20));
Input:	sets A, B;
Output:	Hausdorff distance between sets A, B;
Submodules:	none needed
url:	http://clusters.mathos.unios.hr/modules/HD1.nb

MWKMeans[]	Mahalanobis weighted k-means algorithm (Sect. 6.4.1);
Input:	data set $\mathcal{A} \subset \mathbb{R}^n$, data weights $w \in \mathbb{R}^n$, initial centers $z_1, \ldots, z_k \in \mathbb{R}^n$. Test-examples 9.6–9.10 can be used;
Output:	partition $\Pi = \{\pi_1, \ldots, \pi_k\}$ (each partition consists of clusters with corresponding weights associated with its elements), centers $c_1, \ldots, c_k \in \mathbb{R}^n$, covariance matrices $\Sigma_1, \ldots, \Sigma_k \in \mathbb{R}^{n \times n}$, number of iterations It;
Submodules:	dM[], MWF[], MFmin[], Hdist[], DIRECT[]
url:	http://clusters.mathos.unios.hr/modules/MWKMeans.nb

EM[]	Expectation maximization algorithm for normalized Gaussian mixtures (Sect. 6.4.3);
Input:	data set $\mathcal{A}$, initial centers $z_1, \ldots, z_k \in \mathbb{R}^n$. Test-examples 9.6–9.10 can be used;
Output:	partition $\Pi = \{\pi_1, \ldots, \pi_k\}$, centers $c_1, \ldots, c_k \in \mathbb{R}^n$, covariance matrices $\Sigma_1, \ldots, \Sigma_k \in \mathbb{R}^{n \times n}$, number of iterations It;
Submodules:	dM[], MWF[], MFmin[], Hdist[]
url:	http://clusters.mathos.unios.hr/modules/EM.nb

MInc[]	Mahalanobis weighted incremental algorithm (Sect. 6.4.2). Once the set of optimal partitions is obtained, one should activate the M-Indexes[] module to get the appropriate Mahalanobis indexes;
Input:	data set $\mathcal{A} \subset \mathbb{R}^n$, data weights w, initial center(s) cen, number of iterations K. Each additional center is obtained solving a GOP using Compile. In case when $n > 2$, one has to adapt the minimizing function. Test-examples 9.6–9.10 can also be used;
Output:	set of partitions (each partition consists of clusters with corresponding weights associated with its elements). Graphical representation of each iteration is presented only for $n = 2$;
Submodules:	dM[], MWF[], MFmin[], Hdist[], DIRECT[], MWKMeans[]
url:	http://clusters.mathos.unios.hr/modules/MInc.nb

MCH[]	Mahalanobis Calinski–Harabasz index for optimal Mahalanobis partitions (Sect. 6.5);
Input:	partition $\Pi = \{\pi_1, \ldots, \pi_k\}$, centers $c_1, \ldots, c_k \in \mathbb{R}^n$, covariance matrices $\Sigma_1, \ldots, \Sigma_k \in \mathbb{R}^{n \times n}$;
Output:	MCH indexes for the partition Π;
Submodules:	dM[], MWF[]
url:	http://clusters.mathos.unios.hr/modules/MCH.nb

MDB[]	Mahalanobis Davies–Bouldin index for optimal Mahalanobis partitions (Sect. 6.5);
Input:	partition $\Pi = \{\pi_1, \ldots, \pi_k\}$, centers $c_1, \ldots, c_k \in \mathbb{R}^n$, covariance matrices $\Sigma_1, \ldots, \Sigma_k \in \mathbb{R}^{n \times n}$;
Output:	MDB indexes for the partition Π;
Submodules:	dM[]
url:	http://clusters.mathos.unios.hr/modules/MDB.nb

MSSWC[]	Mahalanobis simplified silhouette width criterion for optimal Mahalanobis partitions (Sect. 6.5);
Input:	partition $\Pi = \{\pi_1, \ldots, \pi_k\}$, centers $c_1, \ldots, c_k \in \mathbb{R}^n$, covariance matrices $\Sigma_1, \ldots, \Sigma_k \in \mathbb{R}^{n \times n}$;
Output:	MSSWC indexes for the partition Π;
Submodules:	dM[]
url:	http://clusters.mathos.unios.hr/modules/MSSWC.nb

MArea[]	Mahalanobis area index for optimal Mahalanobis partitions (Sect. 6.5);
Input:	partition $\Pi = \{\pi_1, \ldots, \pi_k\}$, centers $c_1, \ldots, c_k \in \mathbb{R}^n$, covariance matrices $\Sigma_1, \ldots, \Sigma_k \in \mathbb{R}^{n \times n}$;
Output:	MArea index for the partition Π;
Submodules:	none needed
url:	http://clusters.mathos.unios.hr/modules/MArea.nb

M-Indexes[]	determining the Mahalanobis indexes for optimal Mahalanobis partition (Sect. 6.5);
Input:	set of K partitions obtained using the MInc[] module (each partition consists of clusters with corresponding weights associated with its elements);
Output:	MCH[s], MDB[s], MSSWC[s], and MArea[s] indexes for each $s = 2, \ldots, K$, with graphs;
Submodules:	dM[], MWF[], Hdist[], DIRECT[], MWKMeans[]
url:	http://clusters.mathos.unios.hr/modules/M-Indexes.nb

cmeans[]	c means algorithm (Sect. 7.2.1);
Input:	data set $\mathcal{A} \subset \mathbb{R}^n$, initial centers $z_1, \ldots, z_k \in \mathbb{R}^n$. Test-examples 9.2–9.4 can be used;
Output:	centers $c_1, \ldots, c_k \in \mathbb{R}^n$, membership matrix U, number of iterations;
Submodules:	mu[], Fz[]
url:	http://clusters.mathos.unios.hr/modules/cmeans.nb

FInc[]	fuzzy incremental algorithm (Sect. 7.2.2);
Input:	data set $\mathcal{A}$, initial center(s) cen, number of iterations K. Each additional center is obtained solving a GOP using Compile. In case when $n > 2$, one has to adapt the minimizing function. Test-examples 9.2–9.4 can be used;
Output:	set of partitions and values of FCH, FDB, and FXB indexes (if the original partition is known, one can also obtain the Rand and Jaccard indexes, confusion matrix, and Hausdorff distance between the sets of original and calculated cluster centers). Graphical representation of each iteration is presented only for $n = 2$;
Submodules:	FXB[], FCH[], FDB[], Hdist[], Rand[], DIRECT[], cmeans[]
url:	http://clusters.mathos.unios.hr/modules/FInc.nb

FXB []	Xie–Beni fuzzy index for spherical clustering (Sect. 7.2.3);
Input:	data set $A \subset \mathbb{R}^n$, centers $c_1, \ldots, c_k \in \mathbb{R}^n$, membership matrix U;
Output:	FXB index;
Submodules:	none needed
url:	http://clusters.mathos.unios.hr/modules/FXB.nb

FCH []	fuzzy Calinski–Harabasz index for spherical clustering (Sect. 7.2.3);
Input:	data set $A \subset \mathbb{R}^n$, centers $c_1, \ldots, c_k \in \mathbb{R}^n$, membership matrix U;
Output:	FCH index;
Submodules:	none needed
url:	http://clusters.mathos.unios.hr/modules/FCH.nb

FDB []	fuzzy Davies–Bouldin index for spherical clustering (Sect. 7.2.3);
Input:	data set $A \subset \mathbb{R}^n$, centers $c_1, \ldots, c_k \in \mathbb{R}^n$, membership matrix U;
Output:	FDB index;
Submodules:	none needed
url:	http://clusters.mathos.unios.hr/modules/FDB.nb

GKcmeans []	GKc-means algorithm (Sect. 7.3.1);
Input:	data set $A \subset \mathbb{R}^n$, initial centers $z_1, \ldots, z_k \in \mathbb{R}^n$. Test-examples 9.6–9.10 can be used;
Output:	centers $c_1, \ldots, c_k \in \mathbb{R}^n$, membership matrix U, covariance matrices $\Sigma_1, \ldots, \Sigma_k \in \mathbb{R}^{n \times n}$, number of iterations;
Submodules:	dM [], muGK [], MFz []
url:	http://clusters.mathos.unios.hr/modules/GKcmeans.nb

MFInc []	Mahalanobis fuzzy incremental algorithm (Sect. 7.3.2);
Input:	data set A, initial centers $z_1, \ldots, z_k \in \mathbb{R}^n$, number of iterations K. Each additional center is obtained solving a GOP using Compile. In case when $n > 2$, one has to adapt the minimizing function. Test-examples 9.6–9.10 can be used;
Output:	set of partitions and the values of MFXB, MFCH, MFDB, and MFHV values (if the original partition is known, one can also obtain the Rand and Jaccard indexes, confusion matrix, and Hausdorff distance between the sets of original and calculated cluster centers). Graphical representation of each iteration is presented only for $n = 2$;
Submodules:	dM [], muGK [], MFz [], GKcmeans [], MFXB [], MFCH [], MFDB [], MFHV [], RandFrigue [], DIRECT [], Hdist []
url:	http://clusters.mathos.unios.hr/modules/MInc.nb

MFXB[]	Mahalanobis fuzzy Xie–Beni index (Sect. 7.3.3);
Input:	data set $\mathcal{A} \subset \mathbb{R}^n$, centers $c_1, \ldots, c_k \in \mathbb{R}^n$, membership matrix U, covariance matrices $\Sigma_1, \ldots, \Sigma_k \in \mathbb{R}^{n \times n}$;
Output:	MFXB index;
Submodules:	dM[]
url:	http://clusters.mathos.unios.hr/modules/MFXB.nb

MFCH[]	Mahalanobis fuzzy Calinski–Harabasz index (Sect. 7.3.3);
Input:	data set $\mathcal{A} \subset \mathbb{R}^n$, centers $c_1, \ldots, c_k \in \mathbb{R}^n$, membership matrix U, covariance matrices $\Sigma_1, \ldots, \Sigma_k \in \mathbb{R}^{n \times n}$;
Output:	MFCH index;
Submodules:	dM[]
url:	http://clusters.mathos.unios.hr/modules/MFCH.nb

MFDB[]	Mahalanobis fuzzy Davies–Bouldin index (Sect. 7.3.3);
Input:	data set $\mathcal{A} \subset \mathbb{R}^n$, centers $c_1, \ldots, c_k \in \mathbb{R}^n$, membership matrix U, covariance matrices $\Sigma_1, \ldots, \Sigma_k \in \mathbb{R}^{n \times n}$;
Output:	MFDB index;
Submodules:	dM[]
url:	http://clusters.mathos.unios.hr/modules/MFDB.nb

MFHV[]	Mahalanobis fuzzy hypervolume index (Sect. 7.3.3);
Input:	covariance matrices $\Sigma_1, \ldots, \Sigma_k \in \mathbb{R}^{n \times n}$;
Output:	MFHV index;
Submodules:	dM[]
url:	http://clusters.mathos.unios.hr/modules/MFHV.nb

WKGC[]	weighted k-closest generalized circles algorithm (Sect. 8.1.6 and Algorithm 8.2);
Input:	data set $\mathcal{A} \subset \mathbb{R}^n$, data weights w, initial circle-centers $Circ$;
Output:	partition $\Pi = \{\pi_1, \ldots, \pi_k\}$ of the set $\mathcal{A} \subset \mathbb{R}^n$ (associated with each element of the cluster $\pi_j \in \Pi$ is the corresponding weight), generalized circle-centers $Cir_1, \ldots, Cir_k$, number of iterations It;
Submodules:	DIRECT[], dist[], Dmn[], WGLS[];
url:	http://clusters.mathos.unios.hr/modules/WKGC.nb

WGLS[]	weighted objective function value for generalized circle (Sect. 8.1.6);
Input:	Partition Π of the set $\mathcal{A} \subset \mathbb{R}^n$, circle-center(s) Circ;
Output:	weighted objective function value;
Submodules:	DGcir[]
url:	http://clusters.mathos.unios.hr/modules/WGLS.nb

DBC[]	Density-Based Clustering index for optimal partition with geometric objects (Remark 8.3);
Input:	partition Π of the set $\mathcal{A} \subset \mathbb{R}^n$, geometric objects;
Output:	QD index of the partition Π;
Submodules:	DGcir[]
url:	http://clusters.mathos.unios.hr/modules/DBC.nb

Spherical Clustering is a complete *Mathematica*-program for spherical clustering using the LS distance-like function. All modules needed are listed in the preamble. First one has to load one of the suggested examples which will automatically perform the WKMeans[] algorithm. Then one can call the incremental algorithm Inc[] which will, based on several indexes, propose among obtained partitions the MAPart. If the original partition is known, it will also calculate the Rand and Jaccard indexes, as well as the Hausdorff distance between sets of original and calculated cluster centers.

url: http://clusters.mathos.unios.hr/modules/SphClustering.nb

Mahalanobis Clustering is a complete *Mathematica*-program for Mahalanobis clustering using the normalized Mahalanobis distance-like function. All modules needed are listed in the preamble. First one has to load one of the suggested examples which will automatically perform the MWKMeans[] algorithm. Then one can call the incremental algorithm MInc[] which will produce several partitions sorted in the file PART.txt. Calling these partitions by MCH[], MDB[], MSSWC[], and MArea[] modules, one obtains appropriate indexes, based on which MAPart is suggested.

url: http://clusters.mathos.unios.hr/modules/M-Clustering.nb

Fuzzy Spherical Clustering is a complete *Mathematica*-program for fuzzy spherical clustering using the LS distance-like function. All modules needed are listed in the preamble. First one has to load one of the suggested examples. After that one can perform the fuzzy c-means algorithm cmeans[] or the fuzzy incremental algorithm FInc[], which will, based on several indexes, propose one of the obtained partitions as the MAPart. If the original partition is known, it will also calculate the Rand and Jaccard indexes, as well as the Hausdorff distance between sets of original and calculated cluster centers.

url: http://clusters.mathos.unios.hr/modules/F-SphClustering.nb

Fuzzy Mahalanobis Clustering is a complete *Mathematica*-program for fuzzy Mahalanobis clustering using the normalized Mahalanobis distance-like function. All modules needed are listed in the preamble. First one has to load one of the suggested examples. Then one can perform the GKc-means algorithm GKcmeans [] or the Mahalanobis fuzzy incremental algorithm MFInc [], which will, based on several fuzzy indexes, propose one of the obtained partitions as the MAPart. If the original partition is known, it will also calculate the Rand and Jaccard indexes, the confusion matrix, and the Hausdorff distance between sets of original and calculated cluster centers.

url : http://clusters.mathos.unios.hr/modules/F-MClustering.nb

Oval Recognizing is a complete *Mathematica*-program for recognizing ovals using ℓ_1 or LS distance-like function. All modules needed are listed in the preamble. First one has to input the data file E-ColiA.txt (see Sect. 9.5) and then actuate all cells one by one.

url : http://clusters.mathos.unios.hr/modules/OvalRecognizing.nb

9.3   Data Generating

dist []	distance from the point T to the segment $[\mu, \nu]$ in the plane (see Example 4.39);
Input :	point T, endpoints μ, ν of the segment $[\mu, \nu]$;
Output :	distance;
Submodules:	none needed
url :	http://clusters.mathos.unios.hr/modules/dist.nb

DGcir []	distance from a point T to the oval in the plane (see Example 8.25);
Input :	point T, segment $[\mu, \nu]$, radius r;
Output :	distance;
Submodules:	dist []
url :	http://clusters.mathos.unios.hr/modules/DGcir.nb

darc []	distance from the point T to the circle-arc in the plane (see Example 4.39);
Input :	point T, center C, radius r, angles α_1, α_2;
Output :	distance;
Submodules:	none needed
url :	http://clusters.mathos.unios.hr/modules/darc.nb

`DataDisc[]`	generating data from a disc (see Example 4.37);
`Input:`	center C, disc radius r, number n_p of points per unit square, standard deviation σ, number n_r of points generated from the normal distribution in the neighborhood of each disc point;
`Output:`	data set, circle $K(C, r)$
`Submodules:`	none needed
`url:`	http://clusters.mathos.unios.hr/modules/DataDisc.nb

`DataOval[]`	generating data from an oval (see Example 4.39);
`Input:`	endpoints μ, ν, radius r, number n_p of points per unit square, standard deviation σ, number n_r of points generated from the normal distribution in the neighborhood of each point $a \in \mathsf{Oval}([\mu, \nu], r)$;
`Output:`	data set, oval $\mathsf{Oval}([\mu, \nu], r)$;
`Submodules:`	none needed
`url:`	http://clusters.mathos.unios.hr/modules/DataOval.nb

`DataCAoval[]`	generating data from a circle-arc oval (see Example 4.39);
`Input:`	center C, radius r, number n_p of points per unit square, standard deviation σ, number n_r of points generated from the normal distribution in the neighborhood of each point $a \in \mathsf{CAoval}(\mathrm{arc}(C, R, \alpha_1, \alpha_2)$ with radius R and arcs α_1, α_2;
`Output:`	data set, circle-arc oval $\mathsf{CAoval}(\mathrm{arc}(C, R, \alpha_1, \alpha_2), r)$;
`Submodules:`	none needed
`url:`	http://clusters.mathos.unios.hr/modules/DataCAoval.nb

`DataEllipseGarland[]`	generating data from ellipse garland (see Example 4.41);
`Input:`	ellipse garland $\{x \in \mathbb{R}^2 : 0.8 \le d_m(C, x, \Sigma) \le 1\}$ with center C and covariance matrix Σ;
`Output:`	data set;
`Submodules:`	`dm[]`
`url:`	http://clusters.mathos.unios.hr/modules/DataEllipseGarland. nb

9.4 Test Examples

Test-example 9.1 The modules `KMeansPart[]` and `KMeansCen[]` may be tested on eighteen test-examples with $A \subset \mathbb{R}$ and eleven test-examples with $A \subset \mathbb{R}^2$. One can choose the initial partition Π and call `KMeansPart[]`, or choose initial centers z and call `KMeansCen[]`. For example, choosing the initial partition $\Pi = \{\{1, 2, 3, 8\}, \{9, 10, 25\}\}$ of the set $A = \{1, 2, 3, 8, 9, 10, 25\} \subset \mathbb{R}$, with $p = 2$, Ind $= 0$, num $= 0$, one obtains `LOPart` $\Pi^{\star} = \{\{1, 2, 3, 8, 9, 10\}, \{25\}\}$, centers $c_1 = 11/2$, $c_2 = 25$, the objective function values of $\mathcal{F} = 155/2$, $F_{\min} = 155/2$, and $\mathcal{G} = 4563/14$, indexes CH $= 21.028$, DB $= 0.184$, and the graphical presentation

For data set $A = \{(2, 3), (3, 6), (5, 8), (6, 5), (7, 7), (8, 1), (9, 5), (10, 3)\} \subset \mathbb{R}^2$ and initial centers $z = \{(4, 4), (8, 4)\}$ with $p = 1$, Ind $= 0$, num $= 0$, we obtain `LOPart` $\Pi^{\star} = \{\{(2, 3), (3, 6), (5, 8), (6, 5)\}, \{(7, 7), (8, 1), (9, 5), (10, 3)\}\}$, centers $c_1 = (4, 11/2)$ and $c_2 = (17/2, 4)$, objective function values $\mathcal{F} = 24$ and $F_{\min} = 24$, and graphical presentation

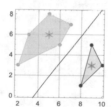

Test-example 9.2 Take five points $C_1 = (2.5, 3)$, $C_2 = (2, 8.5)$, $C_3 = (5, 6)$, $C_4 = (8, 5)$, and $C_5 = (8.5, 1.5)$ in $\Delta = [0, 10]^2 \subset \mathbb{R}^2$. For each point C_j generate 200 random points from the bivariate normal distribution with expectation $C_j \in \mathbb{R}^2$ and covariance matrix $\sigma^2 I$, where $\sigma^2 = 1$ (see Fig. 9.1a). The corresponding data set is shown in Fig. 9.1b.

Mathematica-code: `TExaMethods1` available at http://clusters.mathos.unios. hr/modules/TExaMethods1.nb

Test-example 9.3 Take six points $C_1 = (-8, -6)$, $C_2 = (-3, -8)$, $C_3 = (-1, -1)$, $C_4 = (8, 4)$, $C_5 = (6, 9)$, and $C_6 = (-1, 8)$ in $\Delta = [0, 10]^2 \subset \mathbb{R}^2$. For each point C_j generate 200 random points from the bivariate normal distribution with expectation $C_j \in \mathbb{R}^2$ and covariance matrix $\sigma^2 I$, where $\sigma^2 = 1.5$ (see Fig. 9.2a). The corresponding data set is shown in Fig. 9.2b.

Mathematica-code: `TExaMethods2` available at http://clusters.mathos.unios. hr/modules/TExaMethods2.nb

Test-example 9.4 Take four points $C_1 = (2.5, 3, 2)$, $C_2 = (2, 8.5, 4)$, $C_3 = (5, 6, 4)$, and $C_4 = (8, 5, 6)$ in $\Delta = [0, 10]^3 \subset \mathbb{R}^3$. For each point C_j generate 100 random points from the multivariate normal distribution with expectation $C_j \in \mathbb{R}^3$

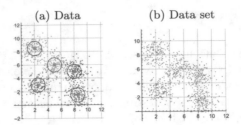

Fig. 9.1 Data set

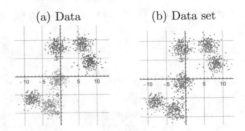

Fig. 9.2 Data set

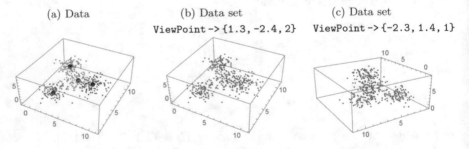

Fig. 9.3 Data set

and covariance matrix $\sigma^2 I$, where $\sigma^2 = 1$ (see Fig. 9.3a). The corresponding data set is shown in Fig. 9.1b (ViewPoint -> {1.3, -2.4, 2}) and inFig. 9.1c (ViewPoint -> {-2.3, 1.4, 1}).

   Mathematica-code: TExaMethods3 available at http://clusters.mathos.unios. hr/modules/TExaMethods3.nb

Test-example 9.5 Figure 9.4b shows the dendrograph of the set $\mathcal{A} = \{(2, 7), (3, 9), (4, 8), (3, 7), (5, 7), (8, 3), (8, 5), (9, 4)\}$.

   Mathematica-code: TExaMethods4 available at http://clusters.mathos.unios. hr/modules/TExaMethods4.nb

Test-example 9.6 Take three points $C_1 = (1, 1)$, $C_2 = (3.5, 3.5)$, and $C_3 = (6, 1)$ in $\Delta = [0, 10]^2 \subset \mathbb{R}^2$. For each point C_j generate 100 random points from the bivariate normal distribution with expectation $C_j \in \mathbb{R}^2$ and covariance matrices (see Fig. 9.5a)

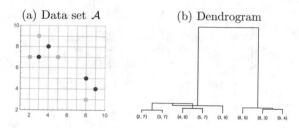

Fig. 9.4 Data set

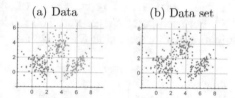

Fig. 9.5 Data set

$$\Sigma_1 = \begin{bmatrix} 1.0 & -0.3 \\ -0.3 & 1.0 \end{bmatrix}, \quad \Sigma_2 = \begin{bmatrix} 1.0 & 0.3 \\ 0.3 & 1.0 \end{bmatrix}, \quad \Sigma_3 = \begin{bmatrix} 1.0 & 0.7 \\ 0.7 & 1.0 \end{bmatrix}$$

Mathematica-code: TExaM1 available at http://clusters.mathos.unios.hr/modules/TExaM1.nb

Test-example 9.7 Take five points $C_1 = (2, 2)$, $C_2 = (4, 6)$, $C_3 = (6, 10)$, $C_4 = (8, 7)$, and $C_5 = (9, 3)$ in $\Delta = [0, 10]^2 \subset \mathbb{R}^2$. For each point C_j generate 100 random points from the bivariate normal distribution with expectation $C_j \in \mathbb{R}^2$ and covariance matrices (see Fig. 9.6a)

$$\Sigma_1 = \begin{bmatrix} 1 & 0 \\ 0 & 1.2 \end{bmatrix}, \; \Sigma_2 = \begin{bmatrix} 2 & -1 \\ -1 & 0.6 \end{bmatrix}, \; \Sigma_3 = \begin{bmatrix} 5 & 0 \\ 0 & 0.5 \end{bmatrix}, \; \Sigma_4 = \begin{bmatrix} 1.2 & 0 \\ 0 & 1 \end{bmatrix}, \; \Sigma_5 = \begin{bmatrix} 2 & 1 \\ 1 & 1 \end{bmatrix}$$

Mathematica-code: TExaM2 available at http://clusters.mathos.unios.hr/modules/TExaM2.nb

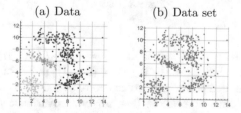

Fig. 9.6 Data set

(a) Data   (b) Data set

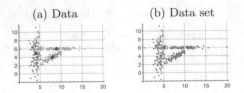

Fig. 9.7 Data set

Test-example 9.8 Take three points $C_1 = (4, 5)$, $C_2 = (8, 4)$, and $C_3 = (9, 6)$ in $\Delta = [0, 10]^2 \subset \mathbb{R}^2$. For each point C_j generate 100 random points from the bivariate normal distribution with expectation $C_j \in \mathbb{R}^2$ and covariance matrices (see Fig. 9.7a)

$$\Sigma_1 = \begin{bmatrix} 0.4 & 0 \\ 0 & 8 \end{bmatrix}, \quad \Sigma_2 = \begin{bmatrix} 2 & 1 \\ 1 & 0.6 \end{bmatrix}, \quad \Sigma_3 = \begin{bmatrix} 10 & 0 \\ 0 & 0.05 \end{bmatrix}$$

Mathematica-code: TExaM3 available at http://clusters.mathos.unios.hr/modules/ TExaM3.nb

Test-example 9.9 Choose three segments $\ell_1 = [(-\sqrt{2}, -\sqrt{2}), (\sqrt{2}, \sqrt{2})]$, $\ell_2 = [(0.5, 2), (2, -2)]$, and $\ell_3 = [(0, -1.5), (3, -1.5)]$ in $\mathbb{R}^2$, on each one generate 100 uniformly distributed points, and then to each point add a random error from the bivariate normal distribution with expectation $0 \in \mathbb{R}^2$ and covariance matrix $\sigma^2 I$, where $\sigma^2 = 0.01$ (see Fig. 9.8a).

Mathematica-code: TExaM4   available   at   http://clusters.mathos.unios.hr/ modules/TExaM4.nb

(a) Data   (b) Data set

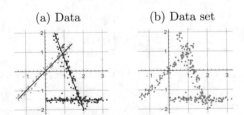

Fig. 9.8 Data set

Test-example 9.10 Take three segments $\ell_1 = [(-\sqrt{2}, \sqrt{2}), (\sqrt{2}, \sqrt{2})]$, $\ell_2 = [(0.5, 2), (1.5, -1.5)]$, and $\ell_3 = [(-1, -0.5), (2, -1)]$ in $\mathbb{R}^2$, on each one generate 100 uniformly distributed points, add to each point a random error from the bivariate normal distribution with expectation $0 \in \mathbb{R}^2$ and covariance matrix $\sigma^2 I$, where $\sigma^2 = 0.0025$ (see Fig. 9.9a).

(a) Data   (b) Data set

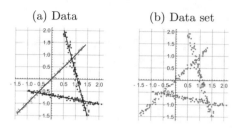

Fig. 9.9 Data set

Mathematica-code: TExaM5 [] available at http://clusters.mathos.unios.hr/ modules/TExaM5.nb

9.5 Data Sets

DataDiscs.txt: data set derived from three discs in Example 4.37, available at http://clusters.mathos.unios.hr/modules/DataDiscs.txt

DataGenCirc.txt: data from ovals in Example 4.39, available at http://clusters. mathos.unios.hr/modules/DataGenCirc.txt

DataEllipseGarland.txt: data set derived from the region between two non-parallel ellipses, and the disc in Example 4.41, available at http://clusters.mathos. unios.hr/modules/DataEllipseGarland.txt

Coral.txt: data set obtained from the real-world image of a red coral in Example 4.43, available at http://clusters.mathos.unios.hr/modules/Coral.txt

E-ColiA.txt: data set obtained from the real-world image of *Escherichia coli* in Sect. 8.6, available at http://clusters.mathos.unios.hr/modules/E-ColiA.txt

Bibliography

1. A.S. Ackleh, E.J. Allen, R.B. Kearfott, P. Seshaiyer, *Classical and Modern Numerical Analysis: Theory, Methods and Practice* (Taylor and Francis Group, London, 2010)
2. C. Akinlar, C. Topal, Edcircles: a real-time circle detector with a false detection control. Pattern Recogn. **46**, 725–740 (2013)
3. G. Andrade, G. Ramos, D. Madeira, R. Sachetto, R. Ferreira, L. Rocha, G-DBSCAN: A GPU accelerated algorithm for density-based clustering. Proc. Comput. Sci. **18**, 369–378 (2013)
4. M. Ankerst, M.M. Breunig, H.-P. Kriegel, J. Sander, OPTICS: ordering points to identify the clustering structure. ACM Sigmod Record **28**, 49–60 (1999)
5. G. Asencio-Cortés, S. Scitovski, R.Scitovski, F. Martínez-Álvarez, Temporal analysis of croatian seismogenic zones to improve earthquake magnitude prediction. Earth Sci. Inf. **10**, 303–320 (2017)
6. F. Aurenhammer, R. Klein, Voronoi diagrams, in *Handbook of Computational Geometry, Chapter V*, ed. by J. Sack, G. Urrutia (Elsevier, Amsterdam, 2000), pp. 201–290
7. R. Babuška, P.J. van der Veen, U. Kaymak, Improved covariance estimation for Gustafson-Kessel clustering, in *IEEE International Conference on Fuzzy Systems*, (2002), pp. 1081–1085
8. A.M. Bagirov, Modified global k-means algorithm for minimum sum-of-squares clustering problems. Pattern Recogn. **41**, 3192–3199 (2008)
9. A.M. Bagirov, An incremental DC algorithm for the minimum sum-of-squares clustering. Iran. J. Oper. Res. **5**, 1–14 (2014)
10. A.M. Bagirov, N. Karmitsa, S. Taheri, *Partitional Clustering via Nonsmooth Optimization. Clustering via Optimization* (Springer, Berlin, 2020)
11. A.M. Bagirov, J. Ugon, An algorithm for minimizing clustering functions. Optimization **54**, 351–368 (2005)
12. A.M. Bagirov, J. Ugon, H. Mirzayeva, Nonsmooth nonconvex optimization approach to clusterwise linear regression problems. Eur. J. Oper. Res. **229**, 132–142 (2013)
13. A.M. Bagirov, J. Ugon, D. Webb, An efficient algorithm for the incremental construction of a piecewise linear classifier. Inf. Syst. **36**, 782–790 (2011)
14. A.M. Bagirov, J. Ugon, D. Webb, Fast modified global k-means algorithm for incremental cluster construction. Pattern Recogn. **44**, 866–876 (2011)
15. A.M. Bagirov, J. Yearwood, A new nonsmooth optimization algorithm for minimum sum-of-squares clustering problems. Eur. J. Oper. Res. **170**, 578–596 (2006)
16. A.M. Bakr, N.M. Ghanem, M.A. Ismail, Efficient incremental density-based algorithm for clustering large datasets. Alexandria Eng. J. **54**, 1147–1154 (2015)

© The Author(s), under exclusive license to Springer Nature Switzerland AG 2021
R. Scitovski et al., *Cluster Analysis and Applications*,
https://doi.org/10.1007/978-3-030-74552-3

17. A. Beck, L. Tetruashvili, On the convergence of block coordinate descent type methods. SIAM J. Optim. 23(4), 2037–2060 (2013)
18. J. Behboodian, On the modes of a mixture of two normal distributions. Technometrics, **12**, 131–139 (1979)
19. L. Beilina, E. Karchevskii, M. Karchevskii, *Numerical Linear Algebra:Theory and Application* (Springer, Berlin, 2017)
20. J.C. Bezdek, R. Ehrlich, W. Full, FCM: the fuzzy c-means clustering algorithm. Comput. Geosci. **10**, 191–203 (1984)
21. J.C. Bezdek, J. Keller, R. Krisnapuram, N. R. Pal, *Fuzzy Models and Algorithms for Pattern Recognition and Image Processing* (Springer, Berlin, 2005)
22. D. Birant, A. Kut, ST-DBSCAN: an algorithm for clustering spatial-temporal data. Data Knowl. Eng. **60**, 208–221 (2007)
23. C.M. Bishop, *Pattern Recognition and Machine Learning* (Springer, Berlin, 2006)
24. D. Blumenfeld, *Operation Research Calculation: Handbook* (CRC Press, Boca Raton, 2001)
25. R.J. Boscovich, De litteraria expeditione per pontificiam ditionem, et synopsis amplioris operis, ac habentur plura eius ex exemplaria etiam sensorum impressa. Bononienci Scientiarum et Artium Znstituto Atque Academia Commentarrii **4**, 353–396 (1757)
26. J.M. Bourjolly, G. Laporte, J.M. Rousseau, Découpage électoral automatisé á i'le de montréal. INFOR **19**, 113–124 (1981)
27. B. Bozkaya, E. Erkut, G. Laporte, A tabu search heuristic and adaptive memory procedure for political districting. Eur. J. Oper. Res. **144**, 12–26 (2003)
28. T. Calinski, J. Harabasz, A dendrite method for cluster analysis. Commun. Stat. **3**, 1–27 (1974)
29. K. Chan, C. Kwong, B. Hu, Market segmentation and ideal point identification for new product design using fuzzy data compression and fuzzy clustering methods. Appl. Soft Comput. **12**, 1371–1378 (2012)
30. W. Cheney, *Analysis for Applied Mathematics*. Graduate Text in Mathematics, vol. 208 (Springer, 2001)
31. N. Chernov, *Circular and Linear Regression: Fitting Circles and Lines by Least Squares*. Monographs on Statistics and Applied Probability, vol. 117 (Chapman & Hall/CRC, London, 2010)
32. C.-H. Chou, M.-C. Su, E. Lai, A new cluster validity measure and its application to image compression. Pattern Anal. Appl. **7**, 205–220 (2004)
33. K.-L. Chung, Y.-H. Huang, S.-M. Shen, A.S. Krylov, D.V. Yurin, E.V. Semeikina, Efficient sampling strategy and refinement strategy for randomized circle detection. Pattern Recogn. **45**, 252–263 (2012)
34. M. Collan, M. Fedrizzi, P. Luukka, A multi-expert system for ranking patents: an approach based on fuzzy pay-off distributions and a TOPSIS-AHP framework. Expert Syst. Appl. **40**, 4749–4759 (2013)
35. E. Cuevas, V. Osuna-Enciso, F. Wario, D. Zaldívar, M. Pérez-Cisneros, Automatic multiple circle detection based on artificial immune systems. Expert Syst. Appl. **39** 713–722 (2012)
36. R. Cupec, R. Grbić, K. Nyarko, K. Sabo, R. Scitovski, Detection of planar surfaces based on RANSAC and LAD plane fitting, in *Proceedings of the 4th European Conference on Mobile Robots, ECMR'09* (2009)
37. R. Cupec, R. Grbić, K. Sabo, R. Scitovski, Three points method for searching the best least absolute deviations plane. Appl. Math. Comput. **215**, 983–994 (2009)
38. H. Darong, W. Peng, Grid-based DBSCAN algorithm with referential parameters. Phys. Proc. **24**, 1166–1170 (2012)
39. J. Dattorro, *Convex Optimization & Euclidean Distance Geometry* (Meboo Publishing, California, 2011). https://meboo.convexoptimization.com/Meboo.html.
40. D. Davies, D. Bouldin, A cluster separation measure. IEEE Trans. Pattern Anal. Mach. Intell. **2**, 224–227 (1979)

41. P.G. de Cortona, C. Manzi, A. Pennisi, F. Ricca, B. Simeone, Evaluation and optimization of electoral systems, in *SIAM Monographs on Discrete Mathematics* (SIAM, Philadelphia, 1999)

42. I.S. Dhillon, Y. Guan, B. Kulis, Kernel k-means, spectral clustering and normalized cuts, in *Proceedings of the 10-th ACM SIGKDD International Conference on Knowledge Discovery and Data Mining (KDD), August 22–25, 2004, Seattle, Washington, USA* (2004), pp. 551–556

43. Y. Dodge (ed.), Statistical data analysis based on the L_1-norm and related methods, in *Proceedings of the Third International Conference on Statistical Data Analysis Based on the L1-norm and Related Methods* (Elsevier, Amsterdam, 1997)

44. Y. Dodge, J. Jurečková, *Adaptive Regression* (Springer, Berlin, 2000)

45. J. Dortet-Bernadet, N. Wicker, Model-based clustering on the unit sphere with an illustration using gene expression profiles. Biostatistics, **9**(1), 66–80 (2008)

46. Z. Drezner, H.W. Hamacher, *Facility Location: Applications and Theory* (Springer, Berlin, 2004)

47. R.O. Duda, P.E. Hart, Use of the Hough Transformation to detect lines and curves in pictures. Commun. ACM **15**, 11–15 (1972)

48. J.C. Dunn, Well separated clusters and optimal fuzzy partitions. J. Cybern. **4**, 95–104 (1974)

49. B. Durak, *A Classification Algorithm Using Mahalanobis Distances Clustering of Data with Applications on Biomedical Data Set*. Ph.D. thesis, The Graduate School of Natural and Applied Sciences of Middle East Technical University, Ankara, 2011

50. L. Ertöz, M. Steinbach, V. Kumar, Finding clusters of different sizes, shapes, and densities in noisy, high dimensional data, in *Proceedings of Second SIAM International Conference on Data Mining, San Francisco*, 2003

51. M. Ester, H. Kriegel, J. Sander, A density-based algorithm for discovering clusters in large spatial databases with noise, in *Second International Conference on Knowledge Discovery and Data Mining (KDD-96)* (Portland 1996), pp. 226–231

52. B.S. Everitt, S. Landau, M. Leese, *Cluster Analysis* (Wiley, London, 2001)

53. L.A. Fernandes, M.M. Oliveira, Real-time line detection through an improved Hough transform voting scheme. Pattern Recogn. **41**, 299–314 (2008)

54. C. Fernández, V. Moreno, B. Curto, J.A. Vicente, Clustering and line detection in laser range measurements. Robot. Auton. Syst. **58**, 720–726 (2010)

55. D.E. Finkel, *DIRECT Optimization Algorithm User Guide* (Center for Research in Scientific Computation. North Carolina State University, 2003). http://www4.ncsu.edu/definkel/research/index.html

56. D.E. Finkel, C.T. Kelley, Additive scaling and the DIRECT algorithm. J. Global Optim. **36**, 597–608 (2006)

57. M. Fischler, R. Bolles, Random sample consensus: a paradigm for model fitting with applications to image analysis and automated cartography. Commun. ACM **24**, 381–395 (1981)

58. R.A. Fisher, The use of multiple measurements in taxonomic problems. Ann. Eugen. **7**, 179–188 (1936)

59. A. Fitzgibbon, M. Pilu, R.B. Fisher, Direct least square fitting of ellipses. IEEE Trans. Pattern Anal. Mach. Intell. **21**, 476–480 (1999)

60. C.A. Floudas, C.E. Gounaris, A review of recent advances in global optimization. J. Global Optim. **45**, 3–38 (2009)

61. H. Frigui, Unsupervised learning of arbitrarily shaped clusters using ensembles of Gaussian models. Pattern Anal. Appl. **8**, 32–49 (2005)

62. H. Frigui, C. Hwang, F.C.-H. Rhee, Clustering and aggregation of relational data with applications to image database categorization. Pattern Recogn. **40**, 3053–3068 (2007)

63. H. Frigui, R. Krishnapuram, Clustering by competitive agglomeration. Pattern Recogn. **30**, 1109–1119 (1997)

64. J.M. Gablonsky, *DIRECT Version 2.0*, Technical report, Center for Research in Scientific Computation. North Carolina State University, 2001

65. J.M. Gablonsky, C.T. Kelley, A locally-biased form of the direct algorithm. J. Global Optim. **21**, 27–37 (2001)

66. W. Gander, G.H. Golub, R. Strebel, Least-squares fitting of circles and ellipses. BIT **34**, 558–578 (1994)

67. I. Gath, A.B. Geva, Unsupervised optimal fuzzy clustering. IEEE Trans. Pattern Anal. Mach. Intell. **11**, 773–781 (1989)

68. R. Grbić, D. Grahovac, R. Scitovski, A method for solving the multiple ellipses detection problem. Pattern Recogn. **60**, 824–834 (2016)

69. R. Grbić, E.K. Nyarko, R. Scitovski, A modification of the DIRECT method for Lipschitz global optimization for a symmetric function. J. Global Optim. **57**, 1193–1212 (2013)

70. J. Guerrero, G. Pajares, M. Montalvo, J. Romeo, M. Guijarro, Support vector machines for crop/weeds identification in maize fields. Expert Syst. Appl. **39**, 11149–11155 (2012)

71. A. Gunawan, *A Faster Algorithm for DBSCAN*, Ph.D. thesis, Technische Universiteit Eindhoven, 2013

72. C. Gurwitz, Weighted median algorithms for l_1 approximation. BIT **30**, 301–310 (1990)

73. D.E. Gustafson, W.C. Kessel, Fuzzy clustering with a fuzzy covariance matrix, in *Proceedings of the IEEE Conference on Decision Control* (San Diego, 1979), pp. 761–766

74. J.P.H. Chen, 0-1 semidefinite programming for graph-cut clustering: modelling and approximation, in *Data Mining and Mathematical Programming*, ed. by P.M. Pardalos, P. Hansen (2008), pp. 15–39

75. P. Hanafizadeh, M. Mirzazadeh, Visualizing market segmentation using self-organizing maps and Fuzzy Delphi method ADSL market of a telecommunication company. Expert Syst. Appl. **38**, 198–205 (2011)

76. J. Harris, J.L. Hirst, M. Mossinghoff, *Combinatorics and Graph Theory*. Undergraduate Texts in Mathematics (Springer, Berlin, 2008)

77. A.E. Hassanien, E. Emary, M.Z. Hossam, *Retinal blood vessel localization approach based on bee colony swarm optimization fuzzy c-means and pattern search*. J. Visual Commun. Image Represent. **31**, 186–196 (2015)

78. E.M.T. Hendrix, B.G. Tóth, *Introduction to Nonlinear and Global Optimization* (Springer, Berlin, 2010)

79. S.W. Hess, J.B. Weaver, H.J. Whelan, P.A. Zitlau, Nonpartisian political redistricting by computer. Oper. Res. **13**, 998–1006 (1965)

80. F. Höppner, F. Klawonn, A contribution to convergence theory of fuzzy c-means and derivatives. IEEE Trans. Fuzzy Syst. **11**, 682–694 (2003)

81. D. Horta, R.J. Campello, Automatic aspect discrimination in data clustering. Pattern Recogn. **45**, 4370–4388 (2012)

82. D. Horta, R.J.G.B. Campello, Comparing hard and overlapping clusterings. J. Mach. Learn. Res. **16**, 2949–2997 (2015)

83. E. Hüllermeier, M. Rifqi, S. Henzgen, R. Senge, Comparing fuzzy partitions: a generalization of the Rand index and related measures. IEEE Trans. Fuzzy Syst. **20**, 546–556 (2012)

84. C. Iyigun, A. Ben-Israel, A generalized Weiszfeld method for the multi-facility location problem. Oper. Res. Lett. **38**(2010) 207–214.

85. F. Jarre, J. Stoer, *Optimierung* (Springer Verlag, Berlin, Heidelberg, 2004)

86. H. Jiang, H. Li, S. Yi, X. Wang, X. Hu, A new hybrid method based on partitioning-based DBSCAN and ant clustering. Expert Syst. Appl. **38**, 9373–9381 (2011)

87. M. Jiang, On the sum of distances along a circle. Discrete Math. **308**, 2038–2045 (2008)

88. D.R. Jones, J.R.R.A. Martins, The DIRECT algorithm—25 years later. J. Global Optim. **79**, 521–566 (2021)

89. D.R. Jones, C.D. Perttunen, B.E. Stuckman, Lipschitzian optimization without the Lipschitz constant. J. Optim. Theory Appl. **79**, 157–181 (1993)

90. D. Jukić, R. Scitovski, H. Späth, Partial linearization of one class of the nonlinear total least squares problem by using the inverse model function. Computing **62**, 163–178 (1999)

91. H. Kälviäinen, P. Hirvonen, L. Xu, E. Oja, Probabilistic and non-probabilistic Hough transforms: overview and comparison. Image Vis. Comput. **13**, 239–252 (1995)

92. A. Karami, R. Johansson, Choosing DBSCAN parameters automatically using differential evolution. Int. J. Comput. Appl. **91**, 1–11 (2014)
93. M.A. Kashiha, C. Bahr, S. Ott, C.P. Moons, T.A. Niewold, F.T.D. Berckmans, Automatic monitoring of pig locomotion using image analysis. Livestock Sci. **159**, 141–148 (2014)
94. L. Kaufman, P.J. Rousseeuw, *Finding Groups in Data: An Introduction to Cluster Analysis* (Wiley, Chichester, 2005)
95. U. Kaymak, M. Setnes, Fuzzy clustering with volume prototype and adaptive cluster merging. IEEE Trans. Fuzzy Syst. **10**, 705–712 (2002)
96. J. Kogan, *Introduction to Clustering Large and High-Dimensional Data* (Cambridge University Press, New York, 2007)
97. M.-C. Körner, J. Brimberg, H. Juel, A. Schöbel, Geometric fit of a point set by generalized circles. J. Global Optim. **51**, 115–132 (2011)
98. K.M. Kumar, A.R.M. Reddy, A fast DBSCAN clustering algorithm by accelerating neighbor searching using groups method. Pattern Recogn. **58**, 39–48 (2016)
99. A. Laha, Building contextual classifiers by integrating fuzzy rule based classification technique and k-nn method for credit scoring. Adv. Eng. Inf. **21**, 281–291 (2007)
100. F. Leisch, A toolbox for k-centroids cluster analysis. Comput. Stat. Data Anal. **51**, 526–544 (2006)
101. C. Liu, H. Shan, B. Wang, Wireless sensor network localization via matrix completion based on Bregman divergence. Sensors **18**, 2974 (2018)
102. H.-C. Liu, B.-C. Jeng, J.-M. Yih, Y.-K. Yu, Fuzzy c-means algorithm based on standard Mahalanobis distances, in *International Symposium on Information Processing (ISIP'09)* (2009), pp. 422–427
103. Y.-S. Liu, K. Ramani, Robust principal axes determination for point-based shapes using least median of squares. Comput. Aided Des. **41**, 293–305 (2009)
104. W. Lu, J. Tan, Detection of incomplete ellipse in images with strong noise by iterative randomized Hough transform (IRHT). Pattern Recogn. **41**, 1268–1279 (2008)
105. A. Manzanera, T.P. Nguyen, X. Xu, Line and circle detection using dense one-to-one Hough transforms on greyscale images. EURASIP J. Image Video Process. **2016**, 46 (2016). https://doi.org/10.1186/s13640-016-0149-y
106. K.V. Mardia, P.E. Jupp, *Directional Statistics* (Wiley, London, 2000)
107. I. Markovsky, S.V. Huffel, Overview of total least squares methods. Signal Process. **87**, 2283–2302 (2007)
108. T. Marošević, Data clustering for circle detection. Croat. Oper. Res. Rev. **5**, 15–24 (2014)
109. T. Marošević, The Hausdorff distance between some sets of points. Math. Commun. **23**, 247–257 (2018)
110. T. Marošević, K. Sabo, P. Taler, A mathematical model for uniform distribution voters per constituencies. Croat. Oper. Res. Rev. **4**, 53–64 (2013)
111. T. Marošević, R. Scitovski, Multiple ellipse fitting by center-based clustering. Croat. Oper. Res. Rev. **6**, 43–53 (2015)
112. F. Martínez-Álvarez, J. Reyes, A. Morales-Esteban, C. Rubio-Escudero, Determining the best set of seismicity indicators to predict earthquakes. two case studies: Chile and the Iberian Peninsula. Knowl. Based Syst. **50** (2013)
113. D.J. Maširević, S. Miodragović, Geometric median in the plane. Elem. Math. **70**, 21–32 (2015)
114. S. Mimaroglu, E. Aksehirli, Improving DBSCAN's execution time by using a pruning technique on bit vectors. Pattern Recogn. Lett. **32**, 1572–1580 (2011)
115. B. Mirkin, *Data clustering for Data Mining* (Chapman & Hall/CRC, 2005)
116. A. Morales-Esteban, F. Martínez-Álvarez, S. Scitovski, R. Scitovski, A fast partitioning algorithm using adaptive Mahalanobis clustering with application to seismic zoning. Comput. Geosci. **73**, 132–141 (2014)
117. A. Morales-Esteban, F. Martínez-Álvarez, S. Scitovski, R. Scitovski, Determination of frequency-magnitude seismic parameters for the Iberian Peninsula and the Republic of Croatia. Comput. Geosci. (2021) Submitted

118. A. Morales-Esteban, F. Martínez-Álvarez, A. Troncoso, J.L. Justo, C. Rubio-Escudero, Pattern recognition to forecast seismic time series. Expert Syst. Appl. **37**, 8333–8342 (2010)
119. M. Moshtaghi, T.C. Havens, J.C. Bezdek, L. Park, C. Leckie, S. Rajasegarar, J.M. Keller, M. Palaniswami, Clustering ellipses for anomaly detection. Pattern Recogn. **44**, 55–69 (2011)
120. P. Mukhopadhyay, B.B. Chaudhuri, A survey of Hough transform. Pattern Recogn. **48**, 993–1010 (2015)
121. Y. Nestorov, Efficiency of coordinate descent methods on huge-scale optimization problems. SIAM J. Optim. **22**, 341–362 (2012)
122. A. Neumaier, Complete search in continuous global optimization and constraint satisfaction. Acta Numer. **13**, 271–369 (2004)
123. M. Ng, A note on constrained k-means algorithms. Pattern Recogn. **33**, 515–519 (2000)
124. Y. Nievergelt, Total least squares: state-of-the-art regression in numerical analysis. SIAM Rev. **36**, 258–264 (1994)
125. Y. Nievergelt, A finite algorithm to fit geometrically all midrange lines, circles, planes, spheres, hyperplanes, and hyperspheres. Numer. Math. **91**, 257–303 (2002)
126. J. Nutini, M. Schmidt, I.H. Laradji, M. Friedlander, H. Koepke, Coordinate descent converges faster with the Gauss-Southwell rule than random selection, in *Proceedings of the 32nd International Conference on Machine Learning (ICML-15)* (2015)
127. A. Okabe, B. Boots, K. Sugihara, *Spatial Tessellations: Concepts and Applications of Voronoi Diagrams* (Wiley, Chichester, 2000)
128. J.M. Ortega, W.C. Rheinboldt, *Iterative Solution of Nonlinear Equations in Several Variables* (SIAM, Philadelphia, 2000)
129. V.M. Panaretos, *Statistics for Mathematicians: A Rigorous First Course*. Compact Textbooks in Mathematics. (Birkhäuser, Basel, 2016)
130. R. Paulavičius, Y. Sergeyev, D. Kvasov, J. Žilinskas, Globally-biased DISIMPL algorithm for expensive global optimization. J. Global Optim. **59**, 545–567 (2014)
131. R. Paulavičius, J. Žilinskas, *Simplicial Global Optimization*. Series: Springer Briefs in Optimization, vol. X (Springer, Berlin, 2014)
132. S.A. Pijavskij, An algorithm for searching for a global minimum of a function. USSR Comput. Math. Math. Phys. **12**, 888–896 (1972) (in Russian)
133. J. Pintér (ed.), *Global Optimization: Scientific and Engineering Case Studies* (Springer, Berlin, 2006)
134. J.D. Pintér, *Global Optimization in Action (Continuous and Lipschitz Optimization: Algorithms, Implementations and Applications)*. Nonconvex Optimization and Its Applications, vol. 6 (Kluwer Academic Publishers, Dordrecht, 1996)
135. D.K. Prasad, M.K.H. Leung, C. Quek, Ellifit: an unconstrained, non-iterative, least squares based geometric ellipse fitting method. Pattern Recogn. **46**, 1449–1465 (2013)
136. D. Reem, S. Reich, A. De Pierro, Re-examination of Bregman functions and new properties of their divergences. Optimization **68**, 279–348 (2019)
137. F. Ricca, A. Scozzari, B. Simeoni, Weighted Voronoi region algorithms for political districting. Math. Comput. Model. **48**, 1468–1477 (2008)
138. F. Ricca, B. Simeoni, Local search algorithms for political districting. Eur. J. Oper. Res. **189**, 1409–1426 (2008)
139. P. Richtarik, M. Takac, Iteration complexity of randomized block-coordinate descent methods for minimizing a composite function. Math. Program. **144**(1–2), 1–38 (2014)
140. P.J. Rousseeuw, M. Hubert, Robust statistics for outlier detection. Wiley Interdiscip. Rev. Data Min. Knowl. Discov. **1**, 73–79 (2011)
141. P.J. Rousseeuw, A.M. Leroy, *Robust Regression and Outlier Detection* (Wiley, New York, 2003)

142. S. Rueda, S. Fathima, C.L. Knight, M. Yaqub, A.T. Papageorghiou, B. Rahmatullah, A. Foi, M. Maggioni, A. Pepe, J. Tohka, R.V. Stebbing, J.E. McManigle, A. Ciurte, X. Bresson, M.B. Cuadra, C. Sun, G.V. Ponomarev, M.S. Gelfand, M.D. Kazanov, C.-W. Wang, H.-C. Chen, C.-W. Peng, C.-M. Hung, J.A. Noble, Evaluation and comparison of current fetal ultrasound image segmentation methods for biometric measurements: a grand challenge. IEEE Trans. Med. Imaging **10**, 1–16 (2013)
143. K. Sabo, D. Grahovac, R. Scitovski, Incremental method for multiple line detection problem—iterative reweighted approach. Math. Comput. Simul. **178**, 588–602 (2020)
144. K. Sabo, R. Scitovski, The best least absolute deviations line—properties and two efficient methods. ANZIAM J. **50**, 185–198 (2008)
145. K. Sabo, R. Scitovski, An approach to cluster separability in a partition. Inf. Sci. **305**, 208–218 (2015)
146. K. Sabo, R. Scitovski, Multiple ellipse detection by using RANSAC and DBSCAN method, in *Proceedings of the 9th International Conference on Pattern Recognition Applications and Methods (ICPRAM)*, vol. 1 (2020), pp. 129–135. https://doi.org/10.5220/0008879301290135
147. K. Sabo, R. Scitovski, P. Taler, Uniform distribution of the number of voters per constituency on the basis of a mathematical model (in Croatian). Hrvatska i komparativna javna uprava **14**, 229–249 (2012)
148. K. Sabo, R. Scitovski, I. Vazler, Grupiranje podataka—klasteri (in Croatian). Osječki matematički list **10**, 149–178 (2010)
149. K. Sabo, R. Scitovski, I. Vazler, One-dimensional center-based l_1-clustering method. Optim. Lett. **7**, 5–22 (2013)
150. K. Sabo, R. Scitovski, I. Vazler, M. Zekić-Sušac, Mathematical models of natural gas consumption. Energy Convers. Manag. **52**, 1721–1727 (2011)
151. J.M. Santos, M. Embrechts, On the use of the adjusted Rand index as a metric for evaluating supervised classification, in *International Conference on Artificial Neural Networks* (2009), pp. 175–184
152. V. Schwämmle, O.N. Jensen, A simple and fast method to determine the parameters for fuzzy c-means cluster analysis. Bioinformatics **26**, 2841–2848 (2010)
153. A. Schöbel, *Locating Lines and Hyperplanes: Theory and Algorithms* (Springer, Berlin, 1999)
154. R. Scitovski, A new global optimization method for a symmetric Lipschitz continuous function and application to searching for a globally optimal partition of a one-dimensional set. J. Global Optim. **68**, 713–727 (2017)
155. R. Scitovski, M.B. Alić, *Grupiranje podataka (In Croatian)*. Odjel za matematiku Sveučilište u Osijeku (2016). http://www.mathos.unios.hr/images/homepages/scitowsk/ASP-2016.pdf
156. R. Scitovski, S. Majstorović, K.Sabo, A combination of RANSAC and DBSCAN methods for solving the multiple geometrical object detection problem. J. Global Optim. **79**, 669–686 (2021). https://doi.org/10.1007/s10898-020-00950-8
157. R. Scitovski, S. Maričić, S. Scitovski, Short-term and long-term water level prediction at one river measurement location. Croat. Oper. Res. Rev. **3**, 80–90 (2012)
158. R. Scitovski, T. Marošević, Multiple circle detection based on center-based clustering. Pattern Recogn. Lett. **52**, 9–16 (2014)
159. R. Scitovski, U. Radojičić, K. Sabo, A fast and efficient method for solving the multiple line detection problem. Rad HAZU, Matematičke znanosti **23** 123–140 (2019)
160. R. Scitovski, K. Sabo, Analysis of the k-means algorithm in the case of data points occurring on the border of two or more clusters. Knowl. Based Syst. **57**, 1–7 (2014)
161. R. Scitovski, K. Sabo, The adaptation of the k-means algorithm to solving the multiple ellipses detection problem by using an initial approximation obtained by the DIRECT global optimization algorithm. Appl. Math. **64**, 663–678 (2019)
162. R. Scitovski, K. Sabo, Application of the DIRECT algorithm to searching for an optimal k-partition of the set A and its application to the multiple circle detection problem. J. Global Optim. **74**(1), 63–77 (2019). https://doi.org/10.1007/s10898-019-00743-8

163. R. Scitovski, K. Sabo, A combination of k-means and dbscan algorithm for solving the multiple generalized circle detection problem. Adv. Data Anal. Classif. (2020). https://doi.org/10.1007/s11634-020-00385-9
164. R. Scitovski, K. Sabo, Klaster analiza i prepoznavanje geometrijskih objekata (In Croatian). Sveučilište u Osijeku, Odjel za matematiku (2020). https://www.mathos.unios.hr/images/homepages/scitowsk/CLUSTERS.pdf
165. R. Scitovski, K. Sabo, D. Grahovac, Globalna optimizacija. Odjel za matematiku (2017) (In Croatian). https://www.mathos.unios.hr/images/homepages/scitowsk/GOP.pdf
166. R. Scitovski, S. Scitovski, A fast partitioning algorithm and its application to earthquake investigation. Comput. Geosci. **59**, 124–131 (2013)
167. R. Scitovski, S.Kosanović, Rate of change in economics research. Econ. Anal. Workers Manag. **19**, 65–75 (1985)
168. R. Scitovski, I. Vidović, D. Bajer, A new fast fuzzy partitioning algorithm. Expert Syst. Appl. **51**, 143–150 (2016)
169. S. Scitovski, A density-based clustering algorithm for earthquake zoning. Comput. Geosci. **110**, 90–95 (2018)
170. Y.D. Sergeyev, D.E. Kvasov, Global search based on efficient diagonal partitions and a set of Lipschitz constants. SIAM J. Optim. **16**, 910–937 (2006)
171. Y.D. Sergeyev, D.E. Kvasov, Lipschitz global optimization, in *Wiley Encyclopedia of Operations Research and Management Science*, ed. by J. Cochran, vol. 4 (Wiley, New York, 2011), pp. 2812–2828
172. Y.D. Sergeyev, R.G. Strongin, D. Lera, *Introduction to Global Optimization Exploiting Space-Filling Curves* Springer Briefs in Optimization (Springer, Berlin, 2013)
173. L. Serir, E. Ramasso, N. Zerhouni, Evidential evolving Gustafson-Kessel algorithm for online data streams partitioning using belief function theory. Int. J. Approx. Reason. **53**, 747–768 (2012)
174. B. Shubert, A sequential method seeking the global maximum of a function. SIAM J. Numer. Anal. **9**, 379–388 (1972)
175. G. Sierksma, *Linear and Integer Programming. Theory and Practice*, 2nd edn. (Marcel Dekker, New York, 2002)
176. H. Späth, *Cluster-Formation und Analyse* (R. Oldenburg Verlag, München, 1983)
177. J.M. Steele, *The Cauchy-Schwarz Master Class: An Introduction to the Art of Mathematical Inequalities* (Mathematical Association of America, Washington, 2004)
178. A. Stetco, X.-J. Zeng, J. Keane, Fuzzy c-means++: Fuzzy c-means with effective seeding initialization. Expert Syst. Appl. **42**, 7541–7548 (2015)
179. E. Süli, D. F. Mayers, *An Introduction to Numerical Analysis* (Cambridge University Press, Cambridge, 2003)
180. P.N. Tan, M. Steinbach, V. Kumar, *Introduction to Data Mining* (Wesley, Reading, 2006)
181. J. Tang, G. Zhang, Y. Wang, H. Wang, F. Liu, A hybrid approach to integrate fuzzy c-means based imputation method with genetic algorithm for missing traffic volume data estimation. Transp. Res. C **51**, 29–40 (2014)
182. M. Teboulle, A unified continuous optimization framework for center-based clustering methods. J. Mach. Learn. Res. **8**, 65–102 (2007)
183. G.R. Terrell, *Mathematical Statistics: A Unified Introduction* (Springer, Berlin, 1999)
184. S. Theodoridis, K. Koutroumbas, *Pattern Recognition*, 4th edn. (Academic Press, Burlington, 2009)
185. J.C.R. Thomas, A new clustering algorithm based on k-means using a line segment as prototype, in *Progress in Pattern Recognition, Image Analysis, Computer Vision, and Applications*, ed. by C.S. Martin, S.-W. Kim (Springer, Berlin, 2011), pp. 638–645
186. A.Y. Uteshev, M.V. Goncharova, Point-to-ellipse and point-to-ellipsoid distance equation analysis. J. Comput. Appl. Math. **328**, 232–251 (2018)
187. I. Vazler, K. Sabo, R. Scitovski, *Weighted median of the data in solving least absolute deviations problems*. Commun. Stat. Theory Methods **41**(8), 1455–1465 (2012)

188. L. Vendramin, R.J.G.B. Campello, E.R. Hruschka, On the comparison of relative clustering validity criteria, in *Proceedings of the SIAM International Conference on Data Mining, SDM 2009, April 30–May 2, 2009, Sparks, Nevada, USA* (SIAM, 2009), pp. 733–744

189. I. Vidović, D. Bajer, R. Scitovski, A new fusion algorithm for fuzzy clustering. Croat. Oper. Res. Rev. **5**, 149–159 (2014)

190. I. Vidović, R. Cupec, v. Hocenski, Crop row detection by global energy minimization. Pattern Recogn. **55**, 68–86 (2016)

191. I. Vidović, R. Scitovski, Center-based clustering for line detection and application to crop rows detection. Comput. Electron. Agric. **109**, 212–220 (2014)

192. P. Viswanath, V.S. Babu, Rough-DBSCAN: a fast hybrid density based clustering method for large data sets. Pattern Recogn. Lett. **30**, 1477–1488 (2009)

193. T. Weise, *Global Optimization Algorithms. Theory and Application* (2008). e-book: http://www.it-weise.de/projects/book.pdf

194. E. Weiszfeld, Sur le point par lequel la somme des distances de n points donnés est minimum. Tohoku Math. J. **43**, 355–386 (1937)

195. I. Wolfram Research, *Mathematica* (Wolfram Research, Champaign, 2020). Version 12.0 edition

196. K.-L. Wu, A derivative parameter selections for fuzzy c-means. Pattern Recogn. **45**, 407–415 (2012)

197. K.-L. Wu, M.-S. Yang, A cluster validity index for fuzzy clustering. Pattern Recogn. Lett. **26**, 1275–1291 (2005)

198. T.J. Wynn, S.A. Stewart, Comparative testing of ellipse-fitting algorithms: implications for analysis of strain and curvature. J. Struct. Geol. **27**, 1973–1985 (2005)

199. J. Xie, H. Gao, W. Xie, X. Liu, P.W. Grant, Robust clustering by detecting density peaks and assigning points based on fuzzy weighted K -nearest neighbors. Inf. Sci. **354**, 19–40 (2016)

200. L. Xu, E. Oja, P. Kultanen, A new curve detection method: Randomized Hough Transform (RHT). Pattern Recogn. Lett. **11**, 331–338 (1990)

201. K.S. Younis, Weighted Mahalanobis distance for hyper-ellipsoidal clustering. Ph.D. thesis, Air Force Institute of Technology, Ohio, 1999

202. Y. Zeng, Z. Xu, Y. He, Y. Rao, Fuzzy entropy clustering by searching local border points for the analysis of gene expression data. Knowl. Based Syst. **129**, 105309 (2019)

203. C. Zhang, Y. Zhou, T. Martin, A validity index for fuzzy and possibilistic c-means algorithm, in *Proceedings of the 12th International Conference on Information Processing and Management of Uncertainty in Knowledge-Based Systems*, ed. by L. Magdalena, M. Ojeda-Aciegoand, J. L. Verdegay (2008), pp. 877–882

204. Y. Zhu, K.M. Ting, M.J. Carman, Density-ratio based clustering for discovering clusters with varying densities. Pattern Recogn. **60**, 983–997 (2016)

205. M. Zuliani, RANSAC for dummies. Technical report, Vision Research Lab, University of California, Santa Barbara, 2009

Index

© The Author(s), under exclusive license to Springer Nature Switzerland AG 2021
R. Scitovski et al., *Cluster Analysis and Applications*,
https://doi.org/10.1007/978-3-030-74552-3

Printed in the United States
by Baker & Taylor Publisher Services